Student's Solutions Manual

Discrete and Combinatorial Mathematics

Fifth Edition

Ralph P. Grimaldi

Rose-Hulman Institute of Technology

Boston San Francisco New York
London Toronto Sydney Tokyo Singapore Madrid
Mexico City Munich Paris Cape Town Hong Kong Montreal

Reproduced by Pearson Addison-Wesley from electronic files supplied by the author.

Publishing as Pearson Addison-Wesley, 75 Arlington Street, Boston, MA 02116

ISBN 0-321-20033-0

3 4 5 6 QEP 06 05 04

Dedicated

to

Mary Lou McCullough

TABLE OF CONTENTS

PART 1

FUNDAMENTALS

OF

DISCRETE MATHEMATICS

CHAPTER 1
FUNDAMENTAL PRINCIPLES OF COUNTING

Sections 1.1 and 1.2

1. (a) By the rule of sum, there are $8 + 5 = 13$ possibilities for the eventual winner.
(b) Since there are eight Republicans and five Democrats, by the rule of product we have $8 \times 5 = 40$ possible pairs of opposing candidates.
(c) The rule of sum in part (a); the rule of product in part (b).

3. By the rule of product there are (a) $4 \times 12 \times 3 \times 2 = 288$ distinct Buicks that can be manufactured. Of these, (b) $4 \times 1 \times 3 \times 2 = 24$ are blue.

5. Based on the evidence supplied by Jennifer and Tiffany, from the rule of product we find that there are $2 \times 2 \times 1 \times 10 \times 10 \times 2 = 800$ different license plates.

7. For each of the nine toppings there are two choices – (1) put the topping on the hamburger; or (2) do not put the topping on the hamburger. Consequently, by the rule of product there are 2^9 possibilities.

9. (a) Since there are 14 different bakery items and 12 different beverages, by the rule of product there are $(14)(12) = 168$ possible orders.
(b) $(14)(12)(6)(18) = 18{,}144$
(c) $(8)(18)(6)(3)(14)(12)(14)(12) = 73{,}156{,}608$

11. (a) There are four roads from town A to town B and three roads from town B to town C, so by the rule of product there are $4 \times 3 = 12$ roads from A to C that pass through B. Since there are two roads from A to C directly, there are $12 + 2 = 14$ ways in which Linda can make the trip from A to C.
(b) Using the result from part (a), together with the rule of product, we find that there are $14 \times 14 = 196$ different round trips (from A to C and back to A).
(c) Here there are $14 \times 13 = 182$ round trips.

13. (a) $8! = P(8,8)$ (b) $7!$ $6!$

15. Here we must place a,b,c,d in the positions denoted by x: e $\underline{x}$ e $\underline{x}$ e $\underline{x}$ e $\underline{x}$ e. By the rule of product there are 4! ways to do this.

17. Class A: $(2^7 - 2)(2^{24} - 2) = 2{,}113{,}928{,}964$
Class B: $2^{14}(2^{16} - 2) = 1{,}073{,}709{,}056$

Class C: $2^{21}(2^8 - 2) = 532,676,608$

19. (a) The number of permutations of seven (distinct) objects is $7! = 5040$.
(b) $4 \times 3 \times 3 \times 2 \times 2 \times 1 \times 1 = (4!)(3!) = 144$
(c) $(3!)(5)(4!) = 720$
(d) There are 3! ways to arrange the C++ books and 4! ways to arrange the Java books. We can have the C++ books first in 3!4! ways – by the rule of product. If the Java books are first we again have 4!3! (or, 3!4!) possible arrangements. So in total there are $(3!)(4!)(2) = 288$ possible arrangements.

21. (a) From the result preceding Example 1.13 there are $12!/(3!2!2!2!)$ possible arrangements.
(b) $[11!/(3!2!2!2!)]$ (for AG) + $[11!/(3!2!2!2!)]$ (for GA)
(c) Consider one case where all the vowels are adjacent: S,C,L,G,C,L, OIOOIA. These seven symbols can be arranged in $(7!)/(2!2!)$ ways. Since O,O,O,I,I,A can be arranged in $(6!)/(3!2!)$ ways, the number of arrangements with all the vowels adjacent is $[7!/(2!2!)][6!/(3!2!)]$.

23. Here the solution is the number of ways we can arrange 12 objects — 4 of the first type, 3 of the second, 2 of the third, and 3 of the fourth. There are $12!/(4!3!2!3!) = 277,200$ ways.

25. (a) $90 = P(n,2) = n(n-1) \Rightarrow n^2 - n - 90 = 0 \Rightarrow (n-10)(n+9) = 0 \Rightarrow n = 10$ or $n = -9 \Rightarrow n = 10$.
(b) $n = 5$
(c) $2n!/(n-2)! + 50 = (2n)!/(2n-2)! \Longrightarrow 2n(n-1) + 50 = (2n)(2n-1) \Longrightarrow n^2 = 25 \Longrightarrow n = 5$.

27. (a) Each path consists of 2 H's, 1 V, and 7 A's. There are $10!/(2!1!7!)$ ways to arrange these 10 letters and this is the number of paths.
(b) $10!/(2!1!7!)$
(c) If $a, b,$ and c are any real numbers and $m, n,$ and p are nonnegative integers, then the number of paths from (a,b,c) to $(a+m, b+n, c+p)$ is $(m+n+p)!/(m!n!p!)$.

29. (a) & (b) By the rule of product the **print** statement is executed $12 \times 6 \times 8 = 576$ times.

31. By the rule of product there are (a) $9 \times 9 \times 8 \times 7 \times 6 \times 5 = 136,080$ six-digit integers with no leading zeros and no repeated digit. (b) When digits may be repeated there are 9×10^5 such six-digit integers.
(i) (a) $(9 \times 8 \times 7 \times 6 \times 5 \times 1)$ (for the integers ending in 0) + $(8 \times 8 \times 7 \times 6 \times 5 \times 4)$ (for the integers ending in 2,4,6, or 8) = 68,800. (b) When the digits may be repeated there are $9 \times 10 \times 10 \times 10 \times 10 \times 5 = 450,000$ six-digit even integers.
(ii) (a) $(9 \times 8 \times 7 \times 6 \times 5 \times 1)$ (for the integers ending in 0) + $(8 \times 8 \times 7 \times 6 \times 5 \times 1)$ (for the integers ending in 5) = 28,560. (b) $9 \times 10 \times 10 \times 10 \times 10 \times 2 = 180,000$.
(iii) We use the fact that an integer is divisible by 4 if and only if the integer formed by

the last two digits is divisible by 4. (a) $(8 \times 7 \times 6 \times 5 \times 6)$ (last two digits are 04, 08, 20, 40, 60, or 80) + $(7 \times 7 \times 6 \times 5 \times 16)$ (last two digits are 12, 16, 24, 28, 32, 36, 48, 52, 56, 64, 68, 72, 76, 84, 92, or 96) = 33,600. (b) $9 \times 10 \times 10 \times 10 \times 25 = 225,000$.

33. (a) With 2 choices per question there are $2^{10} = 1024$ ways to answer the examination. (b) Now there are 3 choices per question and 3^{10} ways.

35. (a) 6! (b) Let A,B denote the two people who insist on sitting next to each other. Then there are 5! (A to the right of B) + 5! (B to the right of A) = 2(5!) seating arrangements.

37. We can select the 10 people to be seated at the table for 10 in $\binom{16}{10}$ ways. For each such selection there are 9! ways of arranging the 10 people around the table. The remaining six people can be seated around the other table in 5! ways. Consequently, there are $\binom{16}{10}9!5!$ ways to seat the 16 people around the two given tables.

39.

```
procedure SumOfFact(i, sum: positive integers; j,k: nonnegative integers;
                              factorial: array [0..9] of ten positive integers)
begin
   factorial [0] := 1
   for i := 1 to 9 do
      factorial [i] := i * factorial [i - 1]

   for i := 1 to 9 do
      for j := 0 to 9 do
         for k := 0 to 9 do
            begin
               sum := factorial [i] + factorial [j] + factorial [k]
               if (100 * i + 10 * j + k) = sum then
                  print (100 * i + 10 * j + k)
            end
end
```

The unique answer is 145 since $(1!) + (4!) + (5!) = 1 + 24 + 120 = 145$.

Section 1.3

1. $\binom{6}{2} = 6!/[2!(6\text{-}2)!] = 6!/(2!4!) = (6)(5)/2 = 15$

a	b	b	c	c	e
a	c	b	d	c	f
a	d	b	e	d	e
a	e	b	f	d	f
a	f	c	d	e	f

3. (a) $C(10,4) = 10!/(4!6!) = (10)(9)(8)(7)/(4)(3)(2)(1) = 210$
(b) $\binom{12}{7} = 12!/(7!5!) = (12)(11)(10)(9)(8)/(5)(4)(3)(2)(1) = 792$
(c) $C(14,12) = 14!/(12!2!) = (14)(13)/(2)(1) = 91$
(d) $\binom{15}{10} = 15!/(10!5!) = (15)(14)(13)(12)(11)/(5)(4)(3)(2)(1) = 3003$

5. (a) There are $P(5,3) = 5!/(5-3)! = 5!/2! = (5)(4)(3) = 60$ permutations of size 3 for the five letters m, r, a, f, and t.
(b) There are $C(5,3) = 5!/[3!(5-3)!] = 5!/(3!2!) = 10$ combinations of size 3 for the five letters m, r, a, f, and t. They are

a,f,m	a,f,r	a,f,t	a,m,r	a,m,t
a,r,t	f,m,r	f,m,t	f,r,t	m,r,t

7. (a) $\binom{20}{12}$ (b) $\binom{10}{6}\binom{10}{6}$
(c) $\binom{10}{2}\binom{10}{10}$(2 women)+$\binom{10}{4}\binom{10}{8}$ (4 women)+...+$\binom{10}{10}\binom{10}{2}$ (10 women) $= \sum_{i=1}^{5} \binom{10}{2i}\binom{10}{12-2i}$
(d) $\binom{10}{7}\binom{10}{5}$ (7 women) + $\binom{10}{8}\binom{10}{4}$ (8 women) + $\binom{10}{9}\binom{10}{3}$ (9 women) +
$\binom{10}{10}\binom{10}{2}$ (10 women) $= \sum_{i=7}^{10} \binom{10}{i}\binom{10}{12-i}$.
(e) $\sum_{i=8}^{10} \binom{10}{i}\binom{10}{12-i}$

9. (a) Since there are eight positions in a byte we can choose two of these positions (for the 1's) in $\binom{8}{2}$ ways.
(b) $\binom{8}{4}$ (c) $\binom{8}{6}$ (d) $\binom{8}{6} + \binom{8}{7} + \binom{8}{8}$.

11. (a) $\binom{10}{7} = 120$ (b) $\binom{8}{5} = 56$ (c) $\binom{6}{4}\binom{4}{3}$ (four of the first six) $+\binom{6}{5}\binom{4}{2}$ (five of the first six) $+\binom{6}{6}\binom{4}{1}$ (all of the first six) $= (15)(4) + (6)(6) + (1)(4) = 100$.

13. The letters M,I,I,I,P,P,I can be arranged in [7!/(4!)(2!)] ways. Each arrangement provides eight locations (one at the start of the arrangement, one at the finish, and six between letters) for placing the four nonconsecutive S's. Four of these locations can be selected in $\binom{8}{4}$ ways. Hence, the total number of these arrangements is $\binom{8}{4}$ [7!/(4!)(2!)].

15. (a) Two distinct points determine a line. With 15 points, no three collinear, there are $\binom{15}{2}$ possible lines.
(b) There are $\binom{25}{3}$ possible triangles or planes, and $\binom{25}{4}$ possible tetrahedra.

17. (a) $\sum_{k=2}^{n} \frac{1}{k!}$ (b) $\sum_{i=1}^{7} i^2$ (c) $\sum_{j=1}^{7}(-1)^{j-1}j^3 = \sum_{k=1}^{7}(-1)^{k+1}k^3$

(d) $\sum_{i=0}^{n} \frac{i+1}{n+i}$ (e) $\sum_{i=0}^{n}(-1)^i \left[\frac{n+i}{(2i)!}\right]$

19. $\binom{10}{3}$ (three 1's, seven 0's) + $\binom{10}{1}\binom{9}{1}$ (one 1, one 2, eight 0's) + $\binom{10}{1}$ (one 3, nine 0's) = 220

$\binom{10}{4} + \binom{10}{2} + \binom{10}{1}\binom{9}{2} + \binom{10}{1}\binom{9}{1} = 705$

$(2^{10})(\sum_{i=0}^{5}\binom{10}{2i})$ – Select an even number of locations for 0,2. This is done in $\binom{10}{2i}$ ways for $0 \leq i \leq 5$. Then for the $2i$ positions selected there are two choices; for the $10-2i$ remaining positions there are also two choices – namely, 1,3.

21. There are $\binom{n}{3}$ triangles if sides of the n-gon may be used. Of these $\binom{n}{3}$ triangles, when $n \geq 4$ there are n triangles that use two sides of the n-gon and $n(n-4)$ triangles that use only one side. So if the sides of the n-gon cannot be used, then there are $\binom{n}{3} - n - n(n-4)$, $n \geq 4$, triangles.

23. (a) $\binom{12}{9}$ (b) $\binom{12}{9}(2^3)$

(c) Let $a = 2x$ and $b = -3y$. By the binomial theorem the coefficient of a^9b^3 in the expansion of $(a+b)^{12}$ is $\binom{12}{9}$. But $\binom{12}{9} a^9b^3 = \binom{12}{9}(2x)^9(-3y)^3 = \binom{12}{9}(2^9)(-3)^3 x^9y^3$, so the coefficient of x^9y^3 is $\binom{12}{9}(2^9)(-3)^3$.

25. (a) $\binom{4}{1,1,2} = 12$ (b) $\binom{4}{0,1,1,2} = 12$

(c) $\binom{4}{1,1,2}(2)(-1)(-1)^2 = -24$ (d) $\binom{4}{1,1,2}(-2)(3)^2 = -216$

(e) $\binom{8}{3,2,1,2}(2)^3(-1)^2(3)(-2)^2 = 161,280$

27. In each of parts (a)–(e) replace the variables by 1 and evaluate the results.

(a) 2^3 (b) 2^{10} (c) 3^{10} (d) 4^5 (e) 4^{10}

29. $n\binom{m+n}{m} = n\frac{(m+n)!}{m!n!} = \frac{(m+n)!}{m!(n-1)!} =$

$(m+1)\frac{(m+n)!}{(m+1)(m!)(n-1)!} = (m+1)\frac{(m+n)!}{(m+1)!(n-1)!} = (m+1)\binom{m+n}{m+1}$

31. (a) $1 = [(1+x)-x]^n = (1+x)^n - \binom{n}{1}x^1(1+x)^{n-1} + \binom{n}{2}x^2(1+x)^{n-2} - \ldots + (-1)^n\binom{n}{n}x^n$.

(b) $1 = [(2+x)-(x+1)]^n$ (c) $2^n = [(2+x)-x]^n$

33. (a) $\sum_{i=1}^{3}(a_i - a_{i-1}) = (a_1 - a_0) + (a_2 - a_1) + (a_3 - a_2) = a_3 - a_0$

(b) $\sum_{i=1}^{n}(a_i - a_{i-1}) = (a_1 - a_0) + (a_2 - a_1) + (a_3 - a_2) + \ldots + (a_{n-1} - a_{n-2}) + (a_n - a_{n-1}) = a_n - a_0$

(c) $\sum_{i=1}^{100}(\frac{1}{i+2}-\frac{1}{i+1}) = (\frac{1}{3}-\frac{1}{2})+(\frac{1}{4}-\frac{1}{3})+(\frac{1}{5}-\frac{1}{4})+\ldots+(\frac{1}{101}-\frac{1}{100})+(\frac{1}{102}-\frac{1}{101}) = \frac{1}{102}-\frac{1}{2} = \frac{1-51}{102} = \frac{-50}{102} = \frac{-25}{51}$.

Section 1.4

1. Let $x_i, 1 \le i \le 5$, denote the amounts given to the five children.
(a) The number of integer solutions of $x_1+x_2+x_3+x_4+x_5 = 10,\ 0 \le x_i,\ 1 \le i \le 5$, is $\binom{5+10-1}{10} = \binom{14}{10}$. Here $n = 5,\ r = 10$.
(b) Giving each child one dime results in the equation $x_1+x_2+x_3+x_4+x_5 = 5,\ 0 \le x_i,\ 1 \le i \le 5$. There are $\binom{5+5-1}{5} = \binom{9}{5}$ ways to distribute the remaining five dimes.
(c) Let x_5 denote the amount for the oldest child. The number of solutions to $x_1 + x_2+x_3+x_4+x_5 = 10,\ 0 \le x_i,\ 1 \le i \le 4,\ 2 \le x_5$ is the number of solutions to $y_1+y_2+y_3+y_4+y_5 = 8,\ 0 \le y_i,\ 1 \le i \le 5$, which is $\binom{5+8-1}{8} = \binom{12}{8}$.

3. Here we want the number of nonnegative integer solutions for $x_1+x_2+x_3+x_4 = 20$. With $n = 20$ and $r = 4$ the number of solutions is $\binom{4+20-1}{20} = \binom{23}{20}$.

5. (a) 2^5
(b) For each of the n distinct objects there are two choices. If an object is not selected, then one of the n identical objects is used in the selection. This results in 2^n possible selections of size n.

7. (a) $\binom{4+32-1}{32} = \binom{35}{32}$
(b) $\binom{4+28-1}{28} = \binom{31}{28}$
(c) $\binom{4+8-1}{8} = \binom{11}{8}$
(d) 1
(e) $x_1+x_2+x_3+x_4 = 32,\ x_i \ge -2,\ 1 \le i \le 4$. Let $y_i = x_i+2,\ 1 \le i \le 4$. The number of solutions to the given problem is then the same as the number of solutions to $y_1+y_2+y_3+y_4 = 40,\ y_i \ge 0,\ 1 \le i \le 4$. This is $\binom{4+40-1}{40} = \binom{43}{40}$.
(f) $\binom{4+28-1}{28} - \binom{4+3-1}{3} = \binom{31}{28} - \binom{6}{3}$, where the term $\binom{6}{3}$ accounts for the solutions where $x_4 \ge 26$.

9. $230,230 = \binom{n+20-1}{20} = \binom{n+19}{20} \Longrightarrow n = 7$

11. (a) Here we want the number of selections of size 5, with repetition, from a collection of ten distinct objects. This is $\binom{10+5-1}{5} = \binom{14}{5}$.
(b) $\binom{7+5-1}{5} + 3\binom{7+4-1}{4} + 3\binom{7+3-1}{3} + \binom{7+2-1}{2} = \binom{11}{5} + 3\binom{10}{4} + 3\binom{9}{3} + \binom{8}{2}$, where the first summand accounts for the case where none of 1,3,7 appears, the second summand for when exactly one of 1,3,7 appears once, the third summand for the case of exactly two of these digits appearing once each, and the last summand for when all three appear.

13. (a) After placing one ball in each container we need to find the number of nonnegative integer solutions of $x_1+x_2+x_3+x_4 = 4$. This is $\binom{4+4-1}{4} = \binom{7}{4}$.

(b) $\binom{3+7-1}{7}$ (container 4 has one marble) $+\binom{3+5-1}{5}$ (container 4 has three marbles) $+\binom{3+3-1}{3}$ (container 4 has five marbles) $+\binom{3+1-1}{1}$ (container 4 has seven marbles) $= \sum_{i=0}^{3} \binom{9-2i}{7-2i}$.

15. Consider one such distribution – the one where there are six books on each of the four shelves. Here there are 24! ways for this to happen. And we see that there are also 24! ways to place the books for any other such distribution.

The number of distributions is the number of positive integer solutions to

$$x_1 + x_2 + x_3 + x_4 = 24.$$

This is the same as the number of nonnegative integer solutions for

$$y_1 + y_2 + y_3 + y_4 = 20.$$

[Here $y_i + 1 = x_i$ for all $1 \leq i \leq 4$.]
So there are $\binom{4+20-1}{20} = \binom{23}{20}$ such distributions of the books, and consequently, $\binom{23}{20}(24!)$ ways in which Beth can arrange the 24 books on the four shelves with at least one book on each shelf.

17. (a) Here we are dealing with selections where repetitions are allowed – and the answer is $\binom{5+12-1}{12} = \binom{16}{12}$.
(b) With five choices for each distinct marble, it follows from the rule of product that there are 5^{12} possible distributions.

19. Here there are $r = 4$ nested **for** loops, so $1 \leq m \leq k \leq j \leq i \leq 20$. We are making selections, with repetition, of size $r = 4$ from a collection of size $n = 20$. Hence the **print** statement is executed $\binom{20+4-1}{4} = \binom{23}{4}$ times.

21. The **begin-end** segment is executed $\binom{10+3-1}{3} = \binom{12}{3} = 220$ times. After the execution of this segment the value of the variable *sum* is $\sum_{i=1}^{220} i = (220)(221)/2 = 24,310$.

23. (a) Put one object into each container. Then there are $m-n$ identical objects to place into n distinct containers. This yields $\binom{n+(m-n)-1}{m-n} = \binom{m-1}{m-n} = \binom{m-1}{n-1}$ distributions.
(b) Place r objects into each container. The remaining $m-rn$ objects can then be distributed among the n distinct containers in $\binom{n+(m-rn)-1}{m-rn} = \binom{m-1+(1-r)n}{m-rn} = \binom{m-1+(1-r)n}{n-1}$ ways.

25. If the summands must all be even, then consider one such composition – say,

$$20 = 10 + 4 + 2 + 4 = 2(5 + 2 + 1 + 2).$$

Here we notice that $5+2+1+2$ provides a composition of 10. Further, each composition of 10, when multiplied through by 2, provides a composition of 20, where each summand is even. Consequently, we see that the number of compositions of 20, where each summand is even, equals the number of compositions of 10 – namely, $2^{10-1} = 2^9$.

27. a) Here we want the number of integer solutions for $x_1 + x_2 + x_3 = 12$, $x_1, x_3 > 0$, $x_2 = 7$. The number of integer solutions for $x_1 + x_3 = 5$, with $x_1, x_3 > 0$, is the same as the number of integer solutions for $y_1 + y_3 = 3$, with $y_1, y_3 \geq 0$. This is $\binom{2+3-1}{3} = \binom{4}{3} = 4$.

b) Now we must also consider the integer solutions for $w_1 + w_2 + w_3 = 12$, $w_1, w_3 > 0$, $w_2 = 5$. The number here is $\binom{2+5-1}{5} = \binom{6}{5} = 6$.

Consequently, there are $4 + 6 = 10$ arrangements that result in three runs.

c) The number of arrangements for four runs requires two cases [as above in part (b)].

If the first run consists of heads, then we need the number of integer solutions for $x_1 + x_2 + x_3 + x_4 = 12$, where $x_1 + x_3 = 5$, $x_1, x_3 > 0$ and $x_2 + x_4 = 7$, $x_2, x_4 > 0$. This number is $\binom{2+3-1}{3}\binom{2+5-1}{5} = \binom{4}{3}\binom{6}{5} = 4 \cdot 6 = 24$. When the first run consists of tails we get $\binom{6}{5}\binom{4}{3} = 6 \cdot 4 = 24$ arrangements.

In all there are $2(24) = 48$ arrangements with four runs.

d) If the first run starts with an H, then we need the number of integer solutions for $x_1 + x_2 + x_3 + x_4 + x_5 = 12$ where $x_1 + x_3 + x_5 = 5$, $x_1, x_3, x_5 > 0$ and $x_2 + x_4 = 7$, $x_2, x_4 > 0$. This is $\binom{3+2-1}{2}\binom{2+5-1}{5} = \binom{4}{2}\binom{6}{5} = 36$. For the case where the first run starts with a T, the number of arrangements is $\binom{3+4-1}{4}\binom{2+3-1}{3} = \binom{6}{4}\binom{4}{3} = 60$.

In total there are $36 + 60 = 96$ ways for these 12 tosses to determine five runs.

e) $\binom{3+4-1}{4}\binom{3+2-1}{2} = \binom{6}{4}\binom{4}{2} = 90$ – the number of arrangements which result in six runs, if the first run starts with an H. But this is also the number when the first run starts with a T. Consequently, six runs come about in $2 \cdot 90 = 180$ ways.

f) $2\binom{1+4-1}{4}\binom{1+6-1}{6} + 2\binom{2+3-1}{3}\binom{2+5-1}{5} + 2\binom{3+2-1}{2}\binom{3+4-1}{4} + 2\binom{4+1-1}{1}\binom{4+3-1}{3} + 2\binom{5+0-1}{0}\binom{5+2-1}{2} = 2\sum_{i=0}^{4}\binom{4}{4-i}\binom{6}{6-i} = 2[1 \cdot 1 + 4 \cdot 6 + 6 \cdot 15 + 4 \cdot 20 + 1 \cdot 15] = 420$.

Section 1.5

1.

$$\binom{2n}{n} - \binom{2n}{n-1} = \frac{(2n)!}{n!n!} - \frac{(2n)!}{(n-1)!(n+1)!} =$$

$$\frac{(2n)!(n+1)}{(n+1)!n!} - \frac{(2n)!n}{n!(n+1)!} = \frac{(2n)![(n+1)-n]}{(n+1)!n!} = \frac{1}{(n+1)}\frac{(2n)!}{n!n!} =$$

$$(\frac{1}{n+1})\binom{2n}{n}$$

3. (a) $5(=b_3)$; $14(=b_4)$

(b) For $n \geq 0$ there are $b_n(=\frac{1}{(n+1)}\binom{2n}{n})$ such paths from (0,0) to (n,n).

(c) For $n \geq 0$ the first move is U and the last is H.

5. Using the results in the third column of Table 1.10 we have:

111000	110010	101010
1 2 3	1 2 5	1 3 5
4 5 6	3 4 6	2 4 6

7. There are $b_5(=42)$ ways.

9. (i) When $n=4$ there are $14(=b_4)$ such diagrams.

(ii) For any $n \geq 0$, there are b_n different drawings of n semicircles on and above a horizontal line, with no two semicircles intersecting. Consider, for instance, the diagram in part (f) of the figure. Going from left to right, write 1 the first time you encounter a semicircle and write 0 the second time that semicircle is encountered. Here we get the list 110100. The list 110010 corresponds with the drawing in part (g). This correspondence shows that the number of such drawings for n semicircles is the same as the number of lists of n 1's and n 0's where, as the list is read from left to right, the number of 0's never exceeds the number of 1's.

11. Consider one of the $(\frac{1}{6+1})\binom{2\cdot 6}{6} = (\frac{1}{7})\binom{12}{6}$ ways in which the \$5 and \$10 bills can be arranged – say,

(*) \$5, \$5, \$10, \$5, \$5, \$10, \$10, \$10, \$5, \$5, \$10, \$10.

Here we consider the six \$5 bills as indistinguishable – likewise, for the six \$10 bills. However, we consider the patrons as distinct. Hence, there are 6! ways for the six patrons, each with a \$5 bill, to occupy positions 1, 2, 4, 5, 9, and 10, in the arrangement (*). Likewise, there are 6! ways to locate the other six patrons (each with a \$10 bill). Consequently, here the number of arrangements is

$$(\frac{1}{7})\binom{12}{6}(6!)(6!) = (\frac{1}{7})(12!) = 68,428,800.$$

Supplementary Exercises

1. There are three cases to consider: (1) one major defect, two minor defects; (2) two major defects, four minor defects; and (3) three major defects, six minor defects. Using the rules of sum and product we find the answer to be $\binom{4}{1}\binom{7}{2} + \binom{4}{2}\binom{7}{4} + \binom{4}{3}\binom{7}{6}$.

3. 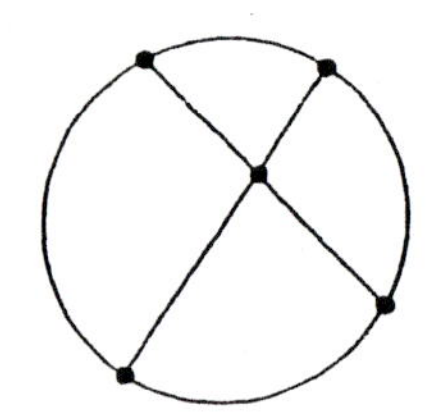

Select any four of these twelve points (on the circumference). As seen in the figure, these points determine a pair of chords that intersect. Consequently, the largest number of points of intersection for all possible chords is $\binom{12}{4} = 495$.

5. (a) 10^{25}
(b) There are 10 choices for the first flag. For the second flag there are 11 choices: The nine poles with no flag, and above or below the first flag on the pole where it is situated. There are 12 choices for the third flag, 13 choices for the fourth,..., and 34 choices for the last (25th). Hence there are $(34!)/(9!)$ possible arrangements.
(c) There are 25! ways to arrange the flags. For each arrangement consider the 24 spaces, one between each pair of flags. Selecting 9 of these spaces provides a distribution among the 10 flagpoles where every flagpole has at least one flag and order is relevant. Hence there are $(25!)\binom{24}{9}$ such arrangements.

7. (a) $C(12, 8)$
(b) Here order is relevant so the answer is $P(12, 8)$.

9. (a) There are two blocks, for example, that differ only in size. There are four that differ only in color, one that differs only in the material used for construction, and five that differ only in shape. In total there are $2 + 4 + 1 + 5 = 12$ blocks that differ from the *small red wooden square* block in exactly one way.
(b) There are $\binom{4}{2} = 6$ ways of selecting the two differing properties. Each such pair must be considered separately.
(i) Material, size: Here there are $1 \times 2 = 2$ such blocks.
(ii) Material, color: This pair yields $1 \times 4 = 4$ such blocks.
(iii) Material, shape: For this pair we obtain $1 \times 5 = 5$ such blocks.
(iv) Size, color: Here we get $2 \times 4 = 8$ of the blocks.
(v) Size, shape: This pair gives us $2 \times 5 = 10$ such blocks.
(vi) Color, shape: For this pair we find $4 \times 5 = 20$ of the blocks we need to count.

In total there are $2 + 4 + 5 + 8 + 10 + 20 = 49$ of Dustin's blocks that differ from the *large blue plastic hexagonal* block in exactly two ways.

11. The number of linear arrangements of the 11 horses is $11!/(5!3!3!)$. Each circular arrangement represents 11 linear arrangements, so there are $(1/11)[11!/(5!3!3!)]$ ways to arrange the horses on the carousel.

13. (a) (i) $\binom{5}{4} + \binom{5}{2}\binom{4}{2} + \binom{4}{4}$ (ii) $\binom{5+4-1}{4} + \binom{5+2-1}{2}\binom{4+2-1}{2} + \binom{4+4-1}{4} = \binom{8}{4} + \binom{6}{2}\binom{5}{2} + \binom{7}{4}$ (iii) $\binom{8}{4} + \binom{6}{2}\binom{5}{2} + \binom{7}{4} - 9$
(b) (i) $\binom{5}{1}\binom{4}{3} + \binom{5}{3}\binom{4}{1}$ (ii) and (iii) $\binom{5}{1}\binom{4+3-1}{3} + \binom{5+3-1}{3}\binom{4}{1} = \binom{5}{1}\binom{6}{3} + \binom{7}{3}\binom{4}{1}$.

15. (a) For each increasing four-digit integer we have four distinct digits, which can only be arranged in one way. These four digits can be chosen in $\binom{9}{4} = 126$ ways. And these same four digits can also be arranged as a decreasing four-digit integer.

To complete the solution we must account for the decreasing four-digit integers where the units digit is 0. There are $\binom{9}{3} = 84$ of these.

Consequently there are $2\binom{9}{4} + \binom{9}{3} = 343$ such four-digit integers.

(b) For each nondecreasing four-digit integer we have four nonzero digits, with repetitions allowed. These four digits can be selected in $\binom{9+4-1}{4} = \binom{12}{4}$ ways. And these same four digits account for a nonincreasing four-digit integer. So at this point we have $2\binom{12}{4} - 9$ of the four-digit integers we want to count. (The reason we subtract 9 is because we have counted the nine integers 1111, 2222, 3333, ..., 9999 twice in $2\binom{12}{4}$.)

We have not accounted for those nonincreasing four-digit integers where the units digit is 0. There are $\binom{10+3-1}{3} - 1 = \binom{12}{3} - 1$ of these four-digit integers. (Here we subtracted 1 since we do not want to include 0000.)

Therefore there are $[2\binom{12}{4} - 9] + [\binom{12}{3} - 1] = [2\binom{12}{4} + \binom{12}{3}] - 10 = 1200$ such four-digit integers.

17. (a) First place person A at the table. There are five distinguishable places available for A (e.g., any of the positions occupied by A,B,C,D,E in Fig. 1.11(a)). Then position the other nine people relative to A. This can be done in 9! ways, so there are (5)(9!) seating arrangements.
(b) There are three distinct ways to position A,B so that they are seated on longer sides of the table across from each other. The other eight people can then be located in 8! different ways, so the total number of arrangements is (3)(8!).

19. (a) Here A must win set 5 and exactly two of the four earlier sets. This can be done in $\binom{4}{2}$ ways. With seven possible scores for each set there are $\binom{4}{2}7^5$ ways for the scores to be recorded.
(b) Here A can win in four sets in $\binom{3}{2}$ ways, and scores can be recorded in $\binom{3}{2}7^4$ ways. So if A wins in four or five sets, then the scores can be recorded in $[\binom{3}{2}7^4 + \binom{4}{2}7^5]$ ways. Since B may be the winner, the final answer is $2[\binom{3}{2}7^4 + \binom{4}{2}7^5]$.

21. For every positive integer n, $0 = (1-1)^n = \binom{n}{0}(1)^0 - \binom{n}{1}(1)^1 + \binom{n}{2}(1)^2 - \binom{n}{3}(1)^3 + \ldots + (-1)^n\binom{n}{n}(1)^n$, and $\binom{n}{0} + \binom{n}{2} + \binom{n}{4} + \ldots = \binom{n}{1} + \binom{n}{3} + \binom{n}{5} + \ldots$

23. (a) There are $P(20,12) = \frac{20!}{8!} = (20)(19)(18)\cdots(11)(10)(9)$ ways in which Francesca can fill her bookshelf.
(b) There are $\binom{17}{9}$ ways in which Francesca can select nine other books. Then she can arrange those nine books and the three books on tennis on her bookshelf in 12! ways. Consequently, among the arrangements in part (a), there are $\binom{17}{9}(12!)$ arrangements that

include Francesca's three books on tennis.

25. (a) For 17 there must be an odd number, between 1 and 17 inclusive, of 1's. For $2k+1$ 1's, where $0 \leq k \leq 8$, there are $2k+2$ locations to select, with repetitions allowed. The selection size is the number of 2's, which is $(1/2)[17-(2k+1)] = 8-k$. The selection can be made in $\binom{2k+2+(8-k)-1}{8-k} = \binom{9+k}{8-k}$ ways, and so the answer is $\sum_{k=0}^{8}\binom{9+k}{8-k} = 2584$.
(b) In the case of 18 the number of 1's must be even: $2k$, for $0 \leq k \leq 9$. If there are $2k$ 1's, there are $2k+1$ locations, with repetitions allowed, for the $(1/2)(18-2k) = 9-k$ 2's. The selection can be made in $\binom{2k+1+(9-k)-1}{9-k} = \binom{9+k}{9-k}$ ways, and the answer is $\sum_{k=0}^{9}\binom{9+k}{9-k} = 4181$.
(c) For n odd, let $n = 2k+1$ for $k \geq 0$. The number of ways to write n as an ordered sum of 1's and 2's is $\sum_{i=0}^{k}\binom{k+1+i}{k-i}$.
For n even, let $n = 2k$ for $k \geq 1$. Here the answer is $\sum_{i=1}^{k}\binom{k+i}{k-i}$.

27. (a) The number of positive integer solutions to the given equation is the same as the number of nonnegative integer solutions for $y_1 + y_2 + \ldots + y_r = n - r$, where $y_i \geq 0$ for all $1 \leq i \leq r$. Here there are $\binom{r+(n-r)-1}{n-r} = \binom{n-1}{n-r} = \binom{n-1}{r-1}$ solutions.
(b) The total is $\sum_{r=1}^{n}\binom{n-1}{r-1} = \binom{n-1}{0} + \binom{n-1}{1} + \ldots + \binom{n-1}{n-1} = 2^{n-1}$.

29. (a) Here we are counting the number of ways to arrange seven R's and 4 U's. This is $11!/(7!4!)$.
(b) There are $[4!/(2!2!)]$ ways to go from $(0,0)$ to $(2,2)$ and there are $[4!/(3!1!)]$ ways to go from $(4,3)$ to $(7,4)$. So there are $[4!/(2!2!)][4!/(3!1!)]$ ways to go from $(0,0)$ to $(7,4)$ and use the path from $(2,2)$ to $(3,2)$ to $(4,2)$ to $(4,3)$. So the paths from $(0,0)$ to $(7,4)$ that do not use the given path is $[11!/(7!4!)] - [4!/(2!2!)][4!/(3!1!)]$.
(c) $[11!/(7!4!)] + [10!/(6!3!1!)] + [9!/(5!2!2!)] + [8!/(4!1!3!)] + [7!/(3!0!4!)]$ (for part (a))
$\{[11!/(7!4!)] + [10!/(6!3!1!)] + [9!/(5!2!2!)] + [8!/(4!1!3!)] + [7!/(3!0!4!)]\} -$
$[\{[4!/(2!2!)] + [3!/(1!1!1!)] + [2!/2!]\} \times \{[4!/(3!1!)] + [3!/(2!1!)]\}]$ (for part (b)).

31. Each rectangle (contained within the 8×5 grid) is determined by four corners of the form $(a,b), (c,b), (c,d), (a,d)$, where a, b, c, d are integers with $0 \leq a < c \leq 8$ and $0 \leq b < d \leq 5$. We can select the pair a, c in $\binom{9}{2}$ ways and the pair b, d in $\binom{6}{2}$ ways. Consequently, the number of rectangles is $\binom{9}{2}\binom{6}{2} = 540$.

33. There are $\binom{6}{4} = 15$ ways to choose the four quarters when Hunter will take these electives. For each of these choices of four quarters, there are $12 \cdot 11 \cdot 10 \cdot 9$ ways to assign the electives. So, in total, there are $\binom{6}{4} \cdot 12 \cdot 11 \cdot 10 \cdot 9 = 178,200$ ways for Hunter to select and schedule these four electives.

CHAPTER 2
FUNDAMENTALS OF LOGIC

Section 2.1

1. The sentences in parts (a), (c), (d), and (f) are statements.

3. Since $p \to q$ is false the truth value for p is 1 and that of q is 0. Consequently, the truth values for the given compound statements are

(a) 0 (b) 0 (c) 1 (d) 0

5. (a) If triangle ABC is equilateral, then it is isosceles.
(b) If triangle ABC is not isosceles, then it is not equilateral.
(c) Triangle ABC is equilateral if and only if it is equiangular.
(d) Triangle ABC is isosceles but it is not equilateral.
(e) If triangle ABC is equiangular, then it is isosceles.

7. (a) If Darci practices her serve daily, then she will have a good chance of winning the tennis tournament.
(b) If you do not fix my air conditioner, then I shall not pay the rent.
or
If I pay the rent, then you will have fixed my air conditioner.
(c) If Mary is to be allowed on Larry's motorcycle, then she must wear her helmet.

9. Propositions (a), (e), (f), and (h) are tautologies. For example, for statements (a) and (e) we find

p	q	$p \lor q$	(a) $\neg(p \lor \neg q) \to \neg p$	$p \to q$	$p \land (p \to q)$	(e) $[p \land (p \to q)] \to q$
0	0	0	1	1	0	1
0	1	1	1	1	0	1
1	0	1	1	0	0	1
1	1	1	1	1	1	1

11. (a) $2^5 = 32$ (b) 2^n

13. Since q has truth value 1, $q \to [(\neg p \lor r) \land \neg s]$ will have truth value 1 when $(\neg p \lor r) \land \neg s$ has truth value 1. This forces s to have truth value 0 and $\neg p \lor r$ to have truth value 1. For $\neg s \to (\neg r \land q)$ to have truth value 1 with s having truth value 0 we must have $\neg r \land q$ with truth value 1. So r has truth value 0 and q truth value 1. Finally, when r has truth value 0, $\neg p \vee r$ will have truth value 1 when p has truth value 0. In summary we have $p: 0;\quad r: 0;\quad s: 0$.

15.

(a) $m = 3,\ n = 6$ (b) $m = 3,\ n = 9$ (c) $m = 18,\ n = 9$
(d) $m = 4,\ n = 9$ (e) $m = 4,\ n = 9$

17. Consider the following possibilities:
(i) Suppose that either the first or the second statement is the true one. Then statements (3) and (4) are false — so their negations are true. And we find from (3) that Tyler did not eat the piece of pie — while from (4) we conclude that Tyler did eat the pie.
(ii) Now we'll suppose that statement (3) is the only true statement. So statements (3) and (4) no longer contradict each other. But now statement (2) is false, and we have Dawn guilty (from statement (2)) and Tyler guilty (from statement (3)).
(iii) Finally, consider the last possibility — that is, statement (4) is the true one. Once again statements (3) and (4) do not contradict each other, and here we learn from statement (2) that Dawn is the vile culprit.

Section 2.2

1. (a)

(i)

p	q	r	$q \wedge r$	$p \rightarrow (q \wedge r)$	$p \rightarrow q$	$p \rightarrow r$	$(p \rightarrow q) \wedge (p \rightarrow r)$
0	0	0	0	1	1	1	1
0	0	1	0	1	1	1	1
0	1	0	0	1	1	1	1
0	1	1	1	1	1	1	1
1	0	0	0	0	0	0	0
1	0	1	0	0	0	1	0
1	1	0	0	0	1	0	0
1	1	1	1	1	1	1	1

(ii)

p	q	r	$p \vee q$	$(p \vee q) \rightarrow r$	$p \rightarrow r$	$q \rightarrow r$	$(p \rightarrow r) \wedge (q \rightarrow r)$
0	0	0	0	1	1	1	1
0	0	1	0	1	1	1	1
0	1	0	1	0	1	0	0
0	1	1	1	1	1	1	1
1	0	0	1	0	0	1	0
1	0	1	1	1	1	1	1
1	1	0	1	0	0	0	0
1	1	1	1	1	1	1	1

(iii)

p	q	r	$q \vee r$	$p \to (q \vee r)$	$p \to q$	$\neg r \to (p \to q)$
0	0	0	0	1	1	1
0	0	1	1	1	1	1
0	1	0	1	1	1	1
0	1	1	1	1	1	1
1	0	0	0	0	0	0
1	0	1	1	1	0	1
1	1	0	1	1	1	1
1	1	1	1	1	1	1

b)

$[p \to (q \vee r)] \iff [\neg r \to (p \to q)]$ From part (iii) of part (a)

$\iff [\neg r \to (\neg p \vee q)]$ By the 2nd Substitution Rule, and $(p \to q) \iff (\neg p \vee q)$

$\iff [\neg(\neg p \vee q) \to \neg\neg r]$ By the lst Substitution Rule, and $(s \to t) \iff (\neg t \to \neg s)$, for primitive statements s, t

$\iff [(\neg\neg p \wedge \neg q) \to r]$ By DeMorgan's Law, Double Negation and the 2nd Substitution Rule

$\iff [(p \wedge \neg q) \to r]$ By Double Negation and the 2nd Substitution Rule

3. a) For a primitive statement s, $s \vee \neg s \iff T_0$. Replace each occurrence of s by $p \vee (q \wedge r)$ and the result follows by the 1st Substitution Rule.

b) For primitive statements s, t we have $(s \to t) \iff (\neg t \to \neg s)$. Replace each occurrence of s by $p \vee q$, and each occurrence of t by r, and the result is a consequence of the 1st Substitution Rule.

5. a) Kelsey placed her studies before her interest in cheerleading, but she (still) did not get a good education.

b) Norma is not doing her mathematics homework or Karen is not practicing her piano lesson.

c) Harold did pass his C++ course and he did finish his data structures project, but he did not graduate at the end of the semester.

7. a)

p	q	$(\neg p \vee q) \wedge (p \wedge (p \wedge q))$	$p \wedge q$
0	0	0	0
0	1	0	0
1	0	0	0
1	1	1	1

b) $(\neg p \wedge q) \vee (p \vee (p \vee q)) \Longleftrightarrow p \vee q$

9. (a) If $0+0=0$, then $2+2=1$.
Let $p : 0+0=0$, $q : 1+1=1$.
(The implication: $p \rightarrow q$) – If $0+0=0$, then $1+1=1$. – False.
(The Converse of $p \rightarrow q$: $q \rightarrow p$) – If $1+1=1$, then $0+0=0$. – True
(The Inverse of $p \rightarrow q$: $\neg p \rightarrow \neg q$) – If $0+0 \neq 0$, then $1+1 \neq 1$. – True
(The Contrapositive of $p \rightarrow q$: $\neg q \rightarrow \neg p$) – If $1+1 \neq 1$, then $0+0 \neq 0$. – False

(b) If $-1 < 3$ and $3+7=10$, then $\sin(\frac{3\pi}{2}) = -1$. (TRUE)
Converse: If $\sin(\frac{3\pi}{2}) = -1$, then $-1 < 3$ and $3+7=10$. (TRUE)
Inverse: If $-1 \geq 3$ or $3+7 \neq 10$, then $\sin(\frac{3\pi}{2}) \neq -1$. (TRUE)
Contrapositive: If $\sin(\frac{3\pi}{2}) \neq -1$, then $-1 \geq 3$ or $3+7 \neq 10$.

11. a) $(q \rightarrow r) \vee \neg p$ b) $(\neg q \vee r) \vee \neg p$

13.

p	q	r	$[(p \leftrightarrow q) \wedge (q \leftrightarrow r) \wedge (r \leftrightarrow p)]$	$[(p \rightarrow q) \wedge (q \rightarrow r) \wedge (r \rightarrow p)]$
0	0	0	1	1
0	0	1	0	0
0	1	0	0	0
0	1	1	0	0
1	0	0	0	0
1	0	1	0	0
1	1	0	0	0
1	1	1	1	1

15. (a) $\neg p \Longleftrightarrow (p \uparrow p)$
(b) $p \vee q \Longleftrightarrow \neg(\neg p \wedge \neg q) \Longleftrightarrow (\neg p \uparrow \neg q) \Longleftrightarrow (p \uparrow p) \uparrow (q \uparrow q)$
(c) $p \wedge q \Longleftrightarrow \neg\neg(p \wedge q) \Longleftrightarrow \neg(p \uparrow q) \Longleftrightarrow (p \uparrow q) \uparrow (p \uparrow q)$
(d) $p \rightarrow q \Longleftrightarrow \neg p \vee q \Longleftrightarrow \neg(p \wedge \neg q) \Longleftrightarrow (p \uparrow \neg q) \Longleftrightarrow p \uparrow (q \uparrow q)$
(e) $p \leftrightarrow q \Longleftrightarrow (p \rightarrow q) \wedge (q \rightarrow p) \Longleftrightarrow t \wedge u \Longleftrightarrow (t \uparrow u) \uparrow (t \uparrow u)$, where t stands for $p \uparrow (q \uparrow q)$ and u for $q \uparrow (p \uparrow p)$.

17.

p	q	$\neg(p \downarrow q)$	$(\neg p \uparrow \neg q)$	$\neg(p \uparrow q)$	$(\neg p \downarrow \neg q)$
0	0	0	0	0	0
0	1	1	1	0	0
1	0	1	1	0	0
1	1	1	1	1	1

19.

		Reasons
(a)	$p \vee [p \wedge (p \vee q)]$	
$\Leftrightarrow$	$p \vee p$	Absorption Law
$\Leftrightarrow$	p	Idempotent Law of $\vee$

		Reasons
(b)	$p \vee q \vee (\neg p \wedge \neg q \wedge r)$	
$\Leftrightarrow$	$(p \vee q) \vee [\neg(p \vee q) \wedge r]$	DeMorgan's Laws
$\Leftrightarrow$	$[(p \vee q) \vee \neg(p \vee q)] \wedge (p \vee q \vee r)$	Distributive Law of $\vee$ over $\wedge$
$\Leftrightarrow$	$T_0 \wedge (p \vee q \vee r)$	Inverse Law
$\Leftrightarrow$	$p \vee q \vee r$	Identity Law

		Reasons
(c)	$[(\neg p \vee \neg q) \rightarrow (p \wedge q \wedge r)]$	
$\Leftrightarrow$	$\neg(\neg p \vee \neg q) \vee (p \wedge q \wedge r)$	$s \rightarrow t \Leftrightarrow \neg s \vee t$
$\Leftrightarrow$	$(\neg\neg p \wedge \neg\neg q) \vee (p \wedge q \wedge r)$	DeMorgan's Laws
$\Leftrightarrow$	$(p \wedge q) \vee (p \wedge q \wedge r)$	Law of Double Negation
$\Leftrightarrow$	$p \wedge q$	Absorption Law

Section 2.3

1. (a)

p	q	r	$p \rightarrow q$	$(p \vee q)$	$(p \vee q) \rightarrow r$
0	0	0	1	0	1
0	0	1	1	0	1
0	1	0	1	1	0
0	1	1	1	1	1
1	0	0	0	1	0
1	0	1	0	1	1
1	1	0	1	1	0
1	1	1	1	1	1

The validity of the argument follows from the results in the last row. (The first seven rows may be ignored.)

(b)

p	q	r	$(p \wedge q) \rightarrow r$	$\neg q$	$p \rightarrow \neg r$	$\neg p \vee \neg q$
0	0	0	1	1	1	1
0	0	1	1	1	1	1
0	1	0	1	0	1	1
0	1	1	1	0	1	1
1	0	0	1	1	1	1
1	0	1	1	1	0	1
1	1	0	0	0	1	0
1	1	1	1	0	0	0

The validity of the argument follows from the results in rows 1, 2, and 5 of the table. The results in the other five rows may be ignored.

(c)

p	q	r	$q \vee r$	$p \vee (q \vee r)$	$[p \vee (q \vee r)] \wedge \neg q$	$p \vee r$
0	0	0	0	0	0	0
0	0	1	1	1	1	1
0	1	0	1	1	0	0
0	1	1	1	1	0	1
1	0	0	0	1	1	1
1	0	1	1	1	1	1
1	1	0	1	1	0	1
1	1	1	1	1	0	1

Consider the last two columns of this truth table. Here we find that whenever the truth value of $[p \vee (q \vee r)] \wedge \neg q$ is 1 then the truth value of $p \vee r$ is also 1. Consequently,

$$[[p \vee (q \vee r)] \wedge \neg q] \Rightarrow p \vee r.$$

(The rows of the table that are crucial for assessing the validity of the argument are rows 2, 5, and 6. Rows 1, 3, 4, 7, and 8 may be ignored.)

3. (a) If p has the truth value 0, then so does $p \wedge q$.
(b) When $p \vee q$ has the truth value 0, then the truth value of p (and that of q) is 0.
(c) If q has truth value 0, then the truth value of $[(p \vee q) \wedge \neg p]$ is 0, regardless of the truth value of p.
(d) The statement $q \vee s$ has truth value 0 only when each of q, s has truth value 0. Then $(p \rightarrow q)$ has truth value 1 when p has truth value 0; $(r \rightarrow s)$ has truth value 1 when r has truth value 0. But then $(p \vee r)$ must have truth value 0, not 1.
(e) For $(\neg p \vee \neg r)$ the truth value is 0 when both p, r have truth value 1. This then forces q, s to have truth value 1, in order for $(p \rightarrow q)$, $(r \rightarrow s)$ to have truth value 1. However, this results in truth value 0 for $(\neg q \vee \neg s)$.

5. (a) Rule of Conjunctive Simplification
(b) Invalid – attempt to argue by the converse

(c) Modus Tollens
(d) Rule of Disjunctive Syllogism
(e) Invalid – attempt to argue by the inverse

7.

(1) & (2)	Premise
(3)	Steps (1), (2) and the Rule of Detachment
(4)	Premise
(5)	Step (4) and $(r \rightarrow \neg q) \Longleftrightarrow (\neg\neg q \rightarrow \neg r) \Longleftrightarrow (q \rightarrow \neg r)$
(6)	Steps (3), (5) and the Rule of Detachment
(7)	Premise
(8)	Steps (6), (7) and the Rule of Disjunctive Syllogism
(9)	Step (8) and the Rule of Disjunctive Amplification

9. (a)

(1)	Premise (The Negation of the Conclusion)
(2)	Step (1) and $\neg(\neg q \rightarrow s) \Longleftrightarrow \neg(\neg\neg q \vee s) \Longleftrightarrow \neg(q \vee s) \Longleftrightarrow \neg q \wedge \neg s$
(3)	Step (2) and the Rule of Conjunctive Simplification
(4)	Premise
(5)	Steps (3), (4) and the Rule of Disjunctive Syllogism
(6)	Premise
(7)	Step (2) and the Rule of Conjunctive Simplification
(8)	Steps (6), (7) and Modus Tollens
(9)	Premise
(10)	Steps (8), (9) and the Rule of Disjunctive Syllogism
(11)	Steps (5), (10) and the Rule of Conjunction
(12)	Step (11) and the Method of Proof by Contradiction

(b)

(1)	$p \rightarrow q$	Premise
(2)	$\neg q \rightarrow \neg p$	Step (1) and $(p \rightarrow q) \Longleftrightarrow (\neg q \rightarrow \neg p)$
(3)	$p \vee r$	Premise
(4)	$\neg p \rightarrow r$	Step (3) and $(p \vee r) \Longleftrightarrow (\neg p \rightarrow r)$
(5)	$\neg q \rightarrow r$	Steps (2), (4) and the Law of the Syllogism
(6)	$\neg r \vee s$	Premise
(7)	$r \rightarrow s$	Step (6) and $(\neg r \vee s) \Longleftrightarrow (r \rightarrow s)$
(8)	$\therefore \neg q \rightarrow s$	Steps (5), (7) and the Law of the Syllogism

(c)

(1)	$\neg p \leftrightarrow q$	Premise
(2)	$(\neg p \rightarrow q) \land (q \rightarrow \neg p)$	Step (1) and $(\neg p \leftrightarrow q) \Longleftrightarrow [(\neg p \rightarrow q) \land (q \rightarrow \neg p)]$
(3)	$\neg p \rightarrow q$	Step (2) and the Rule of Conjunctive Simplification
(4)	$q \rightarrow r$	Premise
(5)	$\neg p \rightarrow r$	Steps (3), (4) and the Law of the Syllogism
(6)	$\neg r$	Premise
(7)	$\therefore p$	Steps (5), (6) and Modus Tollens.

11. (a) Here we want $\neg r$ to have truth value 0 (so r has truth value 1) while the truth value of each of $(p \land \neg q)$ and $[p \rightarrow (q \rightarrow r)]$ is 1. This happens when p has truth value 1 and q has truth value 0. In summary we have $p : 1 \quad q : 0 \quad r : 1$.

(b) $p : 0 \quad q : 0 \quad r : 0$ or 1
$p : 0 \quad q : 1 \quad r : 1$

(c) $p, q, r : 1 \quad s : 0$

(d) $p, q, r : 1 \quad s : 0$

13.

(a)

p	q	r	$p \lor q$	$\neg p \lor r$	t: $(p \lor q) \land (\neg p \lor r)$	$q \lor r$	$t \rightarrow (q \lor r)$
0	0	0	0	1	0	0	1
0	0	1	0	1	0	1	1
0	1	0	1	1	1	1	1
0	1	1	1	1	1	1	1
1	0	0	1	0	0	0	1
1	0	1	1	1	1	1	1
1	1	0	1	0	0	1	1
1	1	1	1	1	1	1	1

From the last column of the truth table it follows that $[(p \lor q) \land (\neg p \lor r)] \rightarrow (q \lor r)$ is a tautology.

Alternately we can try to see if there are truth values that can be assigned to p, q, and r so that $(q \lor r)$ has truth value 0 while $(p \lor q)$, $(\neg p \lor r)$ both have truth value 1.

For $(q \lor r)$ to have truth value 0, it follows that $q : 0$ and $r : 0$. Consequently, for $(p \lor q)$ to have truth value 1, we have $p : 1$ since $q : 0$. Likewise, with $r : 0$ it follows that $\neg p : 1$ if $(\neg p \lor) r$ has truth value 1. But we cannot have $p : 1$ and $\neg p : 1$. So whenever $(p \lor q)$, $(\neg p \lor r)$ have truth value 1, we have $(q \lor r)$ with truth value 1 and it follows that $[(p \lor q) \land (\neg p \lor r)] \rightarrow (q \lor r)$ is a tautology.

Finally we can also argue as follows:

Steps	Reasons
1. $p \vee q$	1. Premise
2. $q \vee p$	2. Step (1) and the Commutative Law of $\vee$
3. $\neg(\neg q) \vee p$	3. Step (2) and the Law of Double Negation
4. $\neg q \rightarrow p$	4. Step (3), $\neg q \rightarrow p \Leftrightarrow \neg(\neg q) \vee p$
5. $\neg p \vee r$	5. Premise
6. $p \rightarrow r$	6. Step (5), $p \rightarrow r \Leftrightarrow \neg p \vee r$
7. $\neg q \rightarrow r$	7. Steps (4), (6), and the Law of the Syllogism
8. $\therefore q \vee r$	8. Step (7), $\neg q \rightarrow r \Leftrightarrow q \vee r$

(b)

(i) Steps	Reasons
1. $p \vee (q \vee r)$	1. Premise
2. $(p \vee q) \wedge (p \vee r)$	2. Step (1) and the Distribution Law of $\vee$ over $\wedge$
3. $p \vee r$	3. Step (2) and the Rule of Conjunctive Simplification
4. $p \rightarrow s$	4. Premise
5. $\neg p \vee s$	5. Step (4), $p \rightarrow s \Leftrightarrow \neg p \vee s$
6. $\therefore r \vee s$	6. Steps (3), (5), the Rule of Conjunction, and Resolution

(ii) Steps	Reasons
1. $p \leftrightarrow q$	1. Premise
2. $(p \rightarrow q) \wedge (q \rightarrow p)$	2. $(p \leftrightarrow q) \Leftrightarrow [(p \rightarrow q) \wedge (q \rightarrow p)]$
3. $p \rightarrow q$	3. Step (2) and the rule of Conjunctive Simplification
4. $\neg p \vee q$	4. Step (3), $p \rightarrow q \Leftrightarrow \neg p \vee q$
5. p	5. Premise
6. $p \vee q$	7. Step (5) and the Rule of Disjunctive Amplification
7. $[(p \vee q) \wedge (\neg p \vee q)]$	7. Steps (6), (4), and the Rule of Conjunction
8. $q \vee q$	8. Step (7) and Resolution
9. $\therefore q$	9. Step (8) and the Idempotent Law of $\vee$.

(iii) Steps	Reasons
1. $p \vee q$	1. Premise
2. $p \rightarrow r$	2. Premise
3. $\neg p \vee r$	3. Step (2), $p \rightarrow r \Leftrightarrow \neg p \vee r$
4. $[(p \vee q) \wedge (\neg p \vee r)]$	4. Steps (1), (3), and the Rule of Conjunction
5. $q \vee r$	5. Step (4) and Resolution
6. $r \rightarrow s$	6. Premise
7. $\neg r \vee s$	7. Step (6), $r \rightarrow s \Leftrightarrow \neg r \vee s$
8. $[(r \vee q) \wedge (\neg r \vee s)]$	8. Steps (5), (7), the Commutative Law of $\vee$, and the Rule of Conjunction
9. $\therefore q \vee s$	9. Step (8) and Resolution

(iv)

	Steps		Reasons
1.	$\neg p \vee q \vee r$	1.	Premise
2.	$q \vee (\neg p \vee r)$	2.	Step (1) and the Commutative and Associative Laws of $\vee$
3.	$\neg q$	3.	Premise
4.	$\neg q \vee (\neg p \vee r)$	4.	Step (3) and the Rule of Disjunctive Amplification
5.	$[[q \vee (\neg \vee r)] \wedge [\neg q \vee (\neg p \vee r)]]$	5.	Steps (2), (4), and the Rule of Conjunction
6.	$(\neg p \vee r)$	6.	Step (5), Resolution, and the Idempotent Law of $\wedge$
7.	$\neg r$	7.	Premise
8.	$\neg r \vee \neg p$	8.	Step (7) and the Rule of Disjunctive Amplification
9.	$[(r \vee \neg p) \wedge (\neg r \vee \neg p)]$	9.	Steps (6), (8), the Commutative Law of $\vee$, and the Rule of Conjunction
10.	$\therefore \neg p$	10.	Step (9), Resolution, and the Idempotent Law of $\vee$

(v)

Steps	Reasons
1. $\neg p \vee s$	1. Premise
2. $p \vee q \vee t$	2. Premise
3. $p \vee (q \vee t)$	3. Step (2) and the Associative Law of $\vee$
4. $[[p \vee (q \vee t)] \wedge (\neg p \vee s)]$	4. Steps (3), (1), and the Rule of Conjunction
5. $(q \vee t) \vee s$	5. Step (4) and Resolution (and the First Substitution Rule)
6. $q \vee (t \vee s)$	6. Step (5) and the Associative Law of $\vee$
7. $\neg q \vee r$	7. Premise
8. $[[q \vee (t \vee s)]$	8. Steps (6), (7), and the Rule of Conjunction
9. $(t \vee s) \vee r$	9. Step (8) and Resolution (and the First Substitution Rule)
10. $t \vee (s \vee r)$	10. Step (9) and the Associative Law of $\vee$
11. $\neg t \vee (s \wedge r)$	11. Premise
12. $(\neg t \vee s) \wedge (\neg t \vee r)$	12. Step (11) and the Distributive Law of $\vee$ over $\wedge$
13. $\neg t \vee s$	13. Step (12) and the Rule of Conjunctive Simplification
14. $[[t \vee (s \vee r)] \wedge (\neg t \vee s)]$	14. Steps (10), (13), and the Rule of Conjunction
15. $(s \vee r) \vee s$	15. Step (14) and Resolution (and the First Substitution Rule)
16. $\therefore\ r \vee s$	16. Step (15) and the Commutative, Associative, and Idempotent Laws of $\vee$

(c) Consider the following assignments.

p: Jonathan has his driver's license.

q: Jonathan's new car is out of gas.

r: Jonathan likes to drive his new car.

Then the given argument can be written in symbolic form as

$$\begin{array}{l} \neg p \vee q \\ p \vee \neg r \\ \underline{\neg q \vee \neg r} \\ \therefore\ \neg r \end{array}$$

Steps		Reasons
1.	$\neg p \vee q$	1. Premise
2.	$p \vee \neg r$	2. Premise
3.	$(p \vee \neg r) \wedge (\neg p \vee q)$	3. Steps (2), (1), and the Rule of Conjunction
4.	$\neg r \vee q$	4. Step (3) and Resolution
5.	$q \vee \neg r$	5. Step (4) and the Commutative Law of $\vee$
6.	$\neg q \vee \neg r$	6. Premise
7.	$(q \vee \neg r) \wedge (\neg q \vee \neg r)$	7. Steps (5), (6), and the Rule of Conjunction
8.	$\neg r \vee \neg r$	8. Step (7) and Resolution
9.	$\therefore \neg r$	9. Step (8) and Idempotent Law of $\vee$

Section 2.4

1.

(a) False (b) False (c) False
(d) True (e) False (f) False

3. Statements (a), (c), and (e) are true, while statements (b), (d), and (f) are false.

5.

(a)	$\exists x\ [m(x) \wedge c(x) \wedge j(x)]$	True
(b)	$\exists x\ [s(x) \wedge c(x) \wedge \neg m(x)]$	True
(c)	$\forall x\ [c(x) \rightarrow (m(x) \veebar p(x))]$	False
(d)	$\forall x\ [(g(x) \wedge c(x)) \rightarrow \neg p(x)]$,	True
	or $\forall x\ [(p(x) \wedge c(x)) \rightarrow \neg g(x)]$,	
	or $\forall x\ [(g(x) \wedge p(x)) \rightarrow \neg c(x)]$	
(e)	$\forall x\ [(c(x) \wedge s(x)) \rightarrow (p(x) \veebar e(x))]$,	True

7. (a)

(i) $\exists x\ q(x)$
(ii) $\exists x\ [p(x) \wedge q(x)]$
(iii) $\forall x\ [q(x) \rightarrow \neg t(x)]$
(iv) $\forall x\ [q(x) \rightarrow \neg t(x)]$
(v) $\exists x\ [q(x) \wedge t(x)]$
(vi) $\forall x\ [(q(x) \wedge r(x)) \rightarrow s(x)]$

(b) Statements (i), (iv), (v), and (vi) are true. Statements (ii) and (iii) are false: $x = 10$ provides a counterexample for either statement.

(c)

(i) If x is a perfect square, then $x > 0$.
(ii) If x is divisible by 4, then x is even.
(iii) If x is divisible by 4, then x is not divisible by 5.
(iv) There exists an integer that is divisible by 4 but it is not a perfect square.

(d) (i) Let $x = 0$. (iii) Let $x = 20$.

9.

(a) (i) True (ii) False – Consider $x = 3$.
(iii) True (iv) True
(b) (i) True (ii) False – Consider $x = 3$.
(iii) True (iv) True
(c) (i) True (ii) True
(iii) True (iv) False – For $x = 2$ or 5, the truth value of $p(x)$ is 1 while that of $r(x)$ is 0.

11. (a) In this case the variable x is free while the variables y, z are bound.
(b) Here the variables x, y are bound; the variable z is free.

13. (a) $p(2,3) \wedge p(3,3) \wedge p(5,3)$
(b) $[p(2,2) \vee p(2,3) \vee p(2,5)] \vee [p(3,2) \vee p(3,3) \vee p(3,5)] \vee [p(5,2) \vee p(5,3) \vee p(5,5)]$
(c) $[p(2,2) \vee p(3,2) \vee p(5,2)] \wedge [p(2,3) \vee p(3,3) \vee p(5,3)] \wedge [p(2,5) \vee p(3,5) \vee p(5,5)]$

15. a) The proposed negation is correct and is a true statement.
b) The proposed negation is wrong. A correct version of the negation is: For all rational numbers x, y, the sum $x + y$ is rational. This correct version of the negation is a true statement.
c) The proposed negation is correct — but false. The (original) statement is true.
d) The proposed negation is wrong. A correct version of the negation is: For all integers x, y, if x, y are both odd, then xy is even.
The (original) statement is true.

17. a) There exists an integer n such that n is not divisible by 2 but n is even (that is, not odd).
b) There exist integers k, m, n such that $k - m$ and $m - n$ are odd, and $k - n$ is odd.
c) For some real number x, $x^2 > 16$ but $-4 \leq x \leq 4$ (that is, $-4 \leq x$ and $x \leq 4$).
d) There exists a real number x such that $|x - 3| < 7$ and either $x \leq -4$ or $x \geq 10$.

19. (a) Statement: For all positive integers m, n, if $m > n$ then $m^2 > n^2$. (TRUE)
Converse: For all positive integers m, n, if $m^2 > n^2$ then $m > n$. (TRUE)
Inverse: For all positive integers m, n, if $m \leq n$ then $m^2 \leq n^2$. (TRUE)
Contrapositive: For all positive integers m, n, if $m^2 \leq n^2$ then $m \leq n$. (TRUE)
(b) Statement: For all integers a, b, if $a > b$ then $a^2 > b^2$. (FALSE — let $a = 1$ and $b = -2$.)
Converse: For all integers a, b, if $a^2 > b^2$ then $a > b$. (FALSE — let $a = -5$ and $b = 3$.)
Inverse: For all integers a, b, if $a \leq b$ then $a^2 \leq b^2$. (FALSE — let $a = -5$ and $b = 3$.)
Contrapositive: For all integers a, b, if $a^2 \leq b^2$ then $a \leq b$. (FALSE — let $a = 1$ and $b = -2$.)
(c) Statement: For all integers m, n, and p, if m divides n and n divides p then m divides p. (TRUE)

Converse: For all integers m and p, if m divides p, then for each integer n it follows that m divides n and n divides p. (FALSE — let $m = 1$, $n = 2$, and $p = 3$.)
Inverse: For all integers m, n, and p, if m does not divide n or n does not divide p, then m does not divide p. (False — let $m = 1$, $n = 2$, and $p = 3$.)
Contrapositive: For all integers m and p, if m does not divide p, then for each integer n it follows that m does not divide n or n does not divide p. (TRUE)
(d) Statement: $\forall x\ [(x > 3) \rightarrow (x^2 > 9)]$ (TRUE)
Converse: $\forall x\ [(x^2 > 9) \rightarrow (x > 3)]$ (FALSE — let $x = -5$.)
Inverse: $\forall x\ [(x \leq 3) \rightarrow (x^2 \leq 9)]$ (FALSE — let $x = -5$.)
Contrapositive: $\forall x\ [(x^2 \leq 9) \rightarrow (x \leq 3)]$ (TRUE)
(e) Statement: $\forall x\ [(x^2 + 4x - 21 > 0) \rightarrow [(x > 3) \vee (x < -7)]]$ (TRUE)
Converse: $\forall x\ [[(x > 3) \vee (x < -7)] \rightarrow (x^2 + 4x - 21 > 0)]$ (TRUE)
Inverse: $\forall x\ [(x^2 + 4x - 21 \leq 0) \rightarrow [(x \leq 3) \wedge (x \geq -7)]]$, or $\forall x\ [(x^2 + 4x - 21 \leq 0) \rightarrow (-7 \leq x \leq 3)]$ (TRUE)
Contrapositive: $\forall x\ [[(x \leq 3) \wedge (x \geq -7)] \rightarrow (x^2 + 4x - 21 \leq 0)]$, or $\forall x\ [(-7 \leq x \leq 3) \rightarrow (x^2 + 4x - 21 \leq 0)]$ (TRUE)

21. (a) True (b) False (c) False (d) True (e) False

23. (a) $\forall a\ \exists b\ [a + b = b + a = 0]$
(b) $\exists u\ \forall a\ [au = ua = a]$
(c) $\forall a \neq 0\ \exists b\ [ab = ba = 1]$
(d) The statement in part (b) remains true but the statement in part (c) is no longer true for this new universe.

25. (a) $\exists x\ \exists y\ [(x > y) \wedge (x - y \leq 0)]$
(b) $\exists x\ \exists y\ [(x < y) \wedge \forall z[x \geq z \vee z \geq y]]$
(c) $\exists x\ \exists y\ [(|x| = |y|) \wedge (y \neq \pm x)]$

Section 2.5

1. Although we may write $28 = 25 + 1 + 1 + 1 = 16 + 4 + 4 + 4$, there is no way to express 28 as the sum of at most three perfect squares.

3. Here we find that

$30 = 25 + 4 + 1$	$40 = 36 + 4$	$50 = 25 + 25$
$32 = 16 + 16$	$42 = 25 + 16 + 1$	$52 = 36 + 16$
$34 = 25 + 9$	$44 = 36 + 4 + 4$	$54 = 25 + 25 + 4$
$36 = 36$	$46 = 36 + 9 + 1$	$56 = 36 + 16 + 4$
$38 = 36 + 1 + 1$	$48 = 16 + 16 + 16$	$58 = 49 + 9$

5. (a) The real number π is not an integer.
(b) Margaret is a librarian.

(c) All administrative directors know how to delegate authority.
(d) Quadrilateral $MNPQ$ is not equiangular.

7. (a) When the statement $\exists x\ [p(x) \vee q(x)]$ is true, there is at least one element c in the prescribed universe where $p(c) \vee q(c)$ is true. Hence at least one of the statements $p(c), q(c)$ has the truth value 1, so at least one of the statements $\exists x\ p(x)$ and $\exists x\ q(x)$ is true. Therefore, it follows that $\exists x\ p(x) \vee \exists x\ q(x)$ is true, and $\exists x\ [p(x) \vee q(x)] \Longrightarrow \exists x\ p(x) \vee \exists x\ q(x)$. Conversely, if $\exists x\ p(x) \vee \exists x\ q(x)$ is true, then at least one of $p(a)$, $q(b)$ has truth value 1, for some a, b in the prescribed universe. Assume without loss of generality that it is $p(a)$. Then $p(a) \vee q(a)$ has truth value 1 so $\exists x\ [p(x) \vee q(x)]$ is a true statement, and $\exists x\ p(x) \vee \exists x\ q(x) \Longrightarrow \exists x\ [p(x) \vee q(x)]$.
(b) First consider when the statement $\forall x\ [p(x) \wedge q(x)]$ is true. This occurs when $p(a) \wedge q(a)$ is true for each a in the prescribed universe. Then $p(a)$ is true (as is $q(a)$) for all a in the universe, so the statements $\forall x\ p(x)$, $\forall x\ q(x)$ are true. Therefore, the statement $\forall x\ p(x) \wedge \forall x\ q(x)$ is true and $\forall x\ [p(x) \wedge q(x)] \Longrightarrow \forall x\ p(x) \wedge \forall x\ q(x)$.
Conversely, suppose that $\forall x\ p(x) \wedge \forall x\ q(x)$ is a true statement. Then $\forall x\ p(x)$, $\forall x\ q(x)$ are both true. So now let c be any element in the prescribed universe. Then $p(c)$, $q(c)$, and $p(c) \wedge q(c)$ are all true. And, since c was chosen arbitrarily, it follows that the statement $\forall x\ [p(x) \wedge q(x)]$ is true, and $\forall x\ p(x) \wedge \forall x\ q(x) \Longrightarrow \forall x\ [p(x) \wedge q(x)]$.

9. (1) Premise
(2) Premise
(3) Step (1) and the Rule of Universal Specification
(4) Step (2) and the Rule of Universal Specification
(5) Step (4) and the Rule of Conjunctive Simplification
(6) Steps (5), (3), and Modus Ponens
(7) Step (6) and the Rule of Conjunctive Simplification
(8) Step (4) and the Rule of Conjunctive Simplification
(9) Steps (7), (8), and the Rule of Conjunction
(10) Step (9) and the Rule of Universal Generalization

11. Consider the open statements
$w(x)$: x works for the credit union
$\ell(x)$: x writes loan applications
$c(x)$: x knows COBOL
$q(x)$: x knows Excel
and let r represent Roxe and i represent Imogene.

In symoblic form the given argument is given as follows:
$\forall x\ [w(x) \rightarrow c(x)]$
$\forall x\ [(w(x) \wedge \ell(x)) \rightarrow q(x)]$
$w(r) \wedge \neg q(r)$
$\underline{q(i) \wedge \neg c(i)}$
$\therefore \neg \ell(r) \wedge \neg w(i)$

The steps (and reasons) needed to verify this argument can now be presented.

	Steps	**Reasons**
(1)	$\forall x\,[w(x) \rightarrow c(x)]$	Premise
(2)	$q(i) \wedge \neg c(i)$	Premise
(3)	$\neg c(i)$	Step (2) and the Rule of Conjunctive Simplification
(4)	$w(i) \rightarrow c(i)$	Step (1) and the Rule of Universal Specification
(5)	$\neg w(i)$	Steps (3), (4), and Modus Tollens
(6)	$\forall x\,[(w(x) \wedge \ell(x)) \rightarrow q(x)]$	Premise
(7)	$w(r) \wedge \neg q(r)$	Premise
(8)	$\neg q(r)$	Step (7) and the Rule of Conjunctive Simplification
(9)	$(w(r) \wedge \ell(r)) \rightarrow q(r)$	Step (6) and the Rule of Universal Specification
(10)	$\neg(w(r) \wedge \ell(r))$	Steps (8), (9), and Modus Tollens
(11)	$w(r)$	Step (7) and the Rule of Conjunctive Simplification
(12)	$\neg w(r) \vee \neg \ell(r)$	Step (10) and DeMorgan's Law
(13)	$\neg \ell(r)$	Steps (11), (12), and the Rule of Disjunctive Syllogism
(14)	$\therefore \neg \ell(r) \wedge \neg w(i)$	Steps (13), (5), and the Rule of Conjunction

13. (a) Contrapositive: For all integers k and ℓ, if k, ℓ are not both odd then $k\ell$ is not odd. — OR, For all integers k and ℓ, if at least one of k, ℓ is even then $k\ell$ is even.
Proof: Let us assume (without loss of generality) that k is even. Then $k = 2c$ for some integer c — because of Definition 2.8. Then $k\ell = (2c)\ell = 2(c\ell)$, by the associative law of multiplication for integers — and $c\ell$ is an integer. Consequently, $k\ell$ is even — once again, by Definition 2.8. [Note that this result does not require anything about the integer ℓ.]

(b) Contrapositive: For all integers k and ℓ, if k and ℓ are not both even or both odd then $k + \ell$ is odd. — OR, For all integers k and ℓ, if one of k, ℓ is odd and the other even then $k + \ell$ is odd.
Proof: Let us assume (without loss of generality) that k is even and ℓ is odd. Then it follows from Definition 2.8 that we may write $k = 2c$ and $\ell = 2d + 1$ for integers c and d. And now we find that $k + \ell = 2c + (2d + 1) = 2(c + d) + 1$, where $c + d$ is an integer — by the associative law of addition and the distributive law of multiplication over addition for integers. From Definition 2.8 we find that $k + \ell = 2(c + d) + 1$ implies that $k + \ell$ is odd.

15. Proof: Assume that for some integer n, n^2 is odd while n is not odd. Then n is even and we may write $n = 2a$, for some integer a — by Definition 2.8. Consequently, $n^2 = (2a)^2 = (2a)(2a) = (2 \cdot 2)(a \cdot a)$, by the commutative and associative laws of multiplication for integers. Hence, we may write $n^2 = 2(2a^2)$, with $2a^2$ an integer — and this means that n^2 is even. Thus we have arrived at a contradiction since we now have n^2 both odd (at the start) and even. This contradiction came about from the false assumption that n is not odd. Therefore, for every integer n, it follows that n^2 odd $\Rightarrow n$ odd.

17. Proof:

(1) Since n is odd we have $n = 2a + 1$ for some integer a. Then $n + 11 = (2a + 1) + 11 = 2a + 12 = 2(a + 6)$, where $a + 6$ is an integer. So by Definition 2.8 it follows that $n + 11$ is even.

(2) If $n + 11$ is not even, then it is odd and we have $n + 11 = 2b + 1$, for some integer b. So $n = (2b + 1) - 11 = 2b - 10 = 2(b - 5)$, where $b - 5$ is an integer, and it follows from Definition 2.8 that n is even — that is, not odd.

(3) In this case we stay with the hypothesis — that n is odd — and also assume that $n + 11$ is not even — hence, odd. So we may write $n + 11 = 2b + 1$, for some integer b. This then implies that $n = 2(b - 5)$, for the integer $b - 5$. So by Definition 2.8 it follows that n is even. But with n both even (as shown) and odd (as in the hypothesis) we have arrived at a contradiction. So our assumption was wrong, and it now follows that $n + 11$ is even for every odd integer n.

19. This result is not true, in general. For example, $m = 4 = 2^2$ and $n = 1 = 1^2$ are two positive integers that are perfect squares, but $m + n = 2^2 + 1^2 = 5$ is not a perfect square.

21. Proof: We shall prove the given result by establishing the truth of its (logically equivalent) contrapositive.

Let us consider the negation of the conclusion — that is, $x < 50$ and $y < 50$. Then with $x < 50$ and $y < 50$ it follows that $x + y < 50 + 50 = 100$, and we have the negation of the hypothesis. The given result now follows by this indirect method of proof (by the contrapositive).

23. Proof: If n is odd, then $n = 2k + 1$ for some (particular) integer k. Then $7n + 8 = 7(2k + 1) + 8 = 14k + 7 + 8 = 14k + 15 = 14k + 14 + 1 = 2(7k + 7) + 1$. It then follows from Definition 2.8 that $7n + 8$ is odd.

To establish the converse, suppose that n is not odd. Then n is even, so we can write $n = 2t$, for some (particular) integer t. But then $7n + 8 = 7(2t) + 8 = 14t + 8 = 2(7t + 4)$, so it follows from Definition 2.8 that $7n+8$ is even – that is, $7n+8$ is not odd. Consequently, the converse follows by contraposition.

Supplementary Exercises

1.

p	q	r	s	$q \wedge r$	$\neg(s \vee r)$	$\overbrace{[(q \wedge r) \rightarrow \neg(s \vee r)]}^{t}$	$p \leftrightarrow t$
0	0	0	0	0	1	1	0
0	0	0	1	0	0	1	0
0	0	1	0	0	0	1	0
0	0	1	1	0	0	1	0
0	1	0	0	0	1	1	0
0	1	0	1	0	0	1	0
0	1	1	0	1	0	0	1
0	1	1	1	1	0	0	1
1	0	0	0	0	1	1	1
1	0	0	1	0	0	1	1
1	0	1	0	0	0	1	1
1	0	1	1	0	0	1	1
1	1	0	0	0	1	1	1
1	1	0	1	0	0	1	1
1	1	1	0	1	0	0	0
1	1	1	1	1	0	0	0

3. (a)

p	q	r	$q \leftrightarrow r$	$p \leftrightarrow (q \leftrightarrow r)$	$(p \leftrightarrow q)$	$(p \leftrightarrow q) \leftrightarrow r$
0	0	0	1	0	1	0
0	0	1	0	1	1	1
0	1	0	0	1	0	1
0	1	1	1	0	0	0
1	0	0	1	1	0	1
1	0	1	0	0	0	0
1	1	0	0	0	1	0
1	1	1	1	1	1	1

It follows from the results in columns 5 and 7 that $[p \leftrightarrow (q \leftrightarrow r)] \Leftrightarrow [(p \leftrightarrow q) \leftrightarrow r]$.

(b) The truth value assignments $p : 0$; $q : 0$; $r : 0$ result in the truth value 1 for $[p \rightarrow (q \rightarrow r)]$ and 0 for $[(p \rightarrow q) \rightarrow r]$. Consequently, these statements are not logically equivalent.

5. Since $p \vee \neg q \Leftrightarrow \neg\neg p \vee \neg q \Leftrightarrow \neg p \rightarrow \neg q$, we can express the given statement as:

(1) If Kaylyn does not practice her piano lessons, then she cannot go to the movies.

But $p \vee \neg q \Leftrightarrow \neg q \vee p \Leftrightarrow q \rightarrow p$, so we can also express the given statement as:
(2) If Kaylyn is to go to the movies, then she will have to practice her piano lessons.

7.

(a) $(\neg p \vee \neg q) \wedge (F_0 \vee p) \wedge p$

(b) $(\neg p \vee \neg q) \wedge (F_0 \vee p) \wedge p$

$\Longleftrightarrow (\neg p \vee \neg q) \wedge (p \wedge p)$	$F_0 \vee p \Longleftrightarrow p$
$\Longleftrightarrow (\neg p \vee \neg q) \wedge p$	Idempotent Law of $\wedge$
$\Longleftrightarrow p \wedge (\neg p \vee \neg q)$	Commutative Law of $\wedge$
$\Longleftrightarrow (p \wedge \neg p) \vee (p \wedge \neg q)$	Distributive Law of $\wedge$ over $\vee$
$\Longleftrightarrow F_0 \vee (p \wedge \neg q)$	$p \wedge \neg p \Longleftrightarrow F_0$
$\Longleftrightarrow p \wedge \neg q$	F_0 is the identity for $\vee$.

9.

(a) contrapositive (b) inverse (c) contrapositive
(d) inverse (e) converse

11. (a)

p	q	r	$p \veebar q$	$(p \veebar q) \veebar r$	$q \veebar r$	$p \veebar (q \veebar r)$
0	0	0	0	0	0	0
0	0	1	0	1	1	1
0	1	0	1	1	1	1
0	1	1	1	0	0	0
1	0	0	1	1	0	1
1	0	1	1	0	1	0
1	1	0	0	0	1	0
1	1	1	0	1	0	1

It follows from the results in columns 5 and 7 that $[(p \veebar q) \veebar r] \Longleftrightarrow [p \veebar (q \veebar r)]$.
(b) The given statements are not logically equivalent. The truth value assignments $p: 1;\ q: 0;\ r: 0$ provide a counterexample.

13.

(a) True (b) False (c) True (d) True
(e) False (f) False (g) False (h) True

15. Suppose that the 62 squares in this 8×8 chessboard (with two opposite missing corners) can be covered with 31 dominos. We agree to place each domino on the board so that the blue part is on top of a blue square (and the white part is then necessarily above a white square). The given chessboard contains 30 blue squares and 32 white ones. Each domino covers one blue and one white square – for a total of 31 blue squares and 31 white ones. This contradiction tells us that we cannot cover this 62 square chessboard with the 31 dominos.

CHAPTER 3
SET THEORY

Section 3.1

1. They are all the same set.

3. All of the statements are true except for parts (b) and (d).

5. (a) $\{0,2\}$
(b) $\{2, 2\frac{1}{2}, 3\frac{1}{3}, 5\frac{1}{5}, 7\frac{1}{7}\}$
(c) $\{0, 2, 12, 36, 80\}$

7. (a) $\forall x\ [x \in A \to x \in B] \land \exists x\ [x \in B \land x \notin A]$
(b) $\exists x\ [x \in A \land x \notin B] \lor \forall x\ [x \notin B \lor x \in A]$

9. (a) $|A| = 6$ (b) $|B| = 7$
(c) If B has 2^n subsets of odd cardinality, then $|B| = n + 1$.

11. (a) There are $2^5 - 1 = 31$ nonempty subsets for the set consisting of one penny, one nickel, one dime, one quarter and one half-dollar.
(b) 30 (c) 28

13. (a) $\binom{30}{5}$
(b) Since the smallest element in A is 5 we must select the other four elements in A from $\{6, 7, 8, \ldots, 29, 30\}$. This can be done in $\binom{25}{4}$ ways.
(c) Let x denote the smallest element in A. Then there are four cases to consider.
($x = 1$) Here we can choose the other four elements in $\binom{29}{4}$ ways.
($x = 2$) Here there are $\binom{28}{4}$ selections.
($x = 3$) There are $\binom{27}{4}$ subsets possible here.
($x = 4$) In this last case we have $\binom{26}{4}$ possibilities.
In total there are $\binom{29}{4} + \binom{28}{4} + \binom{27}{4} + \binom{26}{4}$ subsets A where $|A| = 5$ and the smallest element in A is less than 5.

15. Let $W = \{1\}$, $X = \{\{1\},2\}$, $Y = \{X, 3\}$.

17. (a) Let $x \in A$. Since $A \subseteq B$, $x \in B$. Then with $B \subseteq C$, $x \in C$. So $x \in A \Longrightarrow x \in C$ and $A \subseteq C$.

(b) Since $A \subset B \Longrightarrow A \subseteq B$, by part (a), $A \subseteq C$. With $A \subset B$, there is an element $x \in B$ such that $x \notin A$. Since $B \subseteq C$, $x \in B \Longrightarrow x \in C$, so there is an element $x \in C$ with $x \notin A$ and $A \subset C$.
(c) Since $B \subset C$ it follows that $B \subseteq C$, so by part (a) we have $A \subseteq C$. Also, $B \subset C \Longrightarrow \exists x \in \mathcal{U}\,(x \in C \wedge x \notin B)$. Since $A \subseteq B$, $x \notin B \Longrightarrow x \notin A$. So $A \subseteq C$ and $\exists x \in \mathcal{U}\,(x \in C \wedge x \notin A)$. Hence $A \subset C$.
(d) Since $A \subset B \Longrightarrow A \subseteq B$, the result follows from part (c).

19. (a) For $n, k \in \mathbf{Z}^+$ with $n \geq k+1$, consider the hexagon centered at $\binom{n}{k}$. This has the form

$$\begin{array}{ccccc} & \binom{n-1}{k-1} & & \binom{n-1}{k} & \\ \binom{n}{k-1} & & \binom{n}{k} & & \binom{n}{k+1} \\ & \binom{n+1}{k} & & \binom{n+1}{k+1} & \end{array}$$

where the two alternating triples – namely, $\binom{n-1}{k-1}, \binom{n}{k+1}, \binom{n+1}{k}$ and $\binom{n-1}{k}, \binom{n+1}{k+1}, \binom{n}{k-1}$ – satisfy $\binom{n-1}{k-1}\binom{n}{k+1}\binom{n+1}{k} = \binom{n-1}{k}\binom{n+1}{k+1}\binom{n}{k-1}$.
(b) For $n, k \in \mathbf{Z}^+$ with $n \geq k+1$,

$$\binom{n-1}{k-1}\binom{n}{k+1}\binom{n+1}{k} = \left[\frac{(n-1)!}{(k-1)!(n-k)!}\right]\left[\frac{n!}{(k+1)!(n-k-1)!}\right]\left[\frac{(n+1)!}{k!(n+1-k)!}\right]$$

$$= \left[\frac{(n-1)!}{k!(n-1-k)!}\right]\left[\frac{(n+1)!}{(k+1)!(n-k)!}\right]\left[\frac{n!}{(k-1)!(n-k+1)!}\right] =$$

$$\binom{n-1}{k}\binom{n+1}{k+1}\binom{n}{k-1}.$$

21. $(1/4)\binom{n}{5} = \binom{n-1}{4} \Longrightarrow (1/4)[(n!)/(5!(n-5)!)] = (n-1)!/(4!(n-5)!) \Longrightarrow n! = 20(n-1)! \Longrightarrow n = 20$.

23. For a given $n \in \mathbf{N}$, we need to find $k \in \mathbf{N}$ so that the three consecutive entries $\binom{n}{k}, \binom{n}{k+1}, \binom{n}{k+2}$ are in the ratio $1:2:3$. [Consequently, $n \geq 2$ (and $k \geq 0$).] In order to obtain the given ratio we must have

$$\binom{n}{k+1} = 2\binom{n}{k} \quad \text{and} \quad \binom{n}{k+2} = 3\binom{n}{k}.$$

From $\binom{n}{k+1} = 2\binom{n}{k}$, it follows that $2\frac{n!}{(k!)(n-k)!} = \frac{n!}{(k+1)!(n-k-1)!}$ so $2k+2 = n-k$, or $n = 2+3k$. Likewise, $\binom{n}{k+2} = 3\binom{n}{k}$ implies that $3\frac{n!}{(k!)(n-k)!} = \frac{n!}{(k+2)!(n-k-2)!}$, and we find that $3(k+2)(k+1) = (n-k)(n-k-1)$. Consequently, with $n = 2+3k$, we have $3(k+2)(k+1) = (2+2k)(1+2k)$, or $0 = k^2-3k-4 = (k-4)(k+1)$. Since $k \geq 0$, it follows that $k = 4$

and $n = 14$. So the 5th, 6th, and 7th entries in the row for $n = 14$ provide the unique solution.

25. As an ordered set, $A = \{x, v, w, z, y\}$.

27. (a) If $S \in S$, then since $S = \{A | A \notin A\}$ we have $S \notin S$.
(b) If $S \notin S$, then by the definition of S it follows that $S \in S$.

29.

```
procedure Subsets4(i,j,k,l: positive integers)
begin
    for i := 1 to 4 do
        for j := i+1 to 5 do
            for k := j+1 to 6 do
                for l := k+1 to 7 do
                    print ({i,j,k,l})
end
```

Section 3.2

1.

(a) {1,2,3,5}	(b) A	(c) $\mathcal{U} - \{2\}$
(d) $\mathcal{U} - \{2\}$	(e) {4,8}	(f) {1,2,3,4,5,8}
(g) $\emptyset$	(h) {2,4,8}	(i) {1,3,4,5,8}

3. (a) Since $A = (A - B) \cup (A \cap B)$ we have $A = \{1, 3, 4, 7, 9, 11\}$. Similarly we find $B = \{2, 4, 6, 8, 9\}$.
(b) $C = \{1, 2, 4, 5, 9\}$, $D = \{5, 7, 8, 9\}$.

5.

(a) True	(b) True	(c) True	(d) False	(e) True
(f) True	(g) True	(h) False	(i) False	

7. (a) False. Let $\mathcal{U} = \{1, 2, 3\}$, $A = \{1\}$, $B = \{2\}$, $C = \{3\}$. Then $A \cap C = B \cap C$ but $A \neq B$.
(b) False. Let $\mathcal{U} = \{1, 2\}$, $A = \{1\}$, $B = \{2\}$, $C = \{1, 2\}$. Then $A \cup B = A \cup C$ but $A \neq B$.
(c) $x \in A \Longrightarrow x \in A \cup C \Longrightarrow x \in B \cup C$. So $x \in B$ or $x \in C$. If $x \in B$, then we are finished. If $x \in C$, then $x \in A \cap C = B \cap C$ and $x \in B$. In either case, $x \in B$ so $A \subseteq B$. Likewise, $y \in B \Longrightarrow y \in B \cup C = A \cup C$, so $y \in A$ or $y \in C$. If $y \in C$, then $y \in B \cap C = A \cap C$. In either case, $y \in A$ and $B \subseteq A$. Hence $A = B$.

(d) Let $x \in A$. Consider two cases: (i) $x \in C \Longrightarrow x \notin A\Delta C \Longrightarrow x \notin B\Delta C \Longrightarrow x \in B$. (ii) $x \notin C \Longrightarrow x \in A\Delta C \Longrightarrow x \notin B\Delta C \Longrightarrow x \in B$. In either case $A \subseteq B$. In a similar way we find $B \subseteq A$, so $A = B$.

9. $(A \cap B) \cup C = \{d, x, z\}$ which has $2^3 - 1 = 7$ proper subsets; $A \cap (B \cup C) = \{d\}$ which has 1 proper subset.

11. Here we change each occurrence of $\mathcal{U}$ by $\emptyset$ (and vice versa), and each occurrence of $\cup$ by $\cap$ (and vice versa). This gives us
(a) $\emptyset = (A \cup B) \cap (A \cup \overline{B}) \cap (\overline{A} \cup B) \cap (\overline{A} \cup \overline{B})$
(b) $A = A \cup (A \cap B)$
(c) $A \cap B = (A \cup B) \cap (A \cup \overline{B}) \cap (\overline{A} \cup B)$
(d) $A = (A \cap B) \cup (A \cap \mathcal{U})$

13. (a) False. Let $\mathcal{U} = \{1, 2, 3\}$, $A = \{1\}$, $B = \{2\}$. $P(A) = \{\emptyset, A\}, P(B) = \{\emptyset, B\}$, $P(A \cup B) = \{\emptyset, \{1\}, \{2\}, \{1, 2\}\}$, and $\{1, 2\} \notin P(A) \cup P(B)$.
(b) $X \in P(A) \cap P(B) \Longleftrightarrow X \in P(A)$ and $X \in P(B) \Longleftrightarrow X \subseteq A$ and $X \subseteq B \Longleftrightarrow X \subseteq A \cap B \Longleftrightarrow X \in P(A \cap B)$, so $P(A) \cap P(B) = P(A \cap B)$.

15. (a) $2^6 = 64$ (b) 2^n
(c) In the columns for A, B, whenever a 1 occurs in the column for A, a 1 likewise occurs in the same position in the column for B.
(d)

A	B	C	$A \cup \overline{B}$	$A \cap B$	$\overline{B \cap C}$	$(A \cap B) \cup (\overline{B \cap C})$
0	0	0	1	0	1	1
0	0	1	1	0	1	1
0	1	0	0	0	1	1
0	1	1	0	0	0	0
1	0	0	1	0	1	1
1	0	1	1	0	1	1
1	1	0	1	1	1	1
1	1	1	1	1	0	1

17. (a) $A \cap (B - A) = A \cap (B \cap \overline{A}) = B \cap (A \cap \overline{A}) = B \cap \emptyset = \emptyset$
(b) $[(A \cap B) \cup (A \cap B \cap \overline{C} \cap D)] \cup (\overline{A} \cap B) = (A \cap B) \cup (\overline{A} \cap B)$ by the Absorption Law $= (A \cup \overline{A}) \cap B = \mathcal{U} \cap B = B$
(c) $(A - B) \cup (A \cap B) = (A \cap \overline{B}) \cup (A \cap B) = A \cap (\overline{B} \cup B) = A \cap \mathcal{U} = A$
(d) $\overline{A} \cup \overline{B} \cup (A \cap B \cap \overline{C}) = \overline{(A \cap B)} \cup [(A \cap B) \cap \overline{C}] = [\overline{(A \cap B)} \cup (A \cap B)] \cap [\overline{(A \cap B)} \cup \overline{C}] = [\overline{(A \cap B)} \cup \overline{C}] = \overline{A} \cup \overline{B} \cup \overline{C}$

19.
(a) $[-6, 9]$ (b) $[-8, 12]$ (c) $\emptyset$ (d) $[-8, -6) \cup (9, 12]$
(e) $[-14, 21]$ (f) $[-2, 3]$ (g) $\mathbf{R}$ (h) $[-2, 3]$

Section 3.3

1. Here the universe $\mathcal{U}$ comprises the 600 freshmen. If we let $A, B \subseteq \mathcal{U}$ be the subsets
A: the freshmen who attended the first showing
B: the freshmen who attended the second showing,
then $|\mathcal{U}| = 600$, $|A| = 80$, $|B| = 125$, and $|\overline{A} \cap \overline{B}| = 450$.
Since $|\overline{A} \cap \overline{B}| = |\overline{A \cup B}| = 450$, it follows that $|A \cup B| = 600 - 450 = 150$. Consequently, $|A \cap B| = |A| + |B| - |A \cup B| = 80 + 125 - 150 = 55$ – that is, 55 of the 600 freshmen attended the movie twice.

3. There are 2^9 such strings that start with three 1's and 2^8 that end in four 0's. In addition, 2^5 of these strings start with three 1's and end in four 0's. Consequently, the number that start with three 1's or end in four 0's is

$$2^9 + 2^8 - 2^5 = 512 + 256 - 32 = 736.$$

5. There are 9! permutations that start with 3, 9! permutations that end with 7, and 8! permutations that start with 3 and end with 7. So the number of permutations that start with 3 or end with 7 is $9! + 9! - 8!$

7. (a) There are 24! permutations containing each of the patterns O U T and D I G. There are 22! permutations containing both patterns. Consequently there are $2(24!) - 22!$ permutations containing either O U T or D I G.
(b) There are 26! permutations in total. Of these there are 24! that contain each of the patterns M A N and A N T and 23! that contain both patterns (i.e., contain M A N T). Hence there are $2(24!) - 23!$ permutations that contain either M A N or A N T and $26! - [2(24!) - 23!]$ permutations that contain neither pattern.

9.

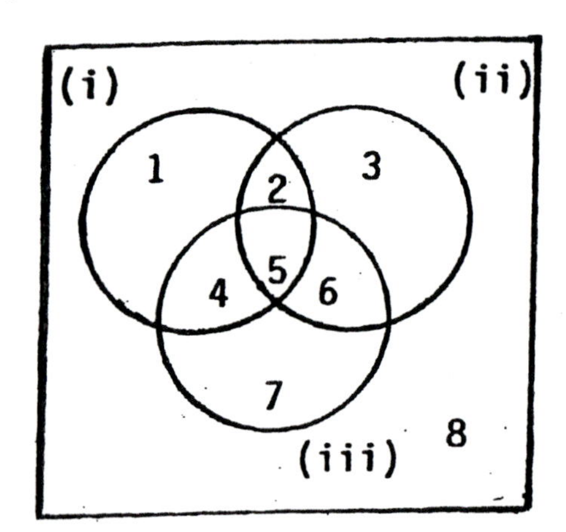

The circle labeled (i) is for the arrangements with consecutive S's; circle (ii) is for consecutive E's; and circle (iii) for consecutive L's. The answer to the problem is the number of arrangements in region 8 which we obtain as follows. For region 5 there are 10! ways to arrange the 10 symbols M,I,C,A,N,O,U,SS,EE,LL. For regions 2,4,6 there are $(11!/2!) - 10!$ arrangements containing exactly two pairs of consecutive letters. Finally each of regions 1,3,7 contains $(12!/(2!2!)) - 2[(11!/2!) - 10!] - 10!$ arrangements, so region 8 contains $[13!/(2!)^3] - 3[12!/(2!2!)] + 3(11!/2!) - 10!$ arrangements.

Section 3.4

1. (a) $Pr(A) = |A|/|S| = 3/8$
(b) $Pr(B) = |B|/|S| = 4/8 = 1/2$
(c) $A \cap B = \{a, c\}$ so $Pr(A \cap B) = 2/8 = 1/4$
(d) $A \cup B = \{a, b, c, e, g\}$ so $Pr(A \cup B) = 5/8$
(e) $\overline{A} = \{d, e, f, g, h\}$ and $Pr(\overline{A}) = 5/8 = 1 - 3/8 = 1 - Pr(A)$
(f) $\overline{A} \cup B = \{a, c, d, e, f, g, h\}$ with $Pr(\overline{A} \cup B) = 7/8$
(g) $A \cap \overline{B} = \{b\}$ so $Pr(A \cap \overline{B}) = 1/8$.

3. Here each equally likely outcome has probability $\frac{1}{25} = 0.04$. Consequently, there are $\frac{0.24}{0.04} = 6$ outcomes in A.

5. (a) $\binom{6}{2}/\binom{12}{2} = 15/66 = 5/22 = 0.2272727\ldots$
(b) $[\binom{1}{1}\binom{10}{1} + \binom{10}{1}\binom{1}{1} + \binom{1}{1}\binom{1}{1}]/\binom{10}{2} = 21/66 = 7/22 = 1 - [\binom{10}{2}/\binom{12}{2}]$

7. $S = \{\{x, y\} | x, y \in \{1, 2, 3, \ldots, 99, 100\}, x \neq y\}$
$A = \{\{x, y\} | \{x, y\} \in S, x + y \text{ is even}\}$
$= \{\{x, y\} | \{x, y\} \in S, x, y \text{ even}\} \cup \{\{x, y\} | \{x, y\} \in S, x, y \text{ odd}\}$
$|S| = \binom{100}{2} = 4950$; $|A| = \binom{50}{2} + \binom{50}{2} = 2450$
$Pr(A) = 2450/4950 = 49/99$

9. The sample space $S = \{(x_1, x_2, x_3, x_4, x_5, x_6) | x_i = H \text{ or } T, 1 \leq i \leq 6\}$. Hence $|S| = 2^6 = 64$.
(a) Here the event $A = \{HHHHHH\}$ and $Pr(A) = 1/64$.
(b) The event $B = \{HHHHHT, HHHHTH, HHHTHH, HHTHHH, HTHHHH, THHHHH\}$ and $Pr(B) = 6/64 = 3/32$.
(c) There are $6!/(4!2!) = 15$ ways to arrange two heads and four tails, so the probability for this event is $15/64$.
(d) 0 heads: 1 arrangement
2 heads: $[6!/(2!4!)] = 15$ arrangements
4 heads: $[6!/(4!2!)] = 15$ arrangements
6 heads: 1 arrangement
The event here includes exactly 32 of the 64 arrangements in S, so the probability for an even number of heads is $32/64 = 1/2$.
(e) 4 heads: $[6!/(4!2!)] = 15$ arrangements
5 heads: $[6!/(5!1!)] = 6$ arrangements
6 heads: 1 arrangement
Here the probability is $22/64 = 11/32$.

11. (a) Let $S =$ the sample space $= \{(x_1, x_2, x_3) | 1 \leq x_i \leq 6,\ i = 1, 2, 3\}$; $|S| = 6^3 = 216$.

Let $A = \{(x_1, x_2, x_3) | x_1 < x_2 \text{ and } x_1 < x_3\} = \bigcup_{n=1}^{5} \{(n, x_2, x_3) | n < x_2 \text{ and } n < x_3\}$.

For $1 \leq n \leq 5$, $|\{(n, x_2, x_3) | n < x_2 \text{ and } n < x_3\}| = (6 - n)^2$.

Consequently, $|A| = 5^2 + 4^2 + 3^2 + 2^2 + 1^2 = 55$.
Therefore, $Pr(A) = 55/216$.
(b) With S as in part (a), let $B = \{(x_1, x_2, x_3)|x_1 < x_2 < x_3\}$.
Then $|\{(1, x_2, x_3)|1 < x_2 < x_3\}| = 10$,
$|\{(2, x_2, x_3)|2 < x_2 < x_3\}| = 6$,
$|\{(3, x_2, x_3)|3 < x_2 < x_3\}| = 3$, and
$|\{(4, x_2, x_3)|4 < x_2 < x_3\}| = 1$,
so $|B| = 20$ and $Pr(B) = 20/216 = 5/54$.

13. (a) $\frac{14!}{15!} = \frac{1}{15}$ (b) $[(14!) + (14!)]/(15!) = 2(14!)/(15!) = 2/15$
(c) $(2)(9)(13!)/(15!) = 3/35$

15. $Pr(A) = 1/3$; $Pr(B) = 7/15$, $Pr(A \cap B) = 2/15$; $Pr(A \cup B) = 2/3$. $Pr(A \cup B) = 2/3 = (1/3) + (7/15) - (2/15) = Pr(A) + Pr(B) - Pr(A \cap B)$.

Section 3.5

1. $Pr(\overline{A}) = 1 - Pr(A) = 1 - 0.4 = 0.6$
$Pr(\overline{B}) = 1 - Pr(B) = 1 - 0.3 = 0.7$
$Pr(A \cup B) = Pr(A) + Pr(B) - Pr(A \cap B) = 0.4 + 0.3 - 0.2 = 0.5$
$Pr(\overline{A \cup B}) = 1 - Pr(A \cup B) = 1 - 0.5 = 0.5$
$Pr(A \cap \overline{B}) = Pr(A) - Pr(A \cap B)$ because $A = (A \cap \overline{B}) \cup (A \cap B)$ with $(A \cap \overline{B}) \cap (A \cap B) = \emptyset$.
So $Pr(A \cap \overline{B}) = 0.4 - 0.2 = 0.2$
$Pr(\overline{A} \cap B) = Pr(B) - Pr(A \cap B) = 0.3 - 0.2 = 0.1$
$Pr(A \cup \overline{B}) = Pr(\overline{\overline{A} \cap B}) = 1 - Pr(\overline{A} \cap B) = 1 - 0.1 = 0.9$
$Pr(\overline{A} \cup B) = Pr(\overline{A \cap \overline{B}}) = 1 - Pr(A \cap \overline{B}) = 1 - 0.2 = 0.8$

3. (a) $S = \{(x, y)|x, y \in \{1, 2, 3, \ldots, 10\}, x \neq y\}$.
(b) For $1 \leq y \leq 9$, if y is the label on the second ball drawn, then there are $10 - y$ possible values for x so that $(x, y) \in S$ and $x > y$. Consequently, if A denotes the event described here, then $|A| = 9 + 8 + 7 + \cdots + 1 = 45$ and $Pr(A) = |A|/|S| = 45/90 = 1/2$.
(c) Let $B = \{(v, w)|v \text{ even}, w \text{ odd}\}$. Then we want $Pr(B \cup C)$ where $B \cap C = \emptyset$. So $Pr(B \cup C) = Pr(B) + Pr(C) = \frac{25}{90} + \frac{25}{90} = \frac{50}{90} = \frac{5}{9}$.

5. Since A, B are disjoint we know that $Pr(A \cup B) = Pr(A) + Pr(B)$, so $Pr(B) = 0.7 - 0.3 = 0.4$.

7. (a) Let p be the probability for the outcome 1. Then for $1 \leq n \leq 6$, the probability for the outcome n is np and $p + 2p + 3p + 4p + 5p + 6p = 1$. Consequently $p = 1/21$.

So the probability for a 5 or 6 is $\frac{5}{21} + \frac{6}{21} = \frac{11}{21}$.
(b) The probability the outcome is even is $\frac{2}{21} + \frac{4}{21} + \frac{6}{21} = \frac{12}{21}$.
(c) $1 - \frac{12}{21} = \frac{9}{21} = \frac{1}{21} + \frac{3}{21} + \frac{5}{21}$.

9. Here the sample space $\mathcal{S} = \{x_1x_2x_3x_4x_5 | x_i \in \{H,T\},\ 1 \le i \le 5\}$. So $|\mathcal{S}| = 2^5 = 32$. The event A of interest here is $A = \{HHHHH, HHHHT, HHHTH, HHHTT, HHTHT, HHTHH\}$ and $Pr(A) = 6/32 = 3/16$.

11. (a) (i) $\frac{18}{38} + \frac{18}{38} - \frac{9}{38} = \frac{27}{38} \doteq 0.710526$
(ii) $\frac{18}{38} + \frac{18}{38} - \frac{9}{38} = \frac{27}{38} \doteq 0.710526$
(b) (i) $\frac{18}{38} \cdot \frac{18}{38} = \frac{81}{361} \doteq 0.224377$
(ii) $\frac{18}{38} \cdot \frac{2}{38} + \frac{2}{38} \cdot \frac{18}{38} = \frac{9}{361} + \frac{9}{361} = \frac{18}{361} \doteq 0.049861$

13. $\frac{6}{14} + \frac{6}{14} - \frac{1}{14} = \frac{11}{14}$

15. Ann selects her seven integers in one of $\binom{80}{7}$ ways. Among these possible selections there are $\binom{11}{7}$ that are winning selections. So the probability Ann is a winner is $\binom{11}{7}/\binom{80}{7} = 330/3,176,716,400 \doteq 0.000000104$. [Using a computer algebra system one gets $0.1038808501 \times 10^{-6}$.]

17. Since $A \cup B \subseteq \mathcal{S}$, it follows from the result of the preceding exercise that $Pr(A \cup B) \le Pr(\mathcal{S}) = 1$. So $1 \ge Pr(A \cup B) = Pr(A) + Pr(B) - Pr(A \cap B)$, and $Pr(A \cap B) \ge Pr(A) + Pr(B) - 1 = 0.7 + 0.5 - 1 = 0.2$.

Section 3.6

1. Let A, B be the events
A: the card drawn is a king
B: the card drawn is an ace or a picture card.
$Pr(A|B) = Pr(A \cap B)/Pr(B) = (\frac{4}{52})/(\frac{16}{52}) = \frac{4}{16} = \frac{1}{4} = 0.25$.

3. Let A, B be the events
A: Coach Mollet works his football team throughout August
B: The team finishes as the division champion.
Here $Pr(B|A) = 0.75$ and $Pr(A) = 0.80$, so $Pr(A \cap B) = Pr(A)Pr(B|A) = (0.80)(0.75) = 0.60$.

5. In general, $Pr(A \cup B) = Pr(A) + Pr(B) - Pr(A \cap B)$. Since A, B are independent, $Pr(A \cap B) = Pr(A)Pr(B)$. So

$$\begin{aligned} Pr(A \cup B) &= Pr(A) + Pr(B) - Pr(A)Pr(B) \\ &= Pr(A) + [1 - Pr(A)]Pr(B) \\ &= Pr(A) + Pr(\overline{A})Pr(B) \end{aligned}$$

and

$$\begin{aligned} Pr(A \cup B) &= Pr(A) + Pr(B) - Pr(A)Pr(B) \\ &= Pr(B) + [1 - Pr(B)]Pr(A) \\ &= Pr(B) + Pr(\overline{B})Pr(A). \end{aligned}$$

7. Let A, B denote the events

A: Bruno selects a gold coin

B: Madeleine selects a gold coin

$$\begin{aligned}
\text{(a) } Pr(B) &= Pr(B\cap A)+Pr(B\cap \overline{A})\\
&= Pr(A)Pr(B|A)+Pr(\overline{A})Pr(B|\overline{A})\\
&= (\tfrac{6}{15})(\tfrac{11}{17})+(\tfrac{9}{15})(\tfrac{10}{17}) = \tfrac{66+90}{255} = \tfrac{156}{255} = \tfrac{52}{85}\\
\text{(b) } Pr(A|B) &= \tfrac{Pr(A\cap B)}{Pr(B)} = \tfrac{Pr(A)Pr(B|A)}{Pr(B)}\\
&= [(\tfrac{6}{15})(\tfrac{11}{17})]/(\tfrac{52}{85}) = \tfrac{66}{156} = \tfrac{11}{26}.
\end{aligned}$$

9.

$$\begin{aligned}
Pr(A\cup B) &= Pr(A)+Pr(B)-Pr(A\cap B)\\
&= Pr(A)+Pr(B)-Pr(A)Pr(B),
\end{aligned}$$

because A, B are independent.

$$\begin{aligned}
0.6 &= 0.3+Pr(B)-(0.3)Pr(B)\\
0.3 &= 0.7Pr(B)
\end{aligned}$$

So $Pr(B)=\frac{3}{7}$.

11.

$$\begin{aligned}
Pr(A) &= \binom{5}{1}(\tfrac{1}{2})^1(\tfrac{1}{2})^4+\binom{5}{3}(\tfrac{1}{2})^3(\tfrac{1}{2})^2+\binom{5}{5}(\tfrac{1}{2})^5\\
&= (\tfrac{1}{2})^5[5+10+1] = \tfrac{16}{32} = \tfrac{1}{2}\\
Pr(B) &= \tfrac{1}{2}\\
Pr(A\cap B) &= (\tfrac{1}{2})[\binom{4}{0}(\tfrac{1}{2})^4+\binom{4}{2}(\tfrac{1}{2})^2(\tfrac{1}{2})^2+\binom{4}{4}(\tfrac{1}{2})^4]\\
&= (\tfrac{1}{2})(\tfrac{1}{2})^4[1+6+1] = \tfrac{8}{32} = \tfrac{1}{4}
\end{aligned}$$

Since $Pr(A\cap B)=\frac{1}{4}=(\frac{1}{2})(\frac{1}{2})=Pr(A)Pr(B)$, the events A, B are independent.

13. Let A, B be the events

A: Paul initially selects a can of lemonade

B: Betty selects two cans of cola.

$$\begin{aligned}
Pr(A|B) &= \tfrac{Pr(A\cap B)}{Pr(B)} = \tfrac{Pr(A)Pr(B|A)}{Pr(B)}\\
Pr(A) &= \tfrac{3}{11}\\
Pr(B) &= Pr(A)Pr(B|A)+Pr(\overline{A})Pr(B|\overline{A})\\
&= (\tfrac{3}{11})(\tfrac{5}{13})(\tfrac{4}{12})+(\tfrac{8}{11})(\tfrac{6}{13})(\tfrac{5}{12})
\end{aligned}$$

So $Pr(A|B)=\frac{(\frac{3}{11})(\frac{5}{13})(\frac{4}{12})}{(\frac{3}{11})(\frac{5}{13})(\frac{4}{12})+(\frac{8}{11})(\frac{6}{13})(\frac{5}{12})}=\frac{60}{60+240}=\frac{6}{30}=\frac{1}{5}$.

15. Let A, B denote the events

A: the first component fails

B: the second component fails.

Here $Pr(A)=0.05$ and $Pr(B|A)=0.02$. The probability the electronic system fails is $Pr(A\cap B)=Pr(A)Pr(B|A)=(0.05)(0.02)=0.001$.

17. In general, $Pr(A\cup B\cup C)=Pr(A)+Pr(B)+Pr(C)-Pr(A\cap B)-Pr(A\cap C)-Pr(B\cap C)+Pr(A\cap B\cap C)$. Since A, B, C are independent we have $\frac{1}{2}=Pr(A\cup B\cup C)=Pr(A)+Pr(B)+Pr(C)-Pr(A)Pr(B)-Pr(A)Pr(C)-Pr(B)Pr(C)+Pr(A)Pr(B)Pr(C)=(\frac{1}{8})+(\frac{1}{4})+Pr(C)-(\frac{1}{8})(\frac{1}{4})-(\frac{1}{8})Pr(C)-(\frac{1}{4})Pr(C)+(\frac{1}{8})(\frac{1}{4})Pr(C)$.
Consequently, $\frac{1}{2}-\frac{1}{8}-\frac{1}{4}+\frac{1}{32}=[1-\frac{1}{8}-\frac{1}{4}+\frac{1}{32}]Pr(C)$ and $Pr(C)=\frac{5}{21}$.

19. Here $A = \{HH, HT\}$ and $Pr(A) = \frac{1}{2}$; $B = \{HT, TT\}$ with $Pr(B) = \frac{1}{2}$; and $C = \{HT, TH\}$ with $Pr(C) = \frac{1}{2}$.

Also $A \cap B = \{HT\}$, so $Pr(A \cap B) = \frac{1}{4} = (\frac{1}{2})(\frac{1}{2}) = Pr(A)Pr(B)$; $A \cap C = \{HT\}$, so $Pr(A \cap C) = \frac{1}{4} = (\frac{1}{2})(\frac{1}{2}) = Pr(A)Pr(C)$; and $B \cap C = \{HT\}$ with $Pr(B \cap C) = (\frac{1}{4}) = (\frac{1}{2})(\frac{1}{2}) = Pr(B)Pr(C)$. Consequently, any two of the events A, B, C are independent.

However, $A \cap B \cap C = \{HT\}$ so $Pr(A \cap B \cap C) = \frac{1}{4} \neq \frac{1}{8} = (\frac{1}{2})(\frac{1}{2})(\frac{1}{2}) = Pr(A)Pr(B)Pr(C)$. Consequently, the events A, B, C are *not* independent.

21. (a) For $0 \leq k \leq 3$, the probability of tossing k heads in three tosses is $\binom{3}{k}(\frac{1}{2})^k(\frac{1}{2})^{3-k} = \binom{3}{k}(\frac{1}{2})^3$. The probability Dustin and Jennifer each toss the same number of heads is $\sum_{k=0}^{3}[\binom{3}{k}\frac{1}{2})^3]^2 = (\frac{1}{2})^6[\binom{3}{0}^2 + \binom{3}{1}^2 + \binom{3}{2}^2 + \binom{3}{3}^2] = (\frac{1}{2})^6[1 + 9 + 9 + 1] = \frac{20}{64} = \frac{5}{16} \doteq 0.3125$.
(b) Let x count the number of heads in Dustin's three tosses and y the number in Jennifer's. Here we consider the cases where $x = 3$: $y = 2, 1$, or 0; $x = 2$: $y = 1$ or 0; $x = 1$: $y = 0$. The probability that Dustin gets more heads than Jennifer is $\binom{3}{3}(\frac{1}{2})^3[\binom{3}{2}(\frac{1}{2})^2(\frac{1}{2}) + \binom{3}{1}(\frac{1}{2})(\frac{1}{2})^2 + \binom{3}{0}(\frac{1}{2})^3] + \binom{3}{2}(\frac{1}{2})^2(\frac{1}{2})[\binom{3}{1}(\frac{1}{2})(\frac{1}{2})^2 + \binom{3}{0}(\frac{1}{2})^3] + \binom{3}{1}(\frac{1}{2})(\frac{1}{2})^2[\binom{3}{0}(\frac{1}{2})^3] = (\frac{1}{2})^6[3+3+1] + (\frac{1}{2})^6(3)[3+1] + (\frac{1}{2})^6(3)(1) = (\frac{1}{2})^6(22) = \frac{11}{32}$.
(c) Here the answer is likewise $\frac{11}{32}$.
[Note: The answers in parts (a), (b), and (c) sum to 1 because the union of the three events for these parts is the entire sample space and the events are disjoint in pairs. Consequently, upon recognizing how the answers in parts (b), (c) are related we see that the answer to part (b) is $(\frac{1}{2})[1 - \frac{5}{16}] = (\frac{1}{2})(\frac{11}{16}) = \frac{11}{32}$.]

23. Let A, B denote the following events:
A: A new airport-security employee has had prior training in weapon detection
B: A new airport-security employee fails to detect a weapon during the first month on the job.

Here $Pr(A) = 0.9$, $Pr(\overline{A}) = 0.1$, $Pr(B|\overline{A}) = 0.03$ and $Pr(B|A) = 0.005$.

The probability a new airport-security employee, who fails to detect a weapon during the first month on the job, has had prior training in weapon detection $= Pr(A|B) = \frac{Pr(A \cap B)}{Pr(B)} = \frac{Pr(A)Pr(B|A)}{Pr(B \cap A) + Pr(B \cap \overline{A})} = \frac{Pr(A)Pr(B|A)}{Pr(A)Pr(B|A) + Pr(\overline{A})Pr(B|\overline{A})} = (0.9)(0.005)/[0.9)(0.005) + (0.1)(0.03)] = 0.0045/[0.0045 + 0.003] = \frac{45}{75} = \frac{3}{5} = 0.6$.

25. (a) There are $\binom{5}{2} = 10$ conditions – one for each pair of events; $\binom{5}{3} = 10$ conditions – one for each triple of events; $\binom{5}{4} = 5$ conditions – one for each quadruple of events; and $\binom{5}{5} = 1$ condition for all five of the events. In total there are 26 [$= 2^5 - \binom{5}{0} - \binom{5}{1}$] conditions to be checked.
(b) $2^n - \binom{n}{0} - \binom{n}{1} = 2^n - 1 - n = 2^n - (n + 1)$ conditions must be checked to establish the independence of n events.

27. Let B_0, B_1, B_2, B_3, and denote A denote the events

B_i: for the three envelopes randomly selected from urn 1 and transferred to urn 2, i

envelopes each contain \$1 while the other $3-i$ envelopes each contain \$5, where $0 \le i \le 3$.
A: Carmen's selection from urn 2 is an envelope that contains \$1.

Here, $Pr(A) = Pr(A \cap B_0) + Pr(A \cap B_1) + Pr(A \cap B_2) + Pr(A \cap B_3)$
$= Pr(B_0)Pr(A|B_0) + Pr(B_1)Pr(A|B_1) + Pr(B_2)Pr(A|B_2) + Pr(B_3)Pr(A|B_3)$
$= [\binom{6}{0}\binom{8}{3}/\binom{14}{3}](\frac{3}{11}) + [\binom{6}{1}\binom{8}{2}/\binom{14}{3}](\frac{4}{11}) + [\binom{6}{2}\binom{8}{1}/\binom{14}{3}](\frac{5}{11}) + [(\binom{6}{3}\binom{8}{0}/\binom{14}{3}](\frac{6}{11}) =$
$(\frac{2}{13})(\frac{3}{11}) + (\frac{6}{13})(\frac{4}{11}) + (\frac{30}{91})(\frac{5}{11}) + (\frac{5}{91})(\frac{6}{11}) =$
$(\frac{1}{91})(\frac{1}{11})[42 + 168 + 150 + 30] = 390/1001 = 30/77.$

29. $0.8 = Pr(A|B) + Pr(B|A) = \frac{Pr(A \cap B)}{Pr(B)} + \frac{Pr(B \cap A)}{Pr(A)} = Pr(A \cap B)[(1/0.3) + (1/0.5)]$, so $(0.15)(0.8) = Pr(A \cap B)[0.5 + 0.3] = (0.8)Pr(A \cap B)$. Consequently, $Pr(A \cap B) = 0.15$.

Section 3.7

1. (a) $Pr(X = 3) = \frac{1}{4}$
(b) $Pr(X \le 4) = \sum_{x=0}^{4} Pr(X = x) = \frac{1}{8} + \frac{1}{4} + \frac{1}{4} + \frac{1}{4} + \frac{1}{8} = 1$
(c) $Pr(X > 0) = \sum_{x=1}^{4} Pr(X = x) = \frac{1}{4} + \frac{1}{4} + \frac{1}{4} + \frac{1}{8} = \frac{7}{8}$
(d) $Pr(1 \le X \le 3) = \sum_{x=1}^{3} Pr(X = x) = \frac{1}{4} + \frac{1}{4} + \frac{1}{4} = \frac{3}{4}$
(e) $Pr(X = 2|X \le 3) = \dfrac{Pr(X = 2 \text{ and } X \le 3)}{Pr(X \le 3)} = \dfrac{Pr(X = 2)}{Pr(X \le 3)} = (\frac{1}{4})/[\frac{1}{8} + \frac{1}{4} + \frac{1}{4} + \frac{1}{4}] =$
$(\frac{1}{4})/(\frac{7}{8}) = (\frac{1}{4})(\frac{8}{7}) = \frac{2}{7}$
(f) $Pr(X \le 1 \text{ or } X = 4) = Pr(X = 0) + Pr(X = 1) + Pr(X = 4) = \frac{1}{8} + \frac{1}{4} + \frac{1}{8} = \frac{1}{2}$

3. (a) $Pr(X = x) = \dfrac{\binom{10}{x}\binom{110}{5-x}}{\binom{120}{5}}$, $x = 0, 1, 2, \ldots, 5$.
(b) $Pr(X = 4) = \dfrac{\binom{10}{4}\binom{110}{1}}{\binom{120}{5}} = \frac{(210)(110)}{190{,}578{,}024} = \frac{23{,}100}{190{,}578{,}024} = \frac{275}{2{,}268{,}786} \doteq 0.000121$
(c) $Pr(X \ge 4) = Pr(X = 4) + Pr(X = 5) = \dfrac{\binom{10}{4}\binom{110}{1}}{\binom{120}{5}} + \dfrac{\binom{10}{5}\binom{110}{0}}{\binom{120}{5}} = \frac{23{,}100+252}{190{,}578{,}024} = \frac{23{,}352}{190{,}578{,}024} =$
$\frac{139}{1{,}134{,}393} \doteq 0.000123.$
(d) $Pr(X = 1|X \le 2) = \dfrac{Pr(X = 1 \text{ and } X \le 2)}{Pr(X \le 2)} = \dfrac{Pr(X = 1)}{Pr(X \le 2)}$
$= \dfrac{\binom{10}{1}\binom{110}{4}/\binom{120}{5}}{[(\binom{10}{0}\binom{110}{5}/\binom{120}{5}) + (\binom{10}{1}\binom{110}{4}/\binom{120}{5}) + (\binom{10}{2}\binom{110}{3}/\binom{120}{5})]}$
$= \binom{10}{1}\binom{110}{4}/[\binom{10}{0}\binom{110}{5} + \binom{10}{1}\binom{110}{4} + \binom{10}{2}\binom{110}{3}]$
$= (10)(5,773,185)/[(1)(122,391,522) + (10)(5,773,185) + (45)(215,820)]$
$= 57,731,850/[122,391,522 + 57,731,850 + 9,711,900]$
$= 57,731,850/189,835,272 = 2675/8796 \doteq 0.304116.$

5. (a) $Pr(X \ge 3) = \sum_{x=3}^{6} Pr(X = x) = Pr(X = 3) + Pr(X = 4) + Pr(X = 5) + Pr(X = 6) = \frac{1}{6} + \frac{1}{6} + \frac{1}{6} + \frac{1}{6} = \frac{4}{6} = \frac{2}{3}$
(b) $Pr(2 \le X \le 5) = \sum_{x=2}^{5} Pr(X = x) = Pr(X = 2) + Pr(X = 3) + Pr(X = 4) + Pr(X = 5) = \frac{1}{6} + \frac{1}{6} + \frac{1}{6} + \frac{1}{6} = \frac{4}{6} = \frac{2}{3}$

(c) $Pr(X=4|X\geq 3)=\dfrac{Pr(X=4 \text{ and } X\geq 3)}{Pr(X\geq 3)}=\dfrac{Pr(X=4)}{Pr(X\geq 3)}=(1/6)/(4/6)=1/4.$
(d) $E(X)=\sum_{x=1}^{6} xPr(X=x)=\sum_{x=1}^{6} x\cdot(\frac{1}{6})=(\frac{1}{6})(1+2+3+4+5+6)=(\frac{1}{6})(21)=7/2.$
(e) $E(X^2)=\sum_{x=1}^{6} x^2Pr(X=x)=(\frac{1}{6})(1+4+9+16+25+36)=(\frac{1}{6})(91)=\frac{91}{6}$

$$\begin{aligned}\text{Var }(X) &= E(X^2)-E(X)^2\\ &= (\tfrac{91}{6})-(\tfrac{7}{2})^2=\tfrac{91}{6}-\tfrac{49}{4}=\tfrac{182-147}{12}=\tfrac{35}{12}\end{aligned}$$

7. (a) $1=\sum_{x=1}^{5} Pr(X=x)=c\sum_{x=1}^{5}(6-x)=c(5+4+3+2+1)=15c$, so $c=1/15$.
(b) $Pr(X\leq 2)=Pr(X=1)+Pr(X=2)=(\frac{1}{15})(6-1)+(\frac{1}{15})(6-2)=\frac{9}{15}=\frac{3}{5}$
(c) $E(X)=\sum_{x=1}^{5} x\cdot Pr(X=x)=\sum_{x=1}^{5} x\cdot(\frac{1}{15})(6-x)$
$=(\frac{1}{15})[1\cdot 5+2\cdot 4+3\cdot 3+4\cdot 2+5\cdot 1]=(\frac{1}{15})(35)=\frac{7}{3}$
(d) $E(X^2)=\sum_{x=1}^{5} x^2\cdot(\frac{1}{15})(6-x)=$
$(\frac{1}{15})[1\cdot 5+4\cdot 4+9\cdot 3+16\cdot 2+25\cdot 1]=(\frac{1}{15})(105)=7$
$\text{Var }(X)=E(X^2)-E(X)^2=7-(\frac{7}{3})^2=\frac{63-49}{9}=\frac{14}{9}$

9. Since X is binomial, $E(X)=70=np$ and $\text{Var }(X)=45.5=npq$. Hence, we find that $45.5=70q$, so $q=45.5/70=0.65$. Consequently, it follows that $p=0.35$ and $n=70/p=70/0.35=200$.

11. Here X is binomial with $n=8$ and $p=0.25$.
(a) $Pr(X=0)=\binom{8}{0}(0.25)^0(0.75)^8\doteq 0.100113$
(b) $Pr(X=3)=\binom{8}{3}(0.25)^3(0.75)^5\doteq 0.207642$
(c) $Pr(X\geq 6)=Pr(X=6)+Pr(X=7)+Pr(X=8)=\binom{8}{6}(0.25)^6(0.75)^2+\binom{8}{7}(0.25)^7(0.75)^1+\binom{8}{8}(0.25)^8(0.75)^0\doteq 0.004227.$
(d) $Pr(X\geq 6|X\geq 4)=\dfrac{Pr(X\geq 6 \text{ and } X\geq 4)}{Pr(X\geq 4)}=\dfrac{Pr(X\geq 6)}{Pr(X\geq 4)}$
$Pr(X\geq 4)=\sum_{x=4}^{8}\binom{8}{x}(0.25)^x(0.75)^{8-x}=\binom{8}{4}(0.25)^4(0.75)^4+\binom{8}{5}(0.25)^5(0.75)^3+\binom{8}{6}(0.25)^6(0.75)^2+\binom{8}{7}(0.25)^7(0.75)^1+\binom{8}{8}(0.25)^8(0.75)^0\doteq 0.113815$
So $Pr(X\geq 6|X\geq 4)\doteq 0.004227/0.113815\doteq 0.037139$.
(e) $E(X)=np=8(0.25)=2$.
(f) $\text{Var }(X)=np(1-p)=8(0.25)(0.75)=1.5$

13. In Chebyshev's Inequality $Pr(|X-E(X)|\leq k\sigma_X)\geq 1-\frac{1}{k^2}$. If $1-\frac{1}{k^2}=0.96$, then $1-0.96=0.04=\frac{1}{k^2}$, and $k^2=\frac{1}{0.04}$. Since $k>0$ we have $k=\frac{1}{0.2}=5$.
Here $\text{Var }(X)=4$ so $\sigma_X=2$ and $c=k\sigma_X=5\cdot 2=10$.

15. Let D denote a defective chip and G a good one. Then the sample space $\mathcal{S}=\{D, GD, GGD, GGG\}$ and $X(D)=1$, $X(GD)=2$, and $X(GGD)=X(GGG)=3$.
(a)
$$\begin{aligned}Pr(X=1) &= \tfrac{4}{20}=\tfrac{1}{5}\\ Pr(X=2) &= (\tfrac{16}{20})(\tfrac{4}{19})=\tfrac{16}{95}\\ Pr(X=3) &= (\tfrac{16}{20})(\tfrac{15}{19})(\tfrac{4}{18})+(\tfrac{16}{20})(\tfrac{15}{19})(\tfrac{14}{18})=\tfrac{12}{19}\end{aligned}$$
(b) $Pr(X\leq 2)=Pr(X=1)+Pr(X=2)=\frac{1}{5}+\frac{16}{95}=\frac{35}{95}=\frac{7}{19}$

(c) $Pr(X=1|X\le 2) = \dfrac{Pr(X=1 \text{ and } X\le 2)}{Pr(X\le 2)} = \dfrac{Pr(X=1)}{Pr(X\le 2)} = (\frac{1}{5})/(\frac{7}{19}) = \frac{19}{35}$

(d) $E(X) = \sum_{x=1}^{3} xPr(X=x) = 1(\frac{1}{5}) + 2(\frac{16}{95}) + 3(\frac{12}{19}) = \frac{1}{5} + \frac{32}{95} + \frac{36}{19} = \frac{19+32+180}{95} = \frac{231}{95} =$ 2.431579

(e) $E(X^2) = \sum_{x=1}^{3} x^2Pr(X=x) = 1(\frac{1}{5}) + 4(\frac{16}{95}) + 9(\frac{12}{19}) = \frac{1}{5} + \frac{64}{95} + \frac{108}{19} = \frac{19+64+540}{95} = \frac{623}{95}$

$$\text{Var }(X) = E(X^2) - E(X)^2 = \frac{623}{95} - \left(\frac{231}{95}\right)^2 = \frac{5824}{(95)^2} = \frac{5824}{9025} \doteq 0.645319$$

17. (a) $E(X(X-1)) = \sum_{x=0}^{n} x(x-1)Pr(X=x)$
$= \sum_{x=2}^{n} x(x-1)Pr(X=x) = \sum_{x=2}^{n} x(x-1)\binom{n}{x}p^x q^{n-x}$
$= \sum_{x=2}^{n} \frac{n!}{x!(n-x)!}x(x-1)p^x q^{n-x}$
$= \sum_{x=2}^{n} \frac{n!}{(x-2)!(n-x)!}p^x q^{n-x} = p^2n(n-1)\sum_{x=2}^{n} \frac{(n-2)!}{(x-2)!(n-x)!}p^{x-2}q^{n-x}$
$= p^2n(n-1)\sum_{y=0}^{n-2} \frac{(n-2)!}{y![n-(y+2)]!}p^y q^{n-(y+2)}$, substituting $x-2=y$,
$= p^2n(n-1)\sum_{y=0}^{n-2} \frac{(n-2)!}{y![(n-2)-y]!}p^y q^{(n-2)-y}$
$= p^2n(n-1)(p+q)^{n-2}$, by the Binomial Theorem
$= p^2n(n-1)(1)^{n-2} = p^2n(n-1) = n^2p^2 - np^2$
(b) $\text{Var }(X) = E(X^2) - E(X)^2 = [E(X(X-1)) + E(X)] - E(X)^2 = [(n^2p^2 - np^2) + np] - (np)^2 = n^2p^2 - np^2 + np - n^2p^2 = np - np^2 = np(1-p) = npq.$

19.

(a)

Word	x, the number of letters and apostrophes in the word
I'll	4
make	4
him	3
an	2
offer	5
he	2
can't	5
refuse	6

x	$Pr(X=x)$
2	$2/8 = 1/4$
3	$1/8$
4	$2/8 = 1/4$
5	$2/8 = 1/4$
6	$1/8$

(b) $E(X) = \sum_{x=2}^{6} x \cdot Pr(X=x)$
$= 2(1/4) + 3(1/8) + 4(1/4) + 5(1/4) + 6(1/8)$
$= (1/8)[4+3+8+10+6] = 31/8$

(c) $E(X^2) = \sum_{x=2}^{6} x^2 \cdot Pr(X = x)$
$= 4(1/4) + 9(1/8) + 16(1/4) + 25(1/4) + 36(1/8)$
$= (1/8)[8 + 9 + 32 + 50 + 36] = 135/8$
$\text{Var}(X) = E(X^2) - E(X)^2 = (135/8) - (31/8)^2 = [1080 - 961]/64 = 119/64$

21. $Pr(X = 2) = [\binom{1}{1}\binom{1}{1}]/\binom{5}{2} = 1/10$ $\quad Pr(X = 3) = [\binom{2}{1}\binom{1}{1}]/\binom{5}{2} = 2/10$
$Pr(X = 4) = [\binom{3}{1}\binom{1}{1}]/\binom{5}{2} = 3/10$ $\quad Pr(X = 5) = [\binom{4}{1}\binom{1}{1}]/\binom{5}{2} = 4/10$
$E(X) = (1/10)(2)+(2/10)(3)+(3/10)(4)+(4/10)(5) = (1/10)[2+6+12+20] = 40/10 = 4$
$E(X^2) = (1/10)(4) + (2/10)(9) + (3/10)(16) + (4/10)(25) = (1/10)[4 + 18 + 48 + 100] = 170/10 = 17$
$\text{Var}(X) = E(X^2) - E(X)^2 = 17 - 16 = 1$, so $\sigma_X = \sqrt{1} = 1$.

Supplementary Exercises

1. Suppose that $(A - B) \subseteq C$ and $x \in A - C$. Then $x \in A$ but $x \notin C$. If $x \notin B$, then $[x \in A \wedge x \notin B] \Longrightarrow x \in (A - B) \subseteq C$. So now we have $x \notin C$ and $x \in C$. This contradiction gives us $x \in B$, so $(A - C) \subseteq B$.
Conversely, if $(A - C) \subseteq B$, let $y \in A - B$. Then $y \in A$ but $y \notin B$. If $y \notin C$, then $[y \in A \wedge y \notin C] \Longrightarrow y \in (A - C) \subseteq B$. This contradiction, i.e., $y \notin B$ and $y \in B$, yields $y \in C$, so $(A - B) \subseteq C$.

3. (a) $\mathcal{U} = \{1,2,3\}$, $A = \{1,2\}$, $B = \{1\}$, $C = \{2\}$ provide a counterexample.
(b) $A = A \cap \mathcal{U} = A \cap (C \cup \overline{C}) = (A \cap C) \cup (A \cap \overline{C}) = (A \cap C) \cup (A - C) = (B \cap C) \cup (B - C) = (B \cap C) \cup (B \cap \overline{C}) = B \cap (C \cup \overline{C}) = B \cap \mathcal{U} = B$
(c) The set assignments for part (a) also provide a counterexample for this situation.

5. (a) 126 (if teams wear different uniforms); 63 (if teams are not distinguishable).
(b) $2^n - 2; (1/2)(2^n - 2)$. $2^n - 2 - 2n; (1/2)(2^n - 2 - 2n)$.

7. (a) 128 $\quad$ (b) $|A| = 8$

9. Suppose that $(A \cap B) \cup C = A \cap (B \cup C)$ and that $x \in C$. Then $x \in C \Longrightarrow x \in (A \cap B) \cup C \Longrightarrow x \in A \cap (B \cup C) \subseteq A$, so $x \in A$, and $C \subseteq A$.
Conversely, suppose that $C \subseteq A$.
(1) If $y \in (A \cap B) \cup C$, then $y \in A \cap B$ or $y \in C$. (i) $y \in A \cap B \Longrightarrow y \in (A \cap B) \cup (A \cap C) \Longrightarrow y \in A \cap (B \cup C)$. (ii) $y \in C \Longrightarrow y \in A$, since $C \subseteq A$. Also, $y \in C \Longrightarrow y \in B \cup C$. So $y \in A \cap (B \cup C)$. In either case ((i) or (ii)) we have $y \in A \cap (B \cup C)$, so $(A \cap B) \cup C \subseteq A \cap (B \cup C)$.
(2) Now let $z \in A \cap (B \cup C)$. Then $z \in A \cap (B \cup C) = (A \cap B) \cup (A \cap C) \subseteq (A \cap B) \cup C$.
From (1) and (2) it follows that $(A \cap B) \cup C = A \cap (B \cup C)$.

11. (a) $[0, 14/3]$ $\quad$ (b) $\{0\} \cup (6, 12]$ $\quad$ (c) $[0, +\infty)$ $\quad$ (d) $\emptyset$

13.

(a)

	A	B	$A \cap B$
$\longrightarrow$	0	0	0
$\longrightarrow$	0	1	0
	1	0	0
$\longrightarrow$	1	1	1

Since $A \subseteq B$, we only consider rows 1,2, and 4 of the table. In these rows A and $A \cap B$ have the same column of results, so $A \subseteq B \Longrightarrow A = A \cap B$.

(c)

	A	B	C	$A \cap \overline{C}$	$A \cap \overline{B}$	$B \cap \overline{C}$	$(A \cap \overline{B}) \cup (B \cap \overline{C})$
$\longrightarrow$	0	0	0	0	0	0	0
	0	0	1	0	0	0	0
	0	1	0	0	0	1	1
	0	1	1	0	0	0	0
$\longrightarrow$	1	0	0	1	1	0	1
	1	0	1	0	1	0	1
$\longrightarrow$	1	1	0	1	0	1	1
$\longrightarrow$	1	1	1	0	0	0	0

We consider only rows 1,5,7, and 8. There $A \cap \overline{C} = (A \cap \overline{B}) \cup (B \cap \overline{C})$.

(b) & (d) The results for these parts are derived in a similar manner.

15. (a) The r 0's determine $r+1$ locations for the m individual 1's. If $r+1 \geq m$, we can select these locations in $\binom{r+1}{m}$ ways.

(b) Using part (a), here we have k 1's (for the elements of A) and $n-k$ 0's (for the elements in $\mathcal{U} - A$). The $n-k$ 0's provide $n-k+1$ locations for the k 1's so that no two are adjacent. These k locations can be selected in $\binom{n-k+1}{k}$ ways if $n-k+1 \geq k$ or $2k \leq n+1$. So there are $\binom{n-k+1}{k}$ subsets A of $\mathcal{U}$ with $|A| = k$ and such that A contains no consecutive integers.

17.

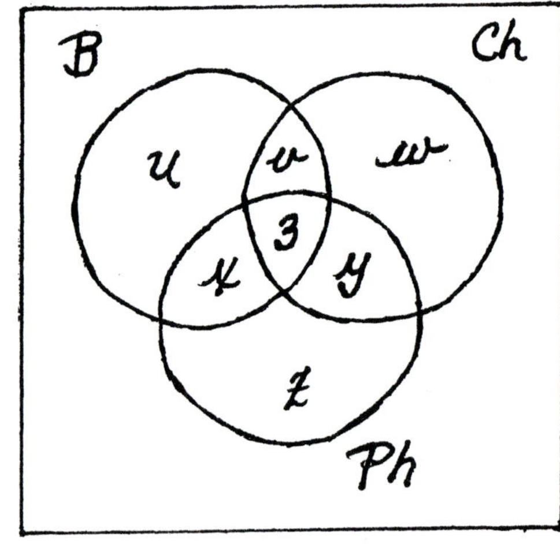

Let u, v, w, x, y, z count the following subsets of students:

$u =$ the number who receive an award only in biology
$w =$ the number who receive an award only in chemistry
$z =$ the number who receive an award only in physics
$v =$ the number who receive awards in biology and chemistry, but not physics
$x =$ the number who receive awards in biology and physics, but not chemistry
$y =$ the number who receive awards in chemistry and physics, but not biology

Then

$u + v + x + 3 = 14$ (Biology)

$$v + w + y + 3 = 13 \qquad \text{(Chemistry)}$$
$$x + y + z + 3 = 21 \qquad \text{(Physics)}$$

Adding these equations we find that $(u + w + z) + 2(v + x + y) = 39$. Also, for the total of 34 we have $34 = u + v + w + x + y + z + 3$, or $(u + w + z) + (v + x + y) = 31$. Let $a = u + w + z$ and $b = v + x + y$. Then from $a + 2b = 39$ and $a + b = 31$ it follows that (a) $a = 23$ and (b) $b = 8$.

19.

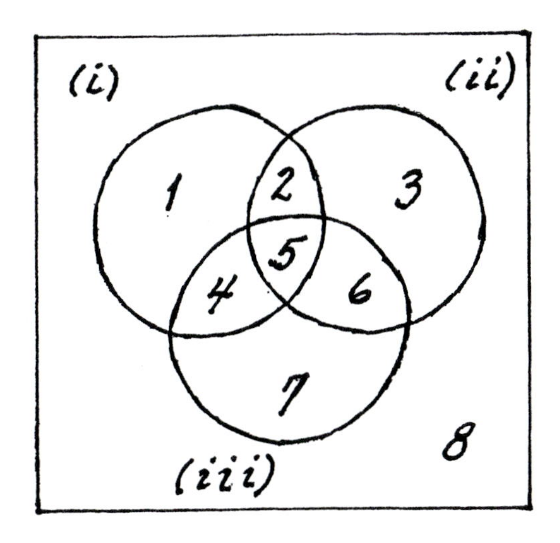

For the given figure let circles (i), (ii), and (iii) denote the subset of assignments where no one is working on experiments 1,2,3, respectively. For each assistant there are seven possibilities: the seven nonempty subsets of $\{1,2,3\}$. So there are 7^{15} possible assignments. To determine the number of assignments in region 8 we need to determine the number of assignments in the union of the three subsets. Region 5 has 0 elements, while regions 2,4,6 each contain 1 element (e.g., for region 2, if all assistants are assigned only to experiment 3 then this is the one way that everyone is working on an experiment, but no one is working on experiments 1 and 2).

In each of regions 1,3,7 there are $3^{15} - 2$ elements (e.g., for regions 1,2,4,5 there are 3 cases to consider where no one is working on experiment 1 – for each assistant can be working on only experiment 2 or only experiment 3 or both experiments 2,3). The number of assignments where at least one person is working on every experiment is $7^{15} - 3[3^{15} - 2] - 3$.

21. Since $|A \cap B| = 0$, $|A \cup B| = 12 + 10 = 22$. There are $\binom{22}{7}$ ways to select seven elements from $A \cup B$. Among these selections $\binom{12}{4}\binom{10}{3}$ contain four elements from A and three from B. Consequently, the probability sought here is $\binom{12}{4}\binom{10}{3}/\binom{22}{7} = (495)(120)/(170,544) \doteq 0.3483$.

23. (a)

$$\underbrace{\binom{16}{8}}_{\text{no diagonal moves}} + \underbrace{\binom{14}{7}}_{\text{one diagonal move}}\binom{15}{1} + \underbrace{\binom{12}{6}}_{\text{two diagonal moves}}\binom{13+2-1}{2}$$

$$+ \underbrace{\binom{10}{5}}_{\text{three diagonal moves}}\binom{11+3-1}{3} + \underbrace{\binom{8}{4}}_{\text{four diagonal moves}}\binom{9+4-1}{4} + \underbrace{\binom{6}{3}}_{\text{five diagonal moves}}\binom{7+5-1}{5}$$

$$+ \underbrace{\binom{4}{2}}_{\text{six diagonal moves}}\binom{5+6-1}{6} + \underbrace{\binom{2}{1}}_{\text{seven diagonal moves}}\binom{3+7-1}{7} + \underbrace{\binom{0}{0}}_{\text{eight diagonal moves}}\binom{1+8-1}{8}$$

$$= \sum_{i=0}^{8} \binom{2i}{i} \binom{(2i+1)+(8-i)-1}{8-i} = \sum_{i=0}^{8} \binom{2i}{i}\binom{8+i}{8-i}$$

(b) (i) $\binom{12}{6}\binom{14}{2} / \sum_{i=0}^{8} \binom{2i}{i}\binom{8+i}{8-i}$

(ii) $\binom{12}{6}\binom{13}{1} / \sum_{i=0}^{8} \binom{2i}{i}\binom{8+i}{8-i}$

(iii) $[\binom{16}{8} + \binom{12}{6}\binom{14}{2} + \binom{8}{4}\binom{12}{4} + \binom{4}{2}\binom{10}{6} + \binom{0}{0}\binom{8}{8}] / \sum_{i=0}^{8} \binom{2i}{i}\binom{8+i}{8-i}$

25. $x^2 - 7x \leq -12 \Rightarrow x^2 - 7x + 12 \leq 0 \Rightarrow (x-3)(x-4) \leq 0 \Rightarrow [(x-3) \leq 0 \text{ and } (x-4) \geq 0]$ or $[(x-3) \geq 0 \text{ and } (x-4) \leq 0] \Rightarrow [x \leq 3 \text{ and } x \geq 4]$ or $[x \geq 3 \text{ and } x \leq 4] \Rightarrow 3 \leq x \leq 4$, so $A = \{x|3 \leq x \leq 4\} = [3,4]$.

$x^2 - x \leq 6 \Rightarrow x^2 - x - 6 \leq 0 \Rightarrow (x-3)(x+2) \leq 0 \Rightarrow [(x-3) \leq 0 \text{ and } (x+2) \geq 0]$ or $[(x-3) \geq 0 \text{ and } (x+2) \leq 0] \Rightarrow [x \leq 3 \text{ and } x \geq -2]$ or $[x \geq 3 \text{ and } x \geq -2] \Rightarrow -2 \leq x \leq 3$, so $B = \{x|-2 \leq x \leq 3\} = [-2,3]$.

Consequently, $A \cap B = \{3\}$ and $A \cup B = [-2,4]$.

27. There are two cases to consider.
(1) The one tail is obtained on the fair coin. The probability for this is $\binom{2}{1}(\frac{1}{2})(\frac{1}{2})\binom{4}{4}(\frac{3}{4})^4$.
(2) The one tail is obtained on the biased coin. The probability in this case is $\binom{2}{2}(\frac{1}{2})^2\binom{4}{3}(\frac{3}{4})^3(\frac{1}{4})$.

Consequently, the answer is the sum of these two probabilities – namely, $\frac{81}{512} + \frac{108}{1024} = \frac{135}{512} \doteq 0.263672$.

29. $Pr(A \cap (B \cup C)) = Pr((A \cap B) \cup (A \cap C)) = Pr(A \cap B) + Pr(A \cap C) - Pr((A \cap B) \cap (A \cap C))$. Since A, B, C are independent and $(A \cap B) \cap (A \cap C) = (A \cap A) \cap (B \cap C) = A \cap B \cap C$,

$Pr(A \cap (B \cup C)) = Pr(A)Pr(B) + Pr(A)Pr(C) - Pr(A)Pr(B)Pr(C) = Pr(A)[Pr(B) + Pr(C) - Pr(B)Pr(C)] = Pr(A)[Pr(B) + Pr(C) - Pr(B \cap C] = Pr(A)Pr(B \cup C)$, so A and $B \cup C$ are independent.

31. (a) The probability that both tires in any single landing gear blow out is $(0.1)(0.1) = 0.01$. So the probability a landing gear will survive even a hard landing with at least one good tire is $1 - 0.01 = 0.99$.
(b) Assuming the landing gears operate independently of each other, the probability that the jet will be able to land safely even on a hard landing is $(0.99)^3 = 0.970299$.

33. Let A, B denote the events
A: the exit door is open
B: Marlo's selection of two keys includes the one key that opens the exit door

The answer then is given as $Pr(A) + Pr(\overline{A} \cap B) = Pr(A) + Pr(\overline{A})Pr(B) = (\frac{1}{2}) + (\frac{1}{2})[\binom{1}{1}\binom{9}{1}/\binom{10}{2}] = (\frac{1}{2}) + (\frac{1}{2})(\frac{1}{5}) = (\frac{1}{2}) + (\frac{1}{10}) = \frac{6}{10} = \frac{3}{5}$

35. $\binom{5}{3}(0.8)^3(0.2)^2 + \binom{5}{4}(0.8)^4(0.2) + \binom{5}{5}(0.8)^5 = 0.2048 + 0.4096 + 0.32768 = 0.94208$

37. Success: one head and two tails – the probability for this is $\binom{3}{1}(\frac{1}{2})^1(\frac{1}{2})^2 = \frac{3}{8}$.
$n = 4$, $p = \frac{3}{8}$
Among the four trials (of tossing three fair coins) we want two successes. The probability for this is $\binom{4}{2}(\frac{3}{8})^2(\frac{5}{8})^2 = (6)(9)(25)/2^{12} = 675/2^{11} = \frac{675}{2048}$.

39. (a) $1 = \sum_{x=0}^{4} Pr(X = x) = c(0+4) + c(1+4) + c(4+4) + c(9+4) + c(16+4) = c[4+5+8+13+20] = 50c$, so $c = \frac{1}{50} = 0.02$
(b) $Pr(X > 1) = Pr(X \geq 2) = Pr(X = 2) + Pr(X = 3) + Pr(X = 4) = (0.02)(8) + (0.02)(13) + (0.02)(20) = 0.02(8 + 13 + 20) = (0.02)(41) = 0.82$
(c) $Pr(X = 3|X \geq 2) = \dfrac{Pr(X = 3 \text{ and } X \geq 2)}{Pr(X \geq 2)} = \dfrac{Pr(X = 3)}{Pr(X \geq 2)} = (0.02)(13)/(0.02)(41) = \dfrac{13}{41} \doteq 0.317073$
(d) $E(X) = \sum_{x=0}^{4} x \cdot Pr(X = x) = 0 \cdot (c)(4) + 1 \cdot (c)(5) + 2 \cdot (c)(8) + 3 \cdot (c)(13) + 4 \cdot (c)(20) = c[5 + 16 + 39 + 80] = 140c = 2.8$
(e) $E(X^2) = \sum_{x=0}^{4} x^2 \cdot Pr(X = x) = 0^2 \cdot (4c) + 1^2 \cdot (5c) + 2^2 \cdot (8c) + 3^2 \cdot (13c) + 4^2 \cdot (20c) = 474c = 9.48$

$\text{Var}(X) = E(X^2) - E(X)^2 = 9.48 - (2.8)^2 = 1.64$

41. (a) To finish with a straight flush, Maureen must draw (i) the 4 and 5 of diamonds; (ii) the 5 and 9 of diamonds; or (iii) the 9 and 10 of diamonds. The probability for each of these three situations is $\binom{2}{2}/\binom{47}{2}$, so the answer is $3/\binom{47}{2}$.
(b) Maureen will finish with a flush if she draws any two of the remaining ten diamonds, which she can do in $\binom{10}{2}$ ways. However, for three choices [as described in part (a)], she actually finishes with a straight flush. Consequently, the answer here is $[\binom{10}{2} - 3]/\binom{47}{2}$.

(c) To finish with a straight from 4 to 8, Maureen must select one of the four 4s and one of the four 5s. This she can do in $\binom{4}{1}\binom{4}{1}$ ways. For the straights from 5 to 9 and 6 to 10 there are likewise $\binom{4}{1}\binom{4}{1}$ possibilities. However, these $3\binom{4}{1}\binom{4}{1}$ straights include three straight flushes, so the answer is $[3\binom{4}{1}\binom{4}{1} - 3]/\binom{47}{2}$.

43. The total number of chips in the grab bag is $1 + 2 + 3 + \cdots + n = n(n + 1)/2$, and the probability a chip with i on it is selected is $2i/[n(n + 1)]$. Let A, B be the events.
A: the chip with 1 on it is selected
B: a red chip is selected.
$Pr(A|B) = Pr(A \cap B)/Pr(B) = Pr(A)/Pr(B)$.
$Pr(A) = 1/[n(n + 1)/2] = 2/[n(n + 1)]$.
$Pr(B) = [1+2+3+\cdots+m]/[n(n+1)/2] = [m(m+1)/2]/[n(n+1)/2] = [m(m+1)]/[n(n+1)]$.
Consequently, $Pr(A|B) = \frac{2/[n(n+1)]}{[m(m+1)]/n(n+1)]} = 2/[m(m + 1)]$.

45.

(a)

Outcome	Probability of Outcome	x, the number of runs
HHH	$(3/4)^3 = 27/64$	1
HHT	$(3/4)^2(1/4) = 9/64$	2
HTH	$(3/4)(1/4)(3/4) = 9/64$	3
THH	$(1/4)(3/4)^2 = 9/64$	2
HTT	$(3/4)(1/4)^2 = 3/64$	2
THT	$(1/4)(3/4)(1/4) = 3/64$	3
TTH	$(1/4)^2(3/4) = 3/64$	2
TTT	$(1/4)^3 = 1/64$	1

The probability distribution for X:

x	$Pr(X = x)$
1	$(27/64) + (1/64) = 28/64 = 7/16$
2	$(9/64) + (9/64) + (3/64) + (3/64) = 24/64 = 3/8$
3	$(9/64) + (3/64) = 12/64 = 3/16$

(b) $E(X) = \sum_{x=1}^{3} x \cdot Pr(X = x) = (1)(7/16) + (2)(3/8) + (3)(3/16)$
$= (1/16)[7 + 12 + 9] = 28/16 = 7/4$

(c) $E(X^2) = \sum_{x=1}^{3} x^2 \cdot Pr(X = x) = (1)(7/16) + (4)(3/8) + (9)(3/16)$
$= (1/16)[7 + 24 + 27] = 58/16 = 29/8$

Var $(X) = E(X^2) - E(X)^2 = (29/8) - (7/4)^2 = (29/8) - (49/16)$
$= (58 - 49)/16 = 9/16$

So $\sigma_X = \sqrt{9/16} = 3/4$.

CHAPTER 4
PROPERTIES OF THE INTEGERS: MATHEMATICAL INDUCTION

Section 4.1

1. (a) $S(n): 1^2 + 3^2 + 5^2 + \ldots + (2n-1)^2 = (n)(2n-1)(2n+1)/3$.
$S(1): 1^2 = (1)(1)(3)/3$. This is true.
Assume $S(k): 1^2 + 3^2 + \ldots + (2k-1)^2 = (k)(2k-1)(2k+1)/3$, for some $k \geq 1$.
Consider $S(k+1)$. $[1^2+3^2+\ldots+(2k-1)^2]+(2k+1)^2 = [(k)(2k-1)(2k+1)/3]+(2k+1)^2 = [(2k+1)/3][k(2k-1)+3(2k+1)] = [(2k+1)/3][2k^2+5k+3] = (k+1)(2k+1)(2k+3)/3$, so $S(k) \Longrightarrow S(k+1)$ and the result follows for all $n \in \mathbf{Z}^+$ by the Principle of Mathematical Induction.
(b) $S(n): 1 \cdot 3 + 2 \cdot 4 + 3 \cdot 5 + \cdots + n(n+2) = \frac{n(n+1)(2n+7)}{6}$.
$S(1): 1 \cdot 3 = (1)(2)(9)/6 = 18/6 = 3$, so $S(1)$ is true. Assume $S(k): 1 \cdot 3 + 2 \cdot 4 + 3 \cdot 5 + \cdots + k(k+2) = \frac{(k)(k+1)(2k+7)}{6}$, for some $k \geq 1$.
Consider $S(k+1)$. $[1 \cdot 3 + 2 \cdot 4 + 3 \cdot 5 + \cdots + k(k+2)] + (k+1)[(k+1)+2] = [k(k+1)(2k+7)/6]+(k+1)(k+3) = [k(k+1)(2k+7)+6(k+1)(k+3)]/6 = (k+1)[2k^2+7k+6k+18]/6 = (k+1)(k+2)(2k+9)/6 = (k+1)[(k+1)+1][2(k+1)+7]/6$, so $S(k) \Rightarrow S(k+1)$ and the result follows for all $n \in \mathbf{Z}^+$ by the Principle of Mathematical Induction.
(c) $S(n): \sum_{i=1}^{n} \frac{1}{i(i+1)} = \frac{n}{n+1}$
$S(1): \sum_{i=1}^{1} \frac{1}{i(i+1)} = \frac{1}{1(2)} = \frac{1}{1+1}$, so $S(1)$ is true.
Assume $S(k): \sum_{i=1}^{k} \frac{1}{i(i+1)} = \frac{k}{k+1}$. Consider $S(k+1)$.
$\sum_{i=1}^{k+1} \frac{1}{i(i+1)} = \sum_{i=1}^{k} \frac{1}{i(i+1)} + \frac{1}{(k+1)(k+2)} = \frac{k}{(k+1)} + \frac{1}{(k+1)(k+2)} = [k(k+2)+1]/[(k+1)(k+2)] = (k+1)/(k+2)$, so $S(k) \Longrightarrow S(k+1)$ and the result follows for all $n \in \mathbf{Z}^+$ by the Principle of Mathematical Induction.
(d) $S(n): \sum_{i=1}^{n} i^3 = \frac{n^2(n+1)^2}{4}$.
$S(1): \sum_{i=1}^{1} i^3 = 1 = 1^2(2)^2/4$, so $S(1)$ is true.
Assume $S(k): \sum_{i=1}^{k} i^3 = k^2(k+1)^2/4$, for some $k \geq 1$. Consider $S(k+1)$. $\sum_{i=1}^{k+1} i^3 = \sum_{i=1}^{k} i^3 + (k+1)^3 = [k^2(k+1)^2/4]+(k+1)^3 = (k+1)^2[k^2+4(k+1)]/4 = (k+1)^2(k+2)^2/4 = (k+1)^2[(k+1)+1]^2/4$, so $S(k) \Rightarrow S(k+1)$ and the result follows for all $n \in \mathbf{Z}^+$ by the Principle of Mathematical Induction.
From Example 4.1 we know that $\sum_{i=1}^{n} i = n(n+1)/2$, so $(\sum_{i=1}^{n} i)^2 = [n(n+1)/2]^2 = n^2(n+1)^2/4$.

3. (a) From $\sum_{i=1}^{n} i^3 + (n+1)^3 = \sum_{i=0}^{n}(i^3+3i^2+3i+1) = \sum_{i=1}^{n} i^3 + 3\sum_{i=1}^{n} i^2 + 3\sum_{i=1}^{n} i + \sum_{i=0}^{n} 1$, we have $(n+1)^3 = 3\sum_{i=1}^{n} i^2 + 3\sum_{i=1} i + (n+1)$. Consequently,

$$\begin{aligned} 3\sum_{i=1}^{n} i^2 &= (n^3+3n^2+3n+1) - 3[(n)(n+1)/2] - n - 1 \\ &= n^3 + (3/2)n^2 + (1/2)n \\ &= (1/2)[2n^3+3n^2+n] = (1/2)n(2n^2+3n+1) \\ &= (1/2)n(n+1)(2n+1), \text{ so} \end{aligned}$$

$\sum_{i=1}^{n} i^2 = (1/6)n(n+1)(2n+1)$ (as shown in Example 4.4).
(b) From $\sum_{i=1}^{n} i^4 + (n+1)^4 = \sum_{i=0}^{n}(i+1)^4 = \sum_{i=0}^{n}(i^4+4i^3+6i^2+4i+1) = \sum_{i=1}^{n} i^4 + 4\sum_{i=1}^{n} i^3 + 6\sum_{i=1}^{n} i^2 + 4\sum_{i=1}^{n} i + \sum_{i=0}^{n} 1$, it follows that $(n+1)^4 = 4\sum_{i=1}^{n} i^3 + 6\sum_{i=1}^{n} i^2 + 4\sum_{i=1}^{n} i + \sum_{i=0}^{n} 1$.
Consequently,
$4\sum_{i=1}^{n} i^3 = (n+1)^4 - 6[n(n+1)(2n+1)/6] - 4[n(n+1)/2] - (n+1) = n^4+4n^3+6n^2+4n+1-(2n^3+3n^2+n)-(2n^2+2n)-(n+1) = n^4+2n^3+n^2 = n^2(n^2+2n+1) = n^2(n+1)^2$.
So $\sum_{i=1}^{n} i^3 = (1/4)n^2(n+1)^2$ [as shown in part (d) of Exercise 1 for this section].

From $\sum_{i=1}^{n} i^5 + (n+1)^5 = \sum_{i=0}^{n}(i+1)^5 = \sum_{i=0}^{n}(i^5+5i^4+10i^3+10i^2+5i+1) = \sum_{i=1}^{n} i^5 + 5\sum_{i=1}^{n} i^4 + 10\sum_{i=1}^{n} i^3 + 10\sum_{i=1}^{n} i^2 + 5\sum_{i=1}^{n} i + \sum_{i=0}^{n} 1$, we have $5\sum_{i=1}^{n} i^4 = (n+1)^5 - (10/4)n^2(n+1)^2 - (10/6)n(n+1)(2n+1) - (5/2)n(n+1) - (n+1)$. So

$$\begin{aligned} 5\sum_{i=1}^{n} i^4 &= n^5+5n^4+10n^3+10n^2+5n+1-(5/2)n^4 \\ &\quad -5n^3-(5/2)n^2-(10/3)n^3-5n^2-(5/3)n-(5/2)n^2-(5/2)n-n-1 \\ &= n^5+(5/2)n^4+(5/3)n^3-(1/6)n. \end{aligned}$$

Consequently, $\sum_{i=1}^{n} i^4 = (1/30)n(n+1)(6n^3+9n^2+n-1)$.

5. (a) $\sum_{i=1}^{123} i = (123)(124)/2 = 7626$.
(b) $\sum_{i=1}^{123} i^2 = (123)(124)(2\cdot 123+1)/6 = 627{,}874$.

7.

$$\begin{aligned} 4n+110 &= 6+8+10+\cdots+[6+(n-1)2] \\ &= 6n+[0+2+4+\cdots+(n-1)2] \\ &= 6n+2[1+2+\cdots+(n-1)] \\ &= 6n+2[(n-1)(n)/2] \\ &= 6n+(n-1)(n) = n^2+5n \\ n^2+n-110 &= (n+11)(n-10) = 0, \end{aligned}$$

so $n = 10$ – the number of layers.

9. (a) $\sum_{i=11}^{33} i = \sum_{i=1}^{33} i - \sum_{i=1}^{10} i = [(33)(34)/2] - [(10)(11)/2] = 561 - 55 = 506$
(b) $\sum_{i=11}^{33} i^2 = \sum_{i=1}^{33} i^2 - \sum_{i=1}^{10} i^2 = [(33)(34)(67)/6] - [(10)(11)(21)/6] = 12144$

11. a) $\sum_{i=1}^{n} t_{2i} = \sum_{i=1}^{n} \frac{(2i)(2i+1)}{2} = \sum_{i=1}^{n}(2i^2+i) = 2\sum_{i=1}^{n} i^2 + \sum_{i=1}^{n} i = 2[(n)(n+1)(2n+1)/6] + [n(n+1)/2] = [n(n+1)(2n+1)/3] + [n(n+1)/2] = n(n+1)[\frac{2n+1}{3} + \frac{1}{2}] = n(n+1)[\frac{4n+5}{6}] = n(n+1)(4n+5)/6$.

b) $\sum_{i=1}^{100} t_{2i} = 100(101)(405)/6 = 681{,}750$.

c) **begin**
sum := 0

```
    for i := 1 to 100 do
        sum := sum + (2 * i) * (2 * i + 1)/2
    print sum
end
```

13. (a) There are $49(= 7^2)$ 2×2 squares and $36(= 6^2)$ 3×3 squares. In total there are $1^2 + 2^2 + 3^2 + \cdots + 8^2 = (8)(8+1)(2 \cdot 8 + 1)/6 = (8)(9)(17)/6 = 204$ squares.

(b) For each $1 \leq k \leq n$ the $n \times n$ chessboard contains $(n-k+1)^2$ $k \times k$ squares. In total there are $1^2 + 2^2 + 3^2 + \cdots + n^2 = n(n+1)(2n+1)/6$ squares.

15. For $n = 5, 2^5 = 32 > 25 = 5^2$. Assume the result for $n = k(\geq 5) : 2^k > k^2$. For $k > 2, k(k-2) > 1$, or $k^2 > 2k + 1$. But $2^k > k^2 \Longrightarrow 2^k + 2^k > k^2 + k^2 \Longrightarrow 2^{k+1} > k^2 + k^2 > k^2 + (2k+1) = (k+1)^2$. Hence the result is true for $n \geq 5$ by the Principle of Mathematical Induction.

17. (a) Once again we start at $n = 0$. Here we find that $1 = 1 + (0/2) \leq H_1 = H_{2^0}$, so this first case is true. Assuming the truth for $n = k(\in \mathbf{N})$ we obtain the induction hypothesis

$$1 + (k/2) \leq H_{2^k}.$$

Turning now to the case where $n = k+1$ we find $H_{2^{k+1}} = H_{2^k} + [1/(2^k+1)] + [1/(2^k+2)] + \ldots + [1/(2^k+2^k)] \geq H_{2^k} + [1/(2^k+2^k)] + [1/(2^k+2^k)] + \ldots + [1/(2^k+2^k)] = H_{2^k} + 2^k[1/2^{k+1}] = H_{2^k} + (1/2) \geq 1 + (k/2) + (1/2) = 1 + (k+1)/2$.

The result now follows for all $n \geq 0$ by the Principle of Mathematical Induction.

(b) Starting with $n = 1$ we find that

$$\sum_{j=1}^{1} jH_j = H_1 = 1 = [(2)(1)/2](3/2) - [(2)(1)/4] = [(2)(1)/2]H_2 - [(2)(1)/4].$$

Assuming the truth of the given statement for $n = k$, we have

$$\sum_{j=1}^{k} jH_j = [(k+1)(k)/2]H_{k+1} - [(k+1)(k)/4].$$

For $n = k+1$ we now find that

$$\sum_{j=1}^{k+1} jH_j = \sum_{j=1}^{k} jH_j + (k+1)H_{k+1}$$

$$= [(k+1)(k)/2]H_{k+1} - [(k+1)(k)/4] + (k+1)H_{k+1}$$

$$= (k+1)[1 + (k/2)]H_{k+1} - [(k+1)(k)/4]$$

$$= (k+1)[1 + (k/2)][H_{k+2} - (1/(k+2))] - [(k+1)(k)/4]$$

$= [(k+2)(k+1)/2]H_{k+2} - [(k+1)(k+2)]/[2(k+2)] - [(k+1)(k)/4]$

$= [(k+2)(k+1)/2]H_{k+2} - [(1/4)[2(k+1) + k(k+1)]]$

$= [(k+2)(k+1)/2]H_{k+2} - [(k+2)(k+1)/4].$

Consequently, by the Principle of Mathematical Induction, it follows that the given statement is true for all $n \in \mathbf{Z}^+$.

19. Assume $S(k)$ true for some $k \geq 1$. For $S(k+1), \sum_{i=1}^{k+1} i = [k + (1/2)]^2/2 + (k+1) = ((k^2+k) + (1/4) + 2k + 2)/2 = [(k+1)^2 + (k+1) + (1/4)]/2 = [(k+1) + (1/2)]^2/2$. S $S(k) \Longrightarrow S(k+1)$. However, we have no first value of k where $S(k)$ is true: For each $k \geq 1, \sum_{i=1}^{k} i = (k)(k+1)/2$ and $(k)(k+1)/2 = [k+(1/2)]^2/2 \Longrightarrow 0 = 1/4$.

21. For $x, n \in \mathbf{Z}^+$, let $S(n)$ denote the statement: If the program reaches the top of the **while** loop, after the two loop instructions are executed $n(> 0)$ times, then the value of the integer variable *answer* is $x(n!)$.
First consider $S(1)$, the statement for the case where $n = 1$. Here the program (if it reaches the top of the **while** loop) will result in one execution of the **while** loop: x will be assigned the value $x \cdot 1 = x(1!)$, and the value of n will be decreased to 0. With the value of n equal to 0 the loop is not processed again and the value of the variable *answer* is $x(1!)$. Hence $S(1)$ is true.
Now assume the truth for $n = k$: For $x, k \in \mathbf{Z}^+$, if the program reaches the top of the **while** loop, then upon exiting the loop, the value of the variable *answer* is $x(k!)$. To establish $S(k+1)$, if the program reaches the top of the **while** loop, then the following occur during the first execution:

The value assigned to the variable x is $x(k+1)$.
The value of n is decreased to $(k+1) - 1 = k$.

But then we can apply the induction hypothesis to the integers $x(k+1)$ and k, and after we exit the **while** loop for these values, the value of the variable *answer* is $(x(k+1))(k!) = x(k+1)!$
Consequently, $S(n)$ is true for all $n \geq 1$, and we have verified the correctness of this program segment by using the Principle of Mathematical Induction.

23. (a) The result is true for $n = 2, 4, 5, 6$. Assume the result is true for all $n = 2, 4, 5, \ldots, k-1, k$, where $k \geq 6$. If $n = k+1$, then $n = 2 + (k-1)$, and since the result is true for $k-1$, it follows by induction that it is true for $k+1$. Consequently, by the Alternative Form of the Principle of Mathematical Induction, every $n \in \mathbf{Z}^+, n \neq 1, 3$, can be written as a sum of 2's and 5's.

(b) $24 = 5+5+7+7$ $\quad 25 = 5+5+5+5+5$ $\quad 26 = 5+7+7+7$
$27 = 5+5+5+5+7$ $\quad 28 = 7+7+7+7$
Hence the result is true for all $24 \leq n \leq 28$. Assume the result true for $24 \leq n \leq 28 \leq k$, and consider $n = k+1$. Since $k+1 \geq 29$, we may write $k+1 = [(k+1)-5]+5 = (k-4)+5$, where $k-4$ can be expressed as a sum of 5's and 7's. Hence $k+1$ can be expressed as

such a sum and the result follows for all $n \geq 24$ by the Alternative Form of the Principle of Mathematical Induction.

25.

$$\begin{aligned}
E(X) &= \sum_x xPr(X=x) = \sum_{x=1}^n x(\tfrac{1}{n}) = (\tfrac{1}{n})\sum_{x=1}^n x = (\tfrac{1}{n})[\tfrac{n(n+1)}{2}] = \tfrac{n+1}{2} \\
E(X^2) &= \sum_x x^2 Pr(X=x) = \sum_{x=1}^n x^2(\tfrac{1}{n}) = (\tfrac{1}{n})\sum_{x=1}^n x^2 = (\tfrac{1}{n})[\tfrac{n(n+1)(2n+1)}{6}] = \tfrac{(n+1)(2n+1)}{6} \\
\text{Var}(X) &= E(X^2) - E(X)^2 = \tfrac{(n+1)(2n+1)}{6} - \tfrac{(n+1)^2}{4} = (n+1)[\tfrac{2n+1}{6} - \tfrac{n+1}{4}] \\
&= (n+1)[\tfrac{4n+2-(3n+3)}{12}] = \tfrac{(n+1)(n-1)}{12} = \tfrac{n^2-1}{12}.
\end{aligned}$$

27. Let $T = \{n \in \mathbf{Z}^+ | n \geq n_0 \text{ and } S(n) \text{ is false}\}$. Since $S(n_0)$, $S(n_0+1)$, $S(n_0+2), \ldots, S(n_1)$ are true, we know that $n_0, n_0+1, n_0+2, \ldots, n_1 \notin T$. If $T \neq \emptyset$, then by the Well-Ordering Principle T has a least element r, because $T \subseteq \mathbf{Z}^+$. However, since $S(n_0)$, $S(n_0+1), \ldots, S(r-1)$ are true, it follows that $S(r)$ is true. Hence $T = \emptyset$ and the result follows.

Section 4.2

1.

(a) $c_1 = 7$; and
$c_{n+1} = c_n + 7$, for $n \geq 1$.

(b) $c_1 = 7$; and
$c_{n+1} = 7c_n$, for $n \geq 1$.

(c) $c_1 = 10$; and
$c_{n+1} = c_n + 3$, for $n \geq 1$.

(d) $c_1 = 7$; and
$c_{n+1} = c_n$, for $n \geq 1$.

(e) $c_1 = 1$; and
$c_{n+1} = c_n + 2n + 1$, for $n \geq 1$.

(f) $c_1 = 3, c_2 = 1$; and
$c_{n+2} = c_n$, for $n \geq 1$.

3. For $n \in \mathbf{Z}^+, n \geq 2$, let $T(n)$ denote the (open) statement: For the statements $p, q_1, q_2, \ldots, q_n$, $p \vee (q_1 \wedge \ldots \wedge q_n) \iff (p \vee q_1) \wedge (p \vee q_2) \wedge \ldots \wedge (p \vee q_n)$.
The statement $T(2)$ is true by virtue of the Distributive Law of $\vee$ over $\wedge$. Assuming $T(k)$, for $k \geq 2$, we now examine the situation for the statements $p, q_1, q_2, \ldots, q_k, q_{k+1}$. We find that $p \vee (q_1 \wedge q_2 \wedge \ldots \wedge q_k \wedge q_{k+1}) \iff p \vee [(q_1 \wedge q_2 \wedge \ldots \wedge q_k) \wedge q_{k+1}] \iff [p \vee (q_1 \wedge q_2 \wedge \ldots \wedge q_k)] \wedge (p \vee q_{k+1}) \iff [(p \vee q_1) \wedge (p \vee q_2) \wedge \ldots \wedge (p \vee q_k)] \wedge (p \vee q_{k+1}) \iff (p \vee q_1) \wedge (p \vee q_2) \wedge \ldots \wedge (p \vee q_k) \wedge (p \vee q_{k+1})$. It then follows by the Principle of Mathematical Induction that the statement $T(n)$ is true for all $n \geq 2$.

5. (a) (i) The intersection of A_1, A_2 is $A_1 \cap A_2$.
(ii) The intersection of $A_1, A_2, \ldots, A_n, A_{n+1}$ is given by $A_1 \cap A_2 \cap \ldots \cap A_n \cap A_{n+1} = (A_1 \cap A_2 \cap \ldots \cap A_n) \cap A_{n+1}$, the intersection of the *two* sets: $A_1 \cap A_2 \cap \ldots \cap A_n$ and A_{n+1}.

(b) Let $S(n)$ denote the given (open) statement. Then the truth of $S(3)$ follows from the Associative Law of $\cap$. Assuming $S(k)$ true for some $k \geq 3$ and all $1 \leq r < k$, consider the case for $k+1$ sets. Then we find that

1) For $r = k$ we have $(A_1 \cap A_2 \cap \ldots \cap A_k) \cap A_{k+1} = A_1 \cap A_2 \cap \ldots \cap A_k \cap A_{k+1}$. This follows from the given recursive definition.

2) For $1 \leq r < k$ we have $(A_1 \cap A_2 \cap \ldots \cap A_r) \cap (A_{r+1} \cap \ldots \cap A_k \cap A_{k+1}) = (A_1 \cap A_2 \cap \ldots \cap A_r) \cap [(A_{r+1} \cap \ldots \cap A_k) \cap A_{k+1}] = [(A_1 \cap A_2 \cap \ldots \cap A_r) \cap (A_{r+1} \cap \ldots \cap A_k)] \cap A_{k+1} = (A_1 \cap A_2 \cap \ldots \cap A_r \cap A_{r+1} \cap \ldots \cap A_k) \cap A_{k+1} = A_1 \cap A_2 \cap \ldots \cap A_r \cap A_{r+1} \cap \ldots \cap A_k \cap A_{k+1}$. So by the Principle of Mathematical Induction, $S(n)$ is true for all $n \geq 3$.

7. For $n = 2$, the truth of the result $A \cap (B_1 \cup B_2) = (A \cap B_1) \cup (A \cap B_2)$ follows by virtue of the Distributive Law of $\cap$ over $\cup$.

Assuming the result for $n = k$, let us examine the case for the sets $A, B_1, B_2, \ldots, B_k, B_{k+1}$. We have $A \cap (B_1 \cup B_2 \cup \ldots \cup B_k \cup B_{k+1}) = A \cap [(B_1 \cup B_2 \cup \ldots \cup B_k) \cup B_{k+1}] = [A \cap (B_1 \cup B_2 \cup \ldots \cup B_k)] \cup (A \cap B_{k+1}) = [(A \cap B_1) \cup (A \cap B_2) \cup \ldots \cup (A \cap B_k)] \cup (A \cap B_{k+1}) = (A \cap B_1) \cup (A \cap B_2) \cup \ldots \cup (A \cap B_k) \cup (A \cap B_{k+1})$.

9. a) (i) For $n = 2$, the expression x_1x_2 denotes the ordinary product of the real numbers x_1 and x_2.
(ii) Let $n \in \mathbf{Z}^+$ with $n \geq 2$. For the real numbers $x_1, x_2, \ldots, x_n, x_{n+1}$, we define

$$x_1x_2\cdots x_nx_{n+1} = (x_1x_2\cdots x_n)x_{n+1},$$

the product of the *two* real numbers $x_1x_2\cdots x_n$ and x_{n+1}.

b) The result holds for $n = 3$ by the Associative Law of Multiplication (for real numbers). So $x_1(x_2x_3) = (x_1x_2)x_3$, and there is no ambiguity in writing $x_1x_2x_3$.
Assuming the result true for some (particular) $k \geq 3$ and all $1 \leq r < k$, let us examine the case for $k+1$ (≥ 4) real numbers. We find that

1) When $r = k$ we have

$$(x_1x_2\cdots x_r)x_{r+1} = x_1x_2\cdots x_rx_{r+1}$$

by virtue of the recursive definition.

2) For $1 \leq r < k$ we have

$$(x_1x_2\cdots x_r)(x_{r+1}\cdots x_kx_{k+1}) = (x_1x_2\cdots x_r)((x_{r+1}\cdots x_k)x_{k+1})$$

$= ((x_1x_2\cdots x_r)(x_{r+1}\cdots x_k))x_{k+1} = (x_1x_2\cdots x_rx_{r+1}\cdots x_k)x_{k+1}$
$= x_1x_2\cdots x_rx_{r+1}\cdots x_kx_{k+1}$,
so the result is true for all $n \geq 3$ and all $1 \leq r < n$, by the Principle of Mathematical Induction.

11. Proof: (By the Alternative Form of the Principle of Mathematical Induction)
For $n = 0, 1, 2$ we have
$(n = 0)$ $a_{0+2} = a_2 = 1 \geq (\sqrt{2})^0$;
$(n = 1)$ $a_{1+2} = a_3 = a_2 + a_0 = 2 \geq \sqrt{2} = (\sqrt{2})^1$; and
$(n = 2)$ $a_{2+2} = a_4 = a_3 + a_1 = 2 + 1 = 3 \geq 2 = (\sqrt{2})^2$.
Therefore the result is true for these first three cases, and this gives us the basis step for the proof.
Next, for some $k \geq 2$ we assume the result true for all $n = 0, 1, 2, \ldots, k$. When $n = k+1$ we find that

$a_{(k+1)+2} = a_{k+3} = a_{k+2} + a_k \geq (\sqrt{2})^k + (\sqrt{2})^{k-2} = [(\sqrt{2})^2 + 1](\sqrt{2})^{k-2} = 3(\sqrt{2})^{k-2}$ $= (3/2)(2)(\sqrt{2})^{k-2} = (3/2)(\sqrt{2})^k \geq (\sqrt{2})^{k+1}$, because $(3/2) = 1.5 > \sqrt{2}$ ($\doteq 1.414$). This provides the inductive step for the proof.

From the basis and inductive steps it now follows by the Alternative Form of the Principle of Mathematical Induction that $a_{n+2} \geq (\sqrt{2})^n$ for all $n \in \mathbf{N}$.

13. Proof: (By Mathematical Induction).

Basis Step: When $n = 1$ we find that

$$\sum_{i=1}^{1} \frac{F_{i-1}}{2^i} = F_0/2 = 0 = 1 - (2/2) = 1 - \frac{F_3}{2} = 1 - \frac{F_{1+2}}{2^1},$$

so the result holds in the first case.

Inductive Step: Assuming the given (open) statement for $n = k$, we have $\sum_{i=1}^{k} \frac{F_{i-1}}{2^i} = 1 - \frac{F_{k+2}}{2^k}$.

When $n = k + 1$, we find that

$$\sum_{i=1}^{k+1} \frac{F_{i-1}}{2^i} = \sum_{i=1}^{k} \frac{F_{i-1}}{2^i} + \frac{F_k}{2^{k+1}} = 1 - \frac{F_{k+2}}{2^k} + \frac{F_k}{2^{k+1}}$$

$$= 1 + (1/2^{k+1})[F_k - 2F_{k+2}] = 1 + (1/2^{k+1})[(F_k - F_{k+2}) - F_{k+2}]$$

$$= 1 + (1/2^{k+1})[-F_{k+1} - F_{k+2}] = 1 - (1/2^{k+1})(F_{k+1} + F_{k+2}) = 1 - (F_{k+3}/2^{k+1}).$$

From the basis and inductive steps it follows from the Principle of Mathematical Induction that

$$\forall n \in \mathbf{Z}^+ \ \sum_{i=1}^{n} (F_{i-1}/2^i) = 1 - (F_{n+2}/2^n).$$

15. Proof: (By the Alternative Form of the Principle of Mathematical Induction)

The result holds for $n = 0$ and $n = 1$ because

$(n = 0)$ $5F_{0+2} = 5F_2 = 5(1) = 5 = 7 - 2 = L_4 - L_0 = L_{0+4} - L_0$; and

$(n = 1)$ $5F_{1+2} = 5F_3 = 5(2) = 10 = 11 - 1 = L_5 - L_1 = L_{1+4} - L_1$.

This establishes the basis step for the proof.

Next we assume the induction hypothesis — that is, that for some k (≥ 1), $5F_{n+2} = L_{n+4} - L_n$ for all $n = 0, 1, 2, \ldots, k-1, k$. It then follows that for $n = k + 1$,

$5F_{(k+1)+2} = 5F_{k+3} = 5(F_{k+2} + F_{k+1}) = 5(F_{k+2} + F_{(k-1)+2})$

$= 5F_{k+2} + 5F_{(k-1)+2} = (L_{k+4} - L_k) + (L_{(k-1)+4} - L_{k-1}) = (L_{k+4} - L_k) + (L_{k+3} - L_{k-1})$

$= (L_{k+4} + L_{k+3}) - (L_k + L_{k-1}) = L_{k+5} - L_{k+1} = L_{(k+1)+4} - L_{k+1}$ — where we have used the recursive definitions of the Fibonacci numbers and Lucas numbers to establish the second and eighth equalities.

It then follows by the Alternative Form of the Principle of Mathematical Induction that

$$\forall n \in \mathbf{N} \ 5F_{n+2} = L_{n+4} - L_n.$$

17.

(a)

Steps	Reasons
(1) p, q, r, T_0	Part (1) of the definition
(2) $(p \vee q)$	Step (1) and Part (2-ii) of the definition
(3) $(\neg r)$	Step (1) and Part (2-i) of the definition
(4) $(T_0 \wedge (\neg r))$	Steps (1), (3), and Part (2-iii) of the definition
(5) $((p \vee q) \to (T_0 \wedge (\neg r)))$	Steps (2), (4), and Part (2-iv) of the definition

(b)

Steps	Reasons
(1) p, q, r, s, F_0	Part (1) of the definition
(2) $(\neg p)$	Step (1) and Part (2-i) of the definition
(3) $((\neg p) \leftrightarrow q)$	Steps (1), (2), and Part (2-v) of the definition
(4) $(s \vee F_0)$	Step (1) and Part (2-ii) of the definition
(5) $(r \wedge (s \vee F_0))$	Steps (1), (4), and Part (2-iii) of the definition
(6) $(((\neg p) \leftrightarrow q) \to (r \wedge (s \vee F_0)))$	Steps (3), (5), and Part (2-iv) of the definition

19. (a) $\binom{k}{2} + \binom{k+1}{2} = [k(k-1)/2] + [(k+1)k/2] = (k^2 - k + k^2 + k)/2 = k^2$.

(c) $\binom{k}{3} + 4\binom{k+1}{3} + \binom{k+2}{3} = [k(k-1)(k-2)/6] + 4[(k+1)(k)(k-1)/6] + [(k+2)(k+1)(k)/6] = (k/6)[(k-1)(k-2) + 4(k+1)(k-1) + (k+2)(k+1)] = (k/6)[6k^2] = k^3$.

(d) $\sum_{k=1}^{n} k^3 = \sum_{k=1}^{n} \binom{k}{3} + 4\sum_{k=1}^{n} \binom{k+1}{3} + \sum_{k=1}^{n} \binom{k+2}{3} = \binom{n+1}{4} + 4\binom{n+2}{4} + \binom{n+3}{4} = (1/24)[(n+1)(n)(n-1)(n-2) + 4(n+2)(n+1)(n)(n-1) + (n+3)(n+2)(n+1)(n)] = [(n+1)(n)/24][(n-1)(n-2) + 4(n+2)(n-1) + (n+3)(n+2)] = [(n+1)(n)/24][6n^2 + 6n] = n^2(n+1)^2/4$.

(e) $k^4 = \binom{k}{4} + 11\binom{k+1}{4} + 11\binom{k+2}{4} + \binom{k+3}{4}$

In general, $k^t = \sum_{r=0}^{t-1} a_{t,r} \binom{k+r}{t}$, where the $a_{t,r}$'s are the Eulerian numbers of Example 4.21. [The given summation formula is known as Worpitzky's identity.]

Section 4.3

1. (a) $a = a \cdot 1$, so $1|a$; $0 = a \cdot 0$, so $a|0$.
(b) $a|b \Longrightarrow b = ac$, for some $c \in \mathbf{Z}$. $b|a \Longrightarrow a = bd$, for some $d \in \mathbf{Z}$. So $b = ac = b(dc)$ and $d = c = 1$ or -1. Hence $a = b$ or $a = -b$.
(c) $a|b \Longrightarrow b = ax$, $b|c \Longrightarrow c = by$, for some $x, y, \in \mathbf{Z}$. So $c = by = a(xy)$ and $a|c$.
(d) $a|b \Longrightarrow ac = b$, for some $c \in \mathbf{Z} \Longrightarrow acx = bx \Longrightarrow a|bx$.
(e) If $a|x, a|y$ then $x = ac, y = ad$ for some $c, d \in \mathbf{Z}$. So $z = x - y = a(c-d)$, and $a|z$. The proofs for the other cases are similar.
(g) Follows from part (f) by the Principle of Mathematical Induction.

3. Since q is prime its only positive divisors are 1 and q. With p a prime, $p > 1$. Hence

$p|q \Longrightarrow p = q$.

5. Proof: (By the Contrapositive)
Suppose that $a \mid b$ or $a \mid c$.
If $a \mid b$, then $ak = b \; \exists k \in \mathbf{Z}$. But $ak = b \Rightarrow (ak)c = a(kc) = bc \Rightarrow a \mid bc$.
A similar result is obtained if $a \mid c$.

7. a) Let $a = 1$, $b = 5$, $c = 2$. Another example is $a = b = 5$, $c = 3$.
b) Proof:
$31|(5a + 7b + 11c) \Longrightarrow 31|(10a + 14b + 22c)$. Also, $31|(31a + 31b + 31c)$, so $31|[(31a + 31b + 31c) - (10a + 14b + 22c)]$. Hence $31|(21a + 17b + 9c)$.

9. $b|a, b|(a + 2) \Longrightarrow b|[ax + (a + 2)y]$ for all $x, y \in \mathbf{Z}$. Let $x = -1, y = 1$. Then $b > 0$ and $b|2$, so $b = 1$ or 2.

11. Let $a = 2m + 1, b = 2n + 1$, for some $m, n \geq 0$. Then $a^2 + b^2 = 4(m^2 + m + n^2 + n) + 2$, so $2|(a^2 + b^2)$ but $4 \not| \, (a^2 + b^2)$.

13. Proof:
For $n = 0$ we have $7^n - 4^n = 7^0 - 4^0 = 1 - 1 = 0$, and $3|0$. So the result is true for this first case. Assuming the truth for $n = k$ we have $3|(7^k - 4^k)$. Turning to the case for $n = k+1$ we find that $7^{k+1} - 4^{k+1} = 7(7^k) - 4(4^k) = (3+4)(7^k) - 4(4^k) = 3(7^k) + 4(7^k - 4^k)$. Since $3|3$ and $3|(7^k - 4^k)$ (by the induction hypothesis), it follows from part (f) of Theorem 4.3 that $3|[3(7^k) + 4(7^k - 4^k)]$, that is, $3|(7^{k+1} - 4^{k+1})$. It now follows by the Principle of Mathematical Induction that $3|(7^n - 4^n)$ for all $n \in \mathbf{N}$.

15.

	Base 10	Base 2	Base 16
(a)	22	10110	16
(b)	527	1000001111	20F
(c)	1234	10011010010	4D2
(d)	6923	1101100001011	1B0B

17.

	Base 2	Base 10	Base 16
(a)	11001110	206	CE
(b)	00110001	49	31
(c)	11110000	240	F0
(d)	01010111	87	57

19. Here n is a divisor of 18 – so $n \in \{1, 2, 3, 6, 9, 18\}$.

21.

	Largest Integer	Smallest Integer
(a)	$7 = 2^3 - 1$	$-8 = -(2^3)$
(b)	$127 = 2^7 - 1$	$-128 = -(2^7)$
(c)	$2^{15} - 1$	$-(2^{15})$
(d)	$2^{31} - 1$	$-(2^{31})$
(e)	$2^{n-1} - 1$	$-(2^{n-1})$

23. $ax = ay \Longrightarrow ax - ay = 0 \Longrightarrow a(x - y) = 0$. In the system of integers, if $b, c \in \mathbf{Z}$ and $bc = 0$, then $b = 0$ or $c = 0$. Since $a(x - y) = 0$ and $a \neq 0$ then $(x - y) = 0$ and $x = y$.

25. (i) If $a = 0$, choose $q = r = 0$.
(ii) Let $a > 0, b < 0$. Then $-b > 0$ so there exist $q, r \in \mathbf{Z}$ with $a = q(-b) + r$, where $0 \leq r < (-b)$. Hence $a = (-q)b + r$ with $0 \leq r < |b|$.
(iii) Finally, consider the case where $a < 0$ and $b < 0$. Then $-a, -b > 0$ so $-a = q'(-b) + r'$ with $0 \leq r' < (-b)$. So $a = q'b - r' = (q' + 1)b + (-r' - b) = qb + r$ with $0 \leq r = -b - r' < -b = |b|$.
For uniqueness, let $a, b, \in \mathbf{Z}, b \neq 0$, and assume $a = q_1b + r_1 = q_2b + r_2$, where $0 \leq r_1, r_2 \leq |b|$. Then $0 = (q_1 - q_2)b + (r_1 - r_2)$ and $|q_1 - q_2||b| = |r_1 - r_2|$. If $r_1 \neq r_2$, then $|r_1 - r_2| > 0$ but $|r_1 - r_2| < |b|$. Hence $|q_1 - q_2||b| < |b|$. This can only happen if $q_1 = q_2$. But then $r_1 = r_2$, so q, r are unique.

27.

```
Program Divisors (input,output);
Var
   N, Divisor: Integer;
Begin
   Write ('The positive integer N whose divisors are sought is N = ');
   Read (N);
   Writeln;
   If N = 1 Then
      Writeln ('The only divisor of 1 is 1.')
   Else
      Begin
      Writeln ('The divisors of ', N:0, 'are :');
      Writeln (1:8);
      If N Mod 2 = 0 Then
         Begin
            For Divisor := 2 to N Div 2 Do
            If N Mod Divisor = 0 Then
               Writeln (Divisor:8)
         End
      Else
         For Divisor := 3 to N Div 3 Do
```

```
            If N Mod Divisor = 0 Then
                Writeln (Divisor:8)
        End;
    Writeln (N:8)
End.
```

29. (a) Since $2|10^t$ for all $t \in \mathbf{Z}^+, 2|n$ iff $2|r_0$. (b) Follows from the fact that $4|10^t$ for $t \geq 2$. (c) Follows from the fact that $8|10^t$ for $t \geq 3$.
In general, $2^{t+1}|n$ iff $2^{t+1}|(r_t \cdot 10^t + \cdots + r_1 \cdot 10 + r_0)$.

Section 4.4

1.

(a) $1820 = 7(231) + 203 \qquad 0 < 208 < 231$
$231 = 1(203) + 28 \qquad 0 < 28 < 203$
$203 = 7(28) + 7 \qquad 0 < 7 < 28$
$28 = 7(4)$, so $\gcd(1820, 23) = 7$

$7 = 203 - 7(28) = 203 - 7[231 - 203] = (-7)(231) + 8(203) = (-7)(231) + 8[1820 - 7(231)]$
$= 8(1820) + (-63)(231) = 231(-63) + 1820(8)$

(b) $2597 = 1(1369) + 1228, \qquad 0 < 1228 < 1369$
$1369 = 1(1228) + 141, \qquad 0 < 141 < 1228$
$1228 = 8(141) + 100, \qquad 0 < 100 < 141$
$141 = 1(100) + 41, \qquad 0 < 41 < 100$
$100 = 2(41) + 18, \qquad 0 < 18 < 41$
$41 = 2(18) + 5, \qquad 0 < 5 < 18$
$18 = 3(5) + 3, \qquad 0 < 3 < 5$
$5 = 1(3) + 2, \qquad 0 < 2 < 3$
$3 = 1(2) + 1, \qquad 0 < 1 < 2$
$2 = 2(1)$, so $\gcd(1369, 2597) = 1$.
$1 = 3 - 2 = 3 - [5 - 3] = (-1)\,5 + 2(3)$
$= (-1)5 + 2[18 - 3(5)] = 2(18) + (-7)(5)$
$= 2(18) + (-7)[41 - 2(18)] = (-7)(41) + (16)(18)$
$= (-7)(41) + 16[100 - 2(41)] = 16(100) + (-39)(41)$
$= 16(100) + (-39)[141 - 100] = (-39)(141) + (55)(100)$
$= (-39)(141) + 55[1228 - 8(141)] = 55(1228) + (-479)(141)$
$= 55(1228) + (-479)[1369 - 1228] = (-479)(1369) + (534)(1228)$
$= (-479)(1369) + (534)[2597 - 1369]$
$= 1369\,(-1013) + 2597(534)$

(c) $\gcd(2689,4001) = 1 = 4001(-1117) + 2689(1662)$

3. $\gcd(a, b) = d \Longrightarrow d = ax + by$, for some $x, y \in \mathbf{Z}$. $\gcd(a, b) = d \Longrightarrow a/d, b/d \in \mathbf{Z}$. $1 = (a/d)x + (b/d)y \Longrightarrow \gcd(a/d, b/d) = 1$.

5. Proof: Since $c = \gcd(a, b)$ we have $a = cx$, $b = cy$ for some $x, y \in \mathbf{Z}^+$. So $ab = (cx)(cy) = c^2(xy)$, and c^2 divides ab.

7. Let $\gcd(a,b) = h$, $\gcd(b,d) = g$. $\gcd(a,b) = h \implies h|a, h|b \implies h|(a \cdot 1 + bc) \implies h|d$. $h|b, h|d \implies h|g$. $\gcd(b,d) = g \implies g|b, g|d \implies g|(d \cdot 1 + b(-c)) \implies g|a$. $g|b, g|a, h = \gcd(a,b) \implies g|h$. $h|g, g|h$, with $g, h \in \mathbf{Z}^+ \implies g = h$.

9. (a) If $c \in \mathbf{Z}^+$, then $c = \gcd(a,b)$ if (and only if)
(1) $c \mid a$ and $c \mid b$; and
(2) $\forall d \in \mathbf{Z}\ [[(d \mid a) \land (d \mid b)] \Rightarrow d \mid c]$

(b) If $c \in \mathbf{Z}^+$, then $c \neq \gcd(a,b)$ if (and only if)
(1) $c \nmid a$ or $c \nmid b$; or
(2) $\exists d \in \mathbf{Z}\ [(d \mid a) \land (d \mid b) \land (d \nmid c)]$.

11. $\gcd(a,b) = 1 \implies ax + by = 1$, for some $a, b \in \mathbf{Z}$. Then $acx + bcy = c$. $a|acx, a|bcy$ (since $a|bc$) $\implies a|c$.

13. Proof: We find that for each $n \in \mathbf{Z}^+$, $(5n+3)(7)+(7n+4)(-5) = (35n+21)-(35n+20) = 1$. Consequently, it follows that the $\gcd(5n+3, 7n+4) = 1$, or $5n+3$ and $7n+4$ are relatively prime.

15. We need to find $x, y \in \mathbf{Z}^+$ where $y > x$ and $20x + 50y = 1020$, or $2x + 5y = 102$. As $\gcd(2,5) = 1$ we start with $2(-2) + 5(1) = 1$ and find that $2(-2) + 5(1) = 1 \Rightarrow 102 = 2(-204) + 5(102) = 2[-204 + 5k] + 5[102 - 2k]$. Since $x = -204 + 5k > 0$, it follows that $k > 204/5 = 40.8$ and $y = 102 - 2k > 0$ implies that $51 > k$. Consequently $k = 41, 42, 43, \ldots, 50$. Since $y > x$ we find the following solutions:

k	$x = -204 + 5k$	$y = 102 - 2k$
41	1	20
42	6	18
43	11	16

17. $\gcd(84, 990) = 6$, so $84x + 990y = c$ has a solution x_0, y_0 in $\mathbf{Z}$ if $6|c$. For $10 < c < 20, 6|c \implies c = 12$ or 18. There is no solution for $c = 11, 13, 14, 15, 16, 17, 19$.

When $c = 12, 84x + 990y = 12$ (or, $14x + 165y = 2$).
$165 = 11(14) + 11$
$14 = 1(11) + 3$
$11 = 3(3) + 2$
$3 = 1(2) + 1$

Therefore $1 = 3 - 2 = 3 - [11 - 3(3)] = 4(3) - 11 = 4[14 - 11] - 11 = 4(14) - 5(11) = 4(14) - 5[165 - 11(14)] = 59(14) - 5(165)$
$1 = 14(59) + 165(-5)$
$2 = 14(118) + 165(-10) = 14(118 - 165k) + 165(-10 + 14k)$.

The solutions for $84x + 990y = 12$ are $x = 118 - 165k, y = -10 + 14k, k \in \mathbf{Z}$.
When $c = 18$, the solutions are $x = 177 - 165k, y = -15 + 14k, k \in \mathbf{Z}$.

19. From Theorem 4.10 we know that $ab = \text{lcm}(a, b) \cdot \gcd(a, b)$. Consequently, $b = [\text{lcm}(a, b) \cdot \gcd(a, b)]/a = (242, 500)(105)/630 = 40, 425$.

21. Since $1 = (n + 1)(1) + n(-1)$, we have $\gcd(n, n + 1) = 1$. From Theorem 4.10 it then follows that $\text{lcm}(n, n + 1) = n(n + 1)$.

Section 4.5

1. (a) $2^2 \cdot 3^3 \cdot 5^3 \cdot 11$ (b) $2^4 \cdot 3 \cdot 5^2 \cdot 7^2 \cdot 11^2$
(c) $3^2 \cdot 5^3 \cdot 7^2 \cdot 11 \cdot 13$

3.

$$m^2 = p_1^{2e_1} p_2^{2e_2} p_3^{2e_3} \cdots p_t^{2e_t}$$
$$m^3 = p_1^{3e_1} p_2^{3e_2} p_3^{3e_3} \cdots p_t^{3e_t}$$

5. Proof: (The proof is similar to that given in Example 4.41.)
If not, we have $\sqrt{p} = a/b$, where $a, b \in \mathbf{Z}^+$ and $\gcd(a, b) = 1$. Then $\sqrt{p} = a/b \Rightarrow p = a^2/b^2 \Rightarrow pb^2 = a^2 \Rightarrow p \mid a^2 \Rightarrow p \mid a$ (by Lemma 4.2). Since $p \mid a$ we know that $a = pk$ $\exists k \in \mathbf{Z}^+$, and $pb^2 = a^2 = (pk)^2 = p^2k^2$, or $b^2 = pk^2$. Hence $p \mid b^2$ and so $p \mid b$. But if $p \mid a$ and $p \mid b$ then $\gcd(a, b) = p > 1$ — contradicting our earlier claim that $\gcd(a, b) = 1$.

7. (a) Since $148, 500 = 2^2 \cdot 3^3 \cdot 5^3 \cdot 11$, any divisor of 148,500 has the form $2^{e_1} \cdot 3^{e_2} \cdot 5^{e_3} \cdot 11^{e_4}$, where $0 \leq e_1 \leq 2$, $0 \leq e_2 \leq 3$, $0 \leq e_3 \leq 3$, and $0 \leq e_4 \leq 1$. So there are three choices for e_1, four choices for each of e_2 and e_3, and two choices for e_4. Consequently, by the rule of product, the number of divisors of 148,500 is $3 \times 4 \times 4 \times 2 = 96$.
(b) 270 (c) 144

9. From Theorem 4.10 we know that $mn = \text{lcm}(m, n) \cdot \gcd(m, n)$, so $\gcd(m, n) = mn/\text{lcm}(m, n) = 2^2 3^1 5^1 11^1 = 660$.

11. $248396544 = 2^8 \cdot 3^6 \cdot 11^3$, so there are $(8 + 1)(6 + 1)(3 + 1) = 252$ possibilities for n.

13. a) Proof: (i) Since $10|a^2$ we have $5|a^2$ and $2|a^2$. Then by Lemma 4.2 it follows that $5|a$ and $2|a$. So $a = 5b$ for some $b \in \mathbf{Z}^+$. Further, since $2|5b$ we have $2|5$ or $2|b$ (by Lemma 4.2). Consequently, $a = 5b = 5(2c) = 10c$, and 10 divides a.
(ii) This result is false — let $a = 2$.

b) We can generalize section (i) of part (a) by replacing 10 by an integer n of the form $p_1 p_2 \cdots p_t$, a product of t distinct primes. (So n is a square-free integer — that is, no square greater than 1 divides n.)

15. Since $7! = 2^4 \cdot 3^2 \cdot 5 \cdot 7$, the smallest perfect square that is divisible by $7!$ is $2^4 \cdot 3^2 \cdot 5^2 \cdot 7^2 = (35) \times (7!) = 176, 400$.

17. For $1260 \times n$ to be a perfect cube, the exponent on each prime divisor must be a multiple of 3. Since $1260 = 2^2 \cdot 3^2 \cdot 5 \cdot 7$, we want $1260 \times n = 2^3 \cdot 3^3 \cdot 5^3 \cdot 7^3$, so $n = 2 \cdot 3 \cdot 5^2 \cdot 7^2 = 7350$.

19. (a) $4 = 2^2$; $8 = 2^3$; $16 = 2^4$; $32 = 2^5$.
Considering the powers of 2, there are 5 different sums of two distinct exponents: $5 = 2+3$; $6 = 2+4$; $7 = 2+5 = 3+4$; $8 = 3+5$; $9 = 4+5$. Hence there are 5 different products that we can form.
(b) Here we have 2^n for $n = 2, 3, 4, 5$ and 6. Now there are 7 different sums of two distinct exponents: $5 = 2+3$; $6 = 2+4$; $7 = 2+5 = 3+4$; $8 = 2+6 = 3+5$; $9 = 3+6 = 4+5$; $10 = 4+6$; $11 = 5+6$. Consequently, we can form 7 different products in this case.
(c) The set here may also be represented as $A \cup B$ where $A = \{2^n | n \in \mathbf{Z}^+, 2 \leq n \leq 6\}$ and $B = \{3^k | k \in \mathbf{Z}^+, 2 \leq k \leq 5\}$.
If the product uses two integers from A then there are 7 possibilities. If both integers are selected from B then we have 5 possibilities. Finally there are $5 \times 4 = 20$ products using one number from each of the sets A, B. In total, the number of different products is $7 + 5 + 20 = 32$.
(d) Consider the set given here as $A \cup B \cup C$ where $A = \{4, 8, 16, 32, 64\}$, $B = \{9, 27, 81, 243, 729\}$ and $C = \{25, 125, 625, 3125\}$.

Here there are six cases to enumerate.

(1) Both elements from A: 7 possibilities.
(2) Both elements from B: 7 possibilities.
(3) Both elements from C: 5 possibilities.
(4) One element from each of A, B: $5 \times 5 = 25$ possibilities.
(5) One element from each of A, C: $5 \times 4 = 20$ possibilities.
(6) One element from each of B, C: $5 \times 4 = 20$ possibilities.
In total there are $7 + 7 + 5 + 25 + 20 + 20 = 84$ possible products.
(e) This case generalizes the result in part (d). Once again there are 84 possible products.

21. The length of $AB = 2^8 = 256$; the length of $AC = 2^9 = 512$. The perimeter of the triangle is 1061.

23. (a) From the Fundamental Theorem of Arithmetic $88,200 = 2^3 \cdot 3^2 \cdot 5^2 \cdot 7^2$. Consider the set $F = \{2^3, 3^2, 5^2, 7^2\}$. Each subset of F determines a factorization ab where $\gcd(a, b) = 1$. There are 2^4 subsets – hence, 2^4 factorizations. Since order is not relevant, this number (of factorizations) reduces to $(1/2)2^4 = 2^3$. And since $1 < a < n$, $1 < b < n$, we remove the case for the empty subset of F (or the subset F itself). This yields $2^3 - 1$ such factorizations.

(b) Here $n = 2^3 \cdot 3^2 \cdot 5^2 \cdot 7^2 \cdot 11$ and there are $2^4 - 1$ such factorizations.

(c) Suppose that $n = p_1^{n_1} p_2^{n_2} \cdots p_k^{n_k}$, where $p_1, p_2, \ldots, p_k$ are k distinct primes and $n_1, n_2, \ldots, n_k \geq 1$. The number of unordered factorizations of n as ab, where $1 < a < n$, $1 < b < n$, and $\gcd(a, b) = 1$, is $2^{k-1} - 1$.

25. Proof: (By Mathematical Induction)

For $n = 2$ we find that $\prod_{i=2}^{2}(1 - \frac{1}{i^2}) = (1 - \frac{1}{2^2}) = (1 - \frac{1}{4}) = 3/4 = (2+1)/(2 \cdot 2)$, so the result is true in this first case and this establishes the basis step for our inductive proof. Next we assume the result true for some (particular) $k \in \mathbf{Z}^+$ where $k \geq 2$. This gives us $\prod_{i=2}^{k}(1 - \frac{1}{i^2}) = (k+1)/(2k)$. When we consider the case for $n = k+1$, using the inductive step, we find that

$$\prod_{i=2}^{k+1}(1 - \frac{1}{i^2})) = \left(\prod_{i=2}^{k}(1 - \frac{1}{i^2})\right)(1 - \frac{1}{(k+1)^2}) = [(k+1)/(2k)][1 - \frac{1}{(k+1)^2}] =$$

$$\left[\frac{k+1}{2k}\right]\left[\frac{(k+1)^2 - 1}{(k+1)^2}\right] = \frac{k^2 + 2k}{(2k)(k+1)} = (k+2)/(2(k+1)) = ((k+1)+1)/(2(k+1)).$$

The result now follows for all positive integers $n \geq 2$ by the Principle of Mathematical Induction.

27. (a) The positive divisors of 28 are 1, 2, 4, 7, 14, and 28, and $1+2+4+7+14+28 = 56 = 2(28)$, so 28 is a perfect integer.

The positive divisors of 496 are 1, 2, 4, 8, 16, 31, 62, 124, 248, and 496, and $1+2+4+8+16+31+62+124+248+496 = 992 = 2(496)$, so 496 is a perfect integer.

(b) It follows from the Fundamental Theorem of Arithmetic that the divisors of $2^{m-1}(2^m - 1)$, for $2^m - 1$ prime, are $1, 2, 2^2, 2^3, \ldots, 2^{m-1}$, and $(2^m - 1)$, $2(2^m - 1)$, $2^2(2^m - 1)$, $2^3(2^m - 1), \ldots$, and $2^{m-1}(2^m - 1)$.

These divisors sum to $[1+2+2^2+2^3+\ldots+2^{m-1}] + (2^m - 1)[1+2+2^2+2^3+\ldots+2^{m-1}] = (2^m - 1) + (2^m - 1)(2^m - 1) = (2^m - 1)[1 + (2^m - 1)] = 2^m(2^m - 1) = 2[2^{m-1}(2^m - 1)]$, so $2^{m-1}(2^m - 1)$ is a perfect integer.

Supplementary Exercises

1. $a + (a+d) + (a+2d) + \ldots + (a + (n-1)d) = na + [(n-1)nd]/2$. For $n = 1, a = a + 0$, and the result is true in this case. Assuming that $\sum_{i=1}^{k}[a + (i-1)d] = ka + [(k-1)kd]/2$, we have $\sum_{i=1}^{k+1}[a + (i-1)d] = (ka + [(k-1)kd]/2) + (a + kd) = (k+1)a + [k(k+1)d]/2$, so the result follows for all $n \in \mathbf{Z}^+$ by the Principle of Mathematical Induction.

3. Conjecture: $\sum_{i=1}^{n}(-1)^{i+1}i^2 = (-1)^{n+1}\sum_{i+1}^{n} i$, for all $n \in \mathbf{Z}^+$.

Proof: (By the Principle of Mathematical Induction)

If $n = 1$ the conjecture provides $\sum_{i=1}^{1}(-1)^{i+1}i^2 = (-1)^{1+1}(1)^2 = 1 = (-1)^{1+1}(1) = (-1)^{1+1}\sum_{i=1}^{1} i$, which is a true statement. This establishes the basis step of the proof.

In order to confirm the inductive step, we shall assume the truth of the result

$$\sum_{i=1}^{k}(-1)^{i+1}i^2 = (-1)^{k+1}\sum_{i=1}^{k} i$$

for some (particular) $k \geq 1$. When $n = k+1$ we find that

$$\sum_{i=1}^{k+1}(-1)^{i+1}i^2 = (\sum_{i=1}^{k}(-1)^{i+1}i^2) + (-1)^{(k+1)+1}(k+1)^2$$

$= (-1)^{k+1}\sum_{i=1}^{k} i + (-1)^{k+2}(k+1)^2 = (-1)^{k+1}(k)(k+1)/2 + (-1)^{k+2}(k+1)^2$
$= (-1)^{k+2}[(k+1)^2 - (k)(k+1)/2]$
$= (-1)^{k+2}(1/2)[2(k+1)^2 - k(k+1)]$
$= (-1)^{k+2}(1/2)[2k^2 + 4k + 2 - k^2 - k]$
$= (-1)^{k+2}(1/2)[k^2 + 3k + 2] = (-1)^{k+2}(1/2)(k+1)(k+2)$
$= (-1)^{k+2}\sum_{i=1}^{k+1} i$, so the truth of the result at $n = k$ implies the truth at $n = k+1$ — and we have the inductive step.
It then follows by the Principle of Mathematical Induction that

$$\sum_{i=1}^{n}(-1)^{i+1}i^2 = (-1)^{n+1}\sum_{i=1}^{n} i,$$

for all $n \in \mathbf{Z}^+$.

5.

(a)

n	n^2+n+41	n	n^2+n+41	n	n^2+n+41
1	43	4	61	7	97
2	47	5	71	8	113
3	53	6	83	9	131

(b) For $n = 39, n^2 + n + 41 = 1601$, a prime. But for $n = 40, n^2 + n + 41 = (41)^2$, so $S(39) \not\Rightarrow S(40)$.

7. (a) For $n = 0, 2^{2n+1} + 1 = 2 + 1 = 3$, so the result is true in this first case. Assuming that 3 divides $2^{2k+1} + 1$ for $n = k \in \mathbf{N}$, consider the case of $n = k+1$. Since $2^{2(k+1)+1} + 1 = 2^{2k+3} + 1 = 4(2^{2k+1}) + 1 = 4(2^{2k+1} + 1) - 3$, and 3 divides both $2^{2k+1} + 1$ and 3, it follows that 3 divides $2^{2(k+1)+1} + 1$. Consequently, the result is true for $n = k+1$ whenever it is true for $n = k$. So by the Principle of Mathematical Induction the result follows for all $n \in \mathbf{N}$.

(b) When $n = 0, 0^3 + (0+1)^3 + (0+2)^3 = 9$, so the statement is true in this case. We assume the truth of the result when $n = k \geq 0$ and examine the result for $n = k+1$. We find that $(k+1)^3 + (k+2)^3 + (k+3)^3 = (k+1)^3 + (k+2)^3 + [k^3 + 9k^2 + 27k + 27] = [k^3 + (k+1)^3 + (k+2)^3] + [9(k^2 + 3k + 3)]$, where the first summand is divisible by 9 because of the induction hypothesis. Consequently, since the result is true for $n = 0$, and since the truth at $n = k$ (≥ 0) implies the truth for $n = k+1$, it follows from the Principle of Mathematical Induction that the statement is true for all integers $n \geq 0$.

9. Converting to base 10 we find that $81x+9y+z=36z+6y+x$, and so $80x+3y-35z=0$. Since $5|(80x-35z)$ and gcd $(3,5)=1$, it follows that $5|y$. Consequently, $y=0$ or $y=5$.

For $y=0$ the equation $80x-35z=0$ leads us to $16x-7z=0$ and $16x=7z \Rightarrow 16|z$. Since $0 \le z \le 5$ we find here that $z=0$ and the solution is $x=y=z=0$.

If $y=5$, then $80x+15-35z=0 \Rightarrow 16x+3-7z=0$. With $0 \le x, z \le 5$, $16x=7z-3 \Rightarrow z$ is odd and $z=1,3$, or 5. Since 16 does *not* divide $4(=7(1)-3)$ or $18(=7(3)-3)$, and since 16 *does* divide $32(=7(5)-3)$ we find that $z=5$ and $x=2$. Hence $x=2$, $y=5$, and $z=5$. [And we see that $(xyz)_9=81x+9y+z=81(2)+9(5)+5=212=36(5)+6(5)+2=36z+6y+x=(zyx)_6$.]

11. For $n=2$ we find that $2^2=4<6=\binom{4}{2}<16=4^2$, so the statement is true in this first case.
Assuming the result true for $n=k\ge 2$ – i.e., $2^k<\binom{2k}{k}<4^k$, we now consider what happens for $n=k+1$. Here we find that
$\binom{2(k+1)}{k+1}=\binom{2k+2}{k+1}=\left[\frac{(2k+2)(2k+1)}{(k+1)(k+1)}\right]\binom{2k}{k}=2[(2k+1)/(k+1)]\binom{2k}{k}>2[(2k+1)/(k+1)]2^k>2^{k+1}$, since $(2k+1)/(k+1)=[(k+1)+k]/(k+1)>1$. In addition, $[(k+1)+k)]/(k+1)<2$,

so $\binom{2k+2}{k+1}=2[(2k+1)/(k+1)]\binom{2k}{k}<(2)(2)\binom{2k}{k}<4^{k+1}$. Consequently the result is true for all $n\ge 2$ by the Principle of Mathematical Induction.

13. First we observe that the statement is true for all $n \in \mathbf{Z}^+$ where $64 \le n \le 68$. This follows from the calculations:
$64=2(17)+6(5)$ $\quad 65=13(5)$ $\quad 66=3(17)+3(5)$ $\quad 67=1(17)+10(5)$ $\quad 68=4(17)$
Now assume the result is true for all n where $68 \le n \le k$ and consider the integer $k+1$. Then $k+1=(k-4)+5$, and since $64 \le k-4 < k$ we can write $k-4=a(17)+b(5)$ for some $a, b \in \mathbf{N}$. Consequently, $k+1=a(17)+(b+1)(5)$, and the result follows for all $n \ge 64$ by the Alternative Form of the Principle of Mathematical Induction.

15. (a) $r=r_0+r_1\cdot 10+r_2\cdot 10^2+\ldots+r_n\cdot 10^n=r_0+r_1(9)+r_1+r_2(99)+r_2+\ldots+r_n\underbrace{(99\ldots 9)}_{n\ 9\text{'s}}+r_n=$
$[9r_1+99r_2+\ldots+(99\ldots 9)r_n]+(r_0+r_1+r_2+\ldots+r_n)$. Hence $9|r$ iff $9|(r_0+r_1+r_2+\ldots+r_n)$.
(c) $3|t$ for $x=1$ or 4 or 7; $9|t$ for $x=7$.

17. (a) Let $n=2^{e_1}\cdot 3^{e_2}\cdot 5^{e_3}\cdot 7^{e_4}\cdot 11^{e_5}$ where $e_1+e_2+e_3+e_4+e_5=9$, with $e_i \ge 0$ for all $1 \le i \le 5$. The number of solutions to this equation is $\binom{5+9-1}{9}=\binom{13}{9}$
(b) $\binom{8}{4}$

19. (a) 1,4,9.
(b) 1,4,9,16, ..., k where k is the largest square less than or equal to n.

21. (a) For all $n \in \mathbf{Z}^+$, $n \ge 3$, $1+2+3+\ldots+n=n(n+1)/2$. If $\{1,2,3,\ldots,n\}=A\cup B$ with $s_A=s_B$, then $2s_A=n(n+1)/2$, or $4s_A=n(n+1)$. Since $4|n(n+1)$ and $\gcd(n,n+1)=1$

then either $4|n$ or $4|(n+1)$.

(b) Here we are verifying the converse of our result in part (a).
(i) If $4|n$ we write $n = 4k$. Here we have $\{1,2,3,\ldots,k,k+1,\ldots,3k,3k+1,\ldots,4k\} = A \cup B$ where $A = \{1,2,3,\ldots,k,3k+1,3k+2,\ldots,4k-1,4k\}$ and $B = \{k+1,k+2,\ldots,2k,2k+1,3k-1,3k\}$, with $s_A = (1+2+3+\ldots+k) + [(3k+1)+(3k+2)+\ldots+(3k+k)] = [k(k+1)/2] + k(3k) + [k(k+1)/2] = k(k+1) + 3k^2 = 4k^2 + k$, and $s_B = [(k+1)+(k+2)+\ldots+(k+k)] + [(2k+1)+(2k+2)+\ldots+(2k+k)] = k(k) + [k(k+1)/2] + k(2k) + [k(k+1)/2] = 3k^2 + k(k+1) = 4k^2 + k$.
(ii) Now we consider the case where $n+1 = 4k$. Then $n = 4k-1$ and we have $\{1,2,3,\ldots,k-1,k,\ldots,3k-1,3k,\ldots,4k-2,4k-1\} = A \cup B$, with $A = \{1,2,3,\ldots,k-1,3k,3k+1,\ldots,4k-1\}$ and $B = \{k,k+1,\ldots,2k-1,2k,2k+1,\ldots,3k-1\}$. Here we find $s_A = [1+2+3+\ldots+(k-1)] + [3k+(3k+1)+\ldots+(3k+(k-1))] = [(k-1)(k)/2] + k(3k) + [(k-1)(k)/2] = 3k^2 + k^2 - k = 4k^2 - k$, and $s_B = [k+(k+1)+\ldots+(k+(k-1))] + [2k+(2k+1)+\ldots+(2k+(k-1))] = k^2 + [(k-1)(k)/2] + k(2k) + [(k-1)(k)/2] = 3k^2 + (k-1)k = 4k^2 - k$.

23. (a) The result is true for $a = 1$, so consider $a > 1$. From the Fundamental Theorem of Arithmetic we can write $a = p_1^{e_1} p_2^{e_2} \cdots p_t^{e_t}$, where $p_1, p_2, \ldots, p_t$, are t distinct primes and $e_i > 0$, for all $1 \le i \le t$. Since $a^2 | b^2$ it follows that $p_i^{2e_i} | b^2$ for all $1 \le i \le t$. So $b^2 = p_1^{2f_1} p_2^{2f_2} \cdots p_t^{2f_t} c^2$, where $f_i \ge e_i$ for all $1 \le i \le t$, and $b = p_1^{f_1} p_2^{f_2} \cdots p_t^{f_t} c = a(p_1^{f_1-e_1} p_2^{f_2-e_2} \cdots p_t^{f_t-e_t})c$, where $f_i - e_i \ge 0$ for all $\le i \le t$. Consequently, $a|b$.

(b) This result is not necessarily true! Let $a = 8$ and $b = 4$. Then $a^2 (= 64)$ divides $b^3 (= 4)$, but a does *not* divide b.

25. (a) Recall that

$$\begin{aligned} a^3 + b^3 &= (a+b)(a^2 - ab + b^2) \\ a^5 + b^5 &= (a+b)(a^4 - a^3 b + a^2 b^2 - ab^3 + b^4) \\ &\vdots \\ a^p + b^p &= (a+b)(a^{p-1} - a^{p-2} b + \cdots + b^{p-1}) \\ &= (a+b)\sum_{i=1}^{p} a^{p-i}(-b)^{i-1}, \end{aligned}$$

for p an odd prime.
Since k is not a power of 2 we write $k = r \cdot p$, where p is an odd prime and $r \ge 1$. Then $a^k + b^k = (a^r)^p + (b^r)^p = (a^r + b^r)\sum_{i=1}^{p} a^{r(p-i)}(-b)^{r(i-1)}$, so $a^k + b^k$ is composite.
(b) Here n is not a power of 2. If, in addition, n is not prime, then $n = r \cdot p$ where p is an odd prime. Then $2^n + 1 = 2^n + 1^n = 2^{r \cdot p} + 1^{r \cdot p} = (2^r + 1^r)\sum_{i=1}^{p} 2^{r(p-i)}(-1)^{r(i-1)} = (2^r + 1)\sum_{i=1}^{p} 2^{r(p-i)}$, so $2^n + 1$ is composite — not prime.

27. Proof: For $n = 0$ we find that $F_0 = 0 \le 1 = (5/3)^0$, and for $n = 1$ we have $F_1 = 1 \le$

$(5/3) = (5/3)^1$. Consequently, the given property is true in these first two cases (and this provides the basis step of the proof).
Assuming that this property is true for $n = 0, 1, 2, \ldots, k-1, k$, where $k \geq 1$, we now examine what happens at $n = k+1$. Here we find that
$F_{k+1} = F_k + F_{k-1} \leq (5/3)^k + (5/3)^{k-1} = (5/3)^{k-1}[(5/3)+1] = (5/3)^{k-1}(8/3)$
$= (5/3)^{k-1}(24/9) \leq (5/3)^{k-1}(25/9) = (5/3)^{k-1}(5/3)^2 = (5/3)^{k+1}$.
It then follows from the Alternative Form of the Principle of Mathematical Induction that $F_n \leq (5/3)^n$ for all $n \in \mathbf{N}$.

29. a) There are $9 \cdot 10 \cdot 10 = 900$ such palindromes and their sum is $\sum_{a=1}^{9}\sum_{b=0}^{9}\sum_{c=0}^{9} abcba = \sum_{a=1}^{9}\sum_{b=0}^{9}\sum_{c=0}^{9}(10001a + 1010b + 100c) = \sum_{a=1}^{9}\sum_{b=0}^{9}[10(10001a + 1010b) + 100(9 \cdot 10/2)] = \sum_{a=1}^{9}\sum_{b=0}^{9}(100010a + 10100b + 4500) = \sum_{a=1}^{9}[10(100010a) + 10100(9 \cdot 10/2) + 10(4500)] = 1000100\sum_{a=1}^{9} a + 9(454500) + 9(45000) = 1000100(9 \cdot 10/2) + 4090500 + 405000 = 49,500,000$.

b)
```
begin
    sum := 0
    for a := 1 to 9 do
        for b := 0 to 9 do
            for c := 0 to 9 do
                sum := sum + 10001 * a + 1010 * b + 100 * c
    print sum
end
```

31. Proof: Suppose that $7|n$. We see that $7|n \Rightarrow 7|(n-21u) \Rightarrow 7|[(n-u)-20u] \Rightarrow 7|[10(\frac{n-u}{10}) - 20u] \Rightarrow 7|[10(\frac{n-u}{10} - 2u)] \Rightarrow 7|(\frac{n-u}{10} - 2u)$, by Lemma 4.2 since $\gcd(7,10) = 1$. [Note: $\frac{n-u}{10} \in \mathbf{Z}^+$ since the units digit of $n-u$ is 0.] Conversely, if $7|(\frac{n-u}{10} - 2u)$, then since $\frac{n-u}{10} - 2u = \frac{n-21u}{10}$ we find that $7|(\frac{n-21u}{10}) \Rightarrow 7 \cdot 10 \cdot x = n - 21u$, for some $x \in \mathbf{Z}^+$. Since $7|7$ and $7|21$, it then follows that $7|n$ – by part (e) of Theorem 4.3.

33. If Catrina's selection includes any of 0,2,4,6,8, then at least two of the resulting three-digit integers will have an even unit's digit, and be even – hence, *not* prime. Should her selection include 5, then two of the resulting three-digit integers will have 5 as their unit's digit; these three-digit integers are then divisible by 5 and so, they are *not* prime. Consequently, to complete the proof we need to consider the four selections of size 3 that Catrina can make from $\{1,3,7,9\}$. The following provides the selections – each with a three-digit integer that is not prime.
(1) $\{1,3,7\} : 713 = 23 \cdot 31$
(2) $\{1,3,9\} : 913 = 11 \cdot 83$
(3) $\{1,7,9\} : 917 = 7 \cdot 131$
(4) $\{3,7,9\} : 793 = 13 \cdot 61$

35. Let x denote the integer Barbara erased. The sum of the integers $1, 2, 3, \ldots, x-1, x+1, x+2, \ldots, n$ is $[n(n+1)/2] - x$, so $[[n(n+1)/2] - x]/(n-1) = 35\frac{7}{17}$. Consequently, $[n(n+1)/2] - x = (35\frac{7}{17})(n-1) = (602/17)(n-1)$. Since $[n(n+1)/2] - x \in \mathbf{Z}^+$, it follows that $(602/17)(n-1) \in \mathbf{Z}^+$. Therefore, from Lemma 4.2, we find that $17|(n-1)$ because 17 does not divide 602. For $n = 1, 18, 35, 52$ we have:

n	$x = [n(n+1)/2] - (602/17)(n-1)$
1	1
18	−431
35	−574
52	−428

When $n = 69$, we find that $x = 7$ [and $(\sum_{i=1}^{69} i - 7)/68 = 602/17 = 35\frac{7}{17}$].

For $n = 69 + 17k$, $k \geq 1$, we have

$$\begin{aligned} x &= [(69+17k)(70+17k)/2] - (602/17)[68+17k] \\ &= 7 + (k/2)[1159 + 289k] \\ &= [7 + (1159k/2)] + (289k^2)/2 > n. \end{aligned}$$

Hence the answer is unique: namely, $n = 69$ and $x = 7$.

37. A common divisor for m, n has the form $p_1^{r_1} p_2^{r_2} p_3^{r_3}$, where $0 \leq r_i \leq \min\{e_i, f_i\}$, for all $1 \leq i \leq 3$. Let $m_i = \min\{e_i, f_i\}$, $1 \leq i \leq 3$. Then the number of common divisors is $(m_1 + 1)(m_2 + 1)(m_3 + 1)$.

CHAPTER 5
RELATIONS AND FUNCTIONS

Section 5.1

1. $A \times B = \{(1,2),(2,2),(3,2),(4,2),(1,5),(2,5),(3,5),(4,5)\}$
$B \times A = \{(2,1),(2,2),(2,3),(2,4),(5,1),(5,2),(5,3),(5,4)\}$
$A \cup (B \times C) = \{1,2,3,4,(2,3),(2,4),(2,7),(5,3),(5,4),(5,7)\}$
$(A \cup B) \times C = (A \times C) \cup (B \times C) = \{(1,3),(2,3),(3,3),(4,3),(5,3),(1,4),(2,4),(3,4),(4,4),$ $(5,4),(1,7),(2,7),(3,7),(4,7),(5,7)\}$

3. (a) $|A \times B| = |A||B| = 9$
(b) Since a relation from A to B is a subset of $A \times B$, there are 2^9 relations from A to B.
(c) Since $|A \times A| = 9$, there are 2^9 relations on A.
(d) For the other seven ordered pairs in $A \times B$ there are two choices: include it in the relation or leave it out. Hence there are 2^7 relations from A to B that contain (1,2) and (1,5).
(e) $\binom{9}{5}$ (f) $\binom{9}{7} + \binom{9}{8} + \binom{9}{9}$

5. (a) Assume that $A \times B \subseteq C \times D$ and let $a \in A$ and $b \in B$. Then $(a,b) \in A \times B$, and since $A \times B \subseteq C \times D$ we have $(a,b) \in C \times D$. But $(a,b) \in C \times D \Rightarrow a \in C$ and $b \in D$. Hence, $a \in A \Rightarrow a \in C$, so $A \subseteq C$, and $b \in B \Rightarrow b \in D$, so $B \subseteq D$.
Conversely, suppose that $A \subseteq C$ and $B \subseteq D$, and that $(x,y) \in A \times B$. Then $(x,y) \in A \times B \Rightarrow x \in A$ and $y \in B \Rightarrow x \in C$ (since $A \subseteq C$) and $y \in D$ (since $B \subseteq D$) $\Rightarrow (x,y) \in C \times D$. Consequently, $A \times B \subseteq C \times D$.

(b) If any of the sets A, B, C, D is empty we still find that

$$[(A \subseteq C) \wedge (B \subseteq D)] \Rightarrow [A \times B \subseteq C \times D].$$

However, the converse need not hold. For example, let $A = \emptyset$, $B = \{1,2\}$, $C = \{1,2\}$ and $D = \{1\}$. Then $A \times B = \emptyset$ — if not, there exists an ordered pair (x,y) in $A \times B$, and this means that the empty set A contains an element x. And so $A \times B = \emptyset \subseteq C \times D$ — but $B = \{1,2\} \not\subseteq \{1\} = D$.

7. (a) Since $|A| = 5$ and $|B| = 4$ we have $|A \times B| = |A||B| = 5 \cdot 4 = 20$. Consequently, $A \times B$ has 2^{20} subsets, so $|\mathcal{P}(A \times B)| = 2^{20}$.

(b) If $|A| = m$ and $|B| = n$, for $m, n \in \mathbf{N}$, then $|A \times B| = mn$. Consequently, $|\mathcal{P}(A \times B)| = 2^{mn}$.

9. (b) $A \times (B \cup C) = \{(x,y)|x \in A \text{ and } y \in (B \cup C)\} = \{(x,y)|x \in A \text{ and } (y \in B \text{ or } y \in C)\} = \{(x,y)|(x \in A \text{ and } y \in B) \text{ or } (x \in A \text{ and } y \in C)\} = \{(x,y)|x \in A \text{ and } y \in B\} \cup \{(x,y)|x \in A \text{ and } y \in C\} = (A \times B) \cup (A \times C)$.
(c) $(A \cap B) \times C = \{(x,y)|x \in A \cap B \text{ and } y \in C\}$
$= \{(x,y)|(x \in A \text{ and } x \in B) \text{ and } y \in C\}$
$= \{(x,y)|(x \in A \text{ and } x \in B) \text{ and } (y \in C \text{ and } y \in C)\}$
$= \{(x,y)|(x \in A \text{ and } y \in C) \text{ and } (x \in B \text{ and } y \in C)\}$
$= \{(x,y)|(x,y) \in A \times C \text{ and } (x,y) \in B \times C\}$
$= (A \times C) \cap (B \times C)$
(d) $(A \cup B) \times C = \{(x,y)|x \in A \cup B \text{ and } y \in C\}$
$= \{(x,y)|(x \in A \text{ or } x \in B) \text{ and } y \in C\}$
$= \{(x,y)|(x \in A \text{ and } y \in C) \text{ or } (x \in B \text{ and } y \in C)\}$
$= \{(x,y)|(x,y) \in A \times C \text{ or } (x,y) \in B \times C\}$
$= (A \times C) \cup (B \times C)$

11. $(x,y) \in A \times (B - C) \Longleftrightarrow x \in A \text{ and } y \in B - C \Longleftrightarrow x \in A \text{ and } (y \in B \text{ and } y \notin C) \Longleftrightarrow (x \in A \text{ and } y \in B) \text{ and } (x \in A \text{ and } y \notin C) \Longleftrightarrow (x,y) \in A \times B \text{ and } (x,y) \notin A \times C \Longleftrightarrow (x,y) \in (A \times B) - (A \times C)$..

13. (a) (1) $(0,2) \in \mathcal{R}$; and
(2) If $(a,b) \in \mathcal{R}$, then $(a+1, b+5) \in \mathcal{R}$.
(b) From part (1) of the definition we have $(0,2) \in \mathcal{R}$. By part (2) of the definition we then find that
(i) $(0,2) \in \mathcal{R} \Rightarrow (0+1, 2+5) = (1,7) \in \mathcal{R}$;
(ii) $(1,7) \in \mathcal{R} \Rightarrow (1+1, 7+5) = (2,12) \in \mathcal{R}$;
(iii) $(2,12) \in \mathcal{R} \Rightarrow (2+1, 12+5) = (3,17) \in \mathcal{R}$; and
(iv) $(3,17) \in \mathcal{R} \Rightarrow (3+1, 17+5) = (4,22) \in \mathcal{R}$.

Section 5.2

1. (a) Function: Range $= \{7, 8, 11, 16, 23, \ldots\}$
(b) Relation, not a function. For example, both (4,2) and $(4,-2)$ are in the relation.
(c) Function: Range = the set of all real numbers.
(d) Relation, not a function. Both (0,1) and $(0,-1)$ are in the relation.
(e) Since $|R| > 5$, R cannot be a function.

3. (a) $\{(1,x),(2,x),(3,x),(4,x)\}, \{(1,y),(2,y),(3,y),(4,y)\}, \{(1,z),(2,z),(3,z),(4,z)\}$
$\{(1,x),(2,y),(3,x),(4,y)\}, \{(1,x),(2,y),(3,z),(4,x)\}$
(b) Since $|B| = 3$, there are three choices for each of the four elements in A. Consequently, there are 3^4 possible functions from A to B.
(c) Since $|A| = 4 > 3 = |B|$, there are no one-to-one functions from A to B.

(d) Since $|A| = 4$, there are four choices for each of the three elements in B. Hence there are 4^3 possible functions from B to A.
(e) For x there are four choices, then for y there are three choices, and finally we have two choices left for z. Consequently, there are $4 \cdot 3 \cdot 2 = 24$ functions from B to A that are one-to-one.
(f) If $f(1) = x$, we still have three choices for each of 2, 3, and 4. So there are $3^3 = 27$ functions $f : A \to B$ where $f(1) = x$.
(g) With $f(1) = f(2) = x$, we still have three choices for each of 3 and 4, and there are 9 possible functions in this case.
(h) Again there are three choices for each of 3 and 4, and 9 possible functions in this case as well [as in (g)].

5. (a) $A \cap B = \{(x, y)|y = 2x + 1 \text{ and } y = 3x\}$
$2x + 1 = 3x \Rightarrow x = 1$
So $A \cap B = \{(1, 3)\}$.
(b) $B \cap C = \{(x, y)|y = 3x \text{ and } y = x - 7\}$
$3x = x - 7 \Rightarrow 2x = -7$, so $x = -7/2$.
Consequently, $B \cap C = \{(-7/2, 3(-7/2)\} = \{(-7/2, -21/2)\}$.
(c) $\overline{\overline{A} \cup \overline{C}} = \overline{\overline{A}} \cap \overline{\overline{C}} = A \cap C = \{(x, y)|y = 2x + 1 \text{ and } y = x - 7\}$
Now $2x + 1 = x - 7 \Rightarrow x = -8$, and so $A \cap C = \{(-8, -15)\}$.
(d) We know that $\overline{B} \cup \overline{C} = \overline{B \cap C}$, and since $B \cap C = \{(-7/2, -21/2)\}$ we have $\overline{B} \cup \overline{C} = \mathbf{R}^2 - \{(-7/2, -21/2)\} = \{(x, y)|x \neq -7/2 \text{ or } y \neq -21/2\}$.

7.
(a) $\lfloor 2.3 - 1.6 \rfloor = \lfloor 0.7 \rfloor = 0$
(b) $\lfloor 2.3 \rfloor - \lfloor 1.6 \rfloor = 2 - 1 = 1$
(c) $\lceil 3.4 \rceil \lfloor 6.2 \rfloor = 4 \cdot 6 = 24$
(d) $\lfloor 3.4 \rfloor \lceil 6.2 \rceil = 3 \cdot 7 = 21$
(e) $\lfloor 2\pi \rfloor = 6$
(f) $2\lceil \pi \rceil = 8$

9. (a) $\ldots [-1, -6/7) \cup [0, 1/7) \cup [1, 8/7) \cup [2, 15/7) \cup \ldots$
(b) $[1, 8/7)$ (c) $\mathbf{Z}$ (d) $\mathbf{R}$

11. (a) $\ldots \cup(-7/3, -2] \cup(-4/3, -1] \cup(-1/3, 0] \cup(2/3, 1] \cup(5/3, 2] \cup \ldots = \bigcup_{m \in \mathbf{Z}^+} (m - 1/3, m]$

(b) $\ldots \cup((-2n-1)/n, -2] \cup((-n-1)/n, -1] \cup(-1/n, 0] \cup((n-1)/n, 1] \cup((2n-1)/n, 2] \cup \ldots$

$= \bigcup_{m \in \mathbf{Z}^+} (m - 1/n, m]$

13. a) Proof (i): If $a \in \mathbf{Z}^+$, then $\lceil a \rceil = a$ and $\lfloor \lceil a \rceil / a \rfloor = \lfloor 1 \rfloor = 1$. If $a \notin \mathbf{Z}^+$, write $a = n + c$, where $n \in \mathbf{Z}^+$ and $0 < c < 1$. Then $\lceil a \rceil / a = (n + 1)/(n + c) = 1 + (1 - c)/(n + c)$, where $0 < (1 - c)/(n + c) < 1$. Hence $\lfloor \lceil a \rceil / a \rfloor = \lfloor 1 + (1 - c)/(n + c) \rfloor = 1$.

Proof (ii): For $a \in \mathbf{Z}^+$, $\lfloor a \rfloor = a$ and $\lceil \lfloor a \rfloor / a \rceil = \lceil 1 \rceil = 1$. When $a \notin \mathbf{Z}^+$, let $a = n + c$, where $n \in \mathbf{Z}^+$ and $0 < c < 1$. Then $\lfloor a \rfloor / a = n/(n + c) = 1 - [c/(n + c)]$, where $0 < c/(n + c) < 1$.

Consequently $\lceil \lfloor a \rfloor / a \rceil = \lceil 1 - (c/(n+c)) \rceil = 1$.

b) Consider $a = 0.1$. Then
(i) $\lfloor \lceil a \rceil / a \rfloor = \lfloor 1/0.1 \rfloor = \lfloor 10 \rfloor = 10 \neq 1$; and
(ii) $\lceil \lfloor a \rfloor / a \rceil = \lceil 0/0.1 \rceil = 0 \neq 1$.

In fact (ii) is false for all $0 < a < 1$, since $\lceil \lfloor a \rfloor / a \rceil = 0$ for all such values of a. In the case of (i), when $0 < a \leq 0.5$, it follows that $\lceil a \rceil / a \geq 2$ and $\lfloor \lceil a \rceil / a \rfloor \geq 2 \neq 1$. However, for $0.5 < a < 1$, $\lceil a \rceil / a = 1/a$ where $1 < 1/a < 2$, and so $\lfloor \lceil a \rceil / a \rfloor = 1$ for $0.5 < a < 1$.

15. (a) Since $f(a) = f(b) \Rightarrow 2a + 1 = 2b + 1 \rightarrow a = b$, we know that f is one-to-one. The range is the set of all odd integers.
(b) One-to-one. Range = $\mathbf{Q}$
(c) Since $f(1) = f(0), f$ is not one-to-one. The range of $f = \{0, \pm 6, \pm 24, \pm 60, \ldots\} = \{n^3 - n | n \in \mathbf{Z}\}$.
(d) Since $f(a) = f(b) \Rightarrow e^a = e^b \Rightarrow \ln(e^a) = \ln(e^b) \Rightarrow a \ln e = b \ln e \Rightarrow a \cdot 1 = b \cdot 1 \Rightarrow a = b$, we have f one-to-one. Range = $(0, +\infty) = \mathbf{R}^+$
(e) One-to-one. Range = $[-1, 1]$
(f) Since $f(\pi/4) = f(3\pi/4), f$ is not one-to-one. The range of $f = [0, 1]$.

17. The extension must include $f(1)$ and $f(4)$. Since $|B| = 4$ there are four choices for each of 1 and 4, so there are $4^2 = 16$ ways to extend the given function g.

19. (a) $f(A_1 \cup A_2) = \{y \in B | y = f(x),\ x \in A_1 \cup A_2\} = \{y \in B | y = f(x),\ x \in A_1$ or $x \in A_2\} = \{y \in B | y = f(x),\ x \in A_1\} \cup \{y \in B | y = f(x),\ x \in A_2\} = f(A_1) \cup f(A_2)$.
(c) $y \in f(A_1) \cap f(A_2) \Longrightarrow y = f(x_1) = f(x_2),\ x_1 \in A_1,\ x_2 \in A_2 \Longrightarrow y = f(x_1)$ with $x_1 = x_2$, since f is one-to-one $\Longrightarrow y \in f(A_1 \cap A_2)$.

21. No. Let $A = \{1, 2\}$, $X = \{1\}$, $Y = \{2\}$, $B = \{3\}$. For $f = \{(1, 3), (2, 3)\}$ we have $f|_X$, $f|_Y$ one-to-one, but f is not one-to-one.

23. (a) $f(a_{ij}) = 12(i-1) + j$ (b) $f(a_{ij}) = 10(i-1) + j$ (c) $f(a_{ij}) = 7(i-1) + j$

25. (a) (i) $f(a_{ij}) = n(i-1) + (k-1) + j$
(ii) $g(a_{ij}) = m(j-1) + (k-1) + i$
(b) $k + (mn - 1) \leq r$

27. (a) $A(1, 3) = A(0, A(1, 2)) = A(1, 2) + 1 = A(0, A(1, 1)) + 1 = [A(1, 1) + 1] + 1 = A(1, 1) + 2 = A(0, A(1, 0)) + 2 = [A(1, 0) + 1] + 2 = A(1, 0) + 3 = A(0, 1) + 3 = (1 + 1) + 3 = 5$

$A(2, 3) = A(1, A(2, 2))$
$A(2, 2) = A(1, A(2, 1))$
$A(2, 1) = A(1, A(2, 0)) = A(1, A(1, 1))$
$A(1, 1) = A(0, A(1, 0)) = A(1, 0) + 1 = A(0, 1) + 1 = (1 + 1) + 1 = 3$
$A(2, 1) = A(1, 3) = A(0, A(1, 2)) = A(1, 2) + 1 = A(0, A(1, 1)) = [A(1, 1) + 1] + 1 = 5$

$A(2,2) = A(1,5) = A(0, A(1,4)) = A(1,4) + 1 = A(0, A(1,3)) + 1 = A(1,3) + 2 = A(0, A(1,2)) + 2 = A(1,2) + 3 = A(0, A(1,1)) + 3 = A(1,1) + 4 = 7$

$A(2,3) = A(1,7) = A(0, A(1,6)) = A(1,6) + 1 = A(0, A(1,5)) + 1 = A(0,7) + 1 = (7+1) + 1 = 9$

(b) Since $A(1,0) = A(0,1) = 2 = 0 + 2$, the result holds for the case where $n = 0$. Assuming the truth of the (open) statement for some k (≥ 0) we have $A(1,k) = k + 2$. Then we find that $A(1, k+1) = A(0, A(1,k)) = A(1,k) + 1 = (k+2) + 1 = (k+1) + 2$, so the truth at $n = k$ implies the truth at $n = k + 1$. Consequently, $A(1,n) = n + 2$ for all $n \in \mathbf{N}$ by the Principle of Mathematical Induction.

(c) Here we find that $A(2,0) = A(1,1) = 1 + 2 = 3$ (by the result in part(b)). So $A(2,0) = 3 + 2 \cdot 0$ and the given (open) statement is true in this first case.
Next we assume the result true for some k (≥ 0) — that is, we assume that $A(2,k) = 3 + 2k$. For $k+1$ we then find that $A(2, k+1) = A(1, A(2,k)) = A(2,k) + 2$ (by part (b)) $= (3 + 2k) + 2$ (by the induction hypothesis) $= 3 + 2(k+1)$. Consequently, for all $n \in \mathbf{N}$, $A(2,n) = 3 + 2n$ — by the Principle of Mathematical Induction.

(d) Once again we consider what happens for $n = 0$. Since $A(3,0) = A(2,1) = 3 + 2(1)$ (by part (c)) $= 5 = 2^{0+3} - 3$, the result holds in this first case.
So now we assume the given (open) statement is true for some k (≥ 0) and this gives us the induction hypothesis: $A(3,k) = 2^{k+3} - 3$. For $n = k + 1$ it then follows that $A(3, k+1) = A(2, A(3,k)) = 3 + 2A(3,k)$ (by part (c)) $= 3 + 2(2^{k+3} - 3)$ (by the induction hypothesis) $= 2^{(k+1)+3} - 3$, so the result holds for $n = k + 1$ whenever it does for $n = k$. Therefore, $A(3,n) = 2^{n+3} - 3$, for all $n \in \mathbf{N}$ — by the Principle of Mathematical Induction.

Section 5.3

1. Let $A = \{1,2,3,4\}, B = \{v,w,x,y,z\}$: (a) $f = \{(1,v),(2,v),(3,w),(4,x)\}$
(b) $f = \{(1,v),(2,x),(3,y),(4,z)\}$
(c) Let $A = \{1,2,3,4,5\}, B = \{w,x,y,z\}, f = \{(1,w),(2,w),(3,x),(4,y),(5,z)\}$.
(d) Let $A = \{1,2,3,4\}, B = \{w,x,y,z\}, f = \{(1,w),(2,x),(3,y),(4,z)\}$.

3. (a) $g(a) = g(b) \Rightarrow a + 7 = b + 7 \Rightarrow a = b$, so g is one-to-one. For any $r \in \mathbf{R}$, $r - 7 \in \mathbf{R}$ and $g(r-7) = (r-7) + 7 = r$, so g is also an onto function.
(b) $g(a) = g(b) \Rightarrow 2a - 3 = 2b - 3 \Rightarrow 2a = 2b \Rightarrow a = b$, so g is one-to-one. For any $r \in \mathbf{R}$, $(\frac{1}{2})(r+3) \in \mathbf{R}$ and $g(\frac{r+3}{2}) = 2(\frac{r+3}{2}) - 3 = r + 3 - 3 = r$, so g is also an onto function.
(c) This function is one-to-one and onto.
(d) Neither one-to-one nor onto. Range $= [0, +\infty)$
(e) Neither one-to-one nor onto. Range $= [-1/4, +\infty)$
(f) $g(a) = g(b) \Rightarrow a^3 = b^3 \Rightarrow a = b$, so g is one-to-one. For any $r \in \mathbf{R}$, $\sqrt[3]{r} \in \mathbf{R}$ and $g(\sqrt[3]{r}) = (\sqrt[3]{r})^3 = r$, so g is also an onto function.

5. For $n = 5$, $m = 2$, $\sum_{k=0}^{5}(-1)^k\binom{5}{5-k}(5-k)^2$

$= (-1)^0\binom{5}{5}(5)^2 + (-1)^1\binom{5}{4}(4)^2 + (-1)^2\binom{5}{3}(3)^2 + (-1)^3\binom{5}{2}(2)^2 + (-1)^4\binom{5}{1}(1)^2 + (-1)^5\binom{5}{0}(0)^2$
$= 25 - (5)(16) + (10)(9) - (10)(4) + (5)(1)$
$= 25 - 80 + 90 - 40 + 5 = 0$

For $n = 5, m = 3, \sum_{k=0}^{5}(-1)^k\binom{5}{5-k}(5-k)^3 = (-1)^0\binom{5}{5}(5)^3 + (-1)^1\binom{5}{4}(4)^3 + (-1)^2\binom{5}{3}(3)^3 + (-1)^3\binom{5}{2}(2)^3 + (-1)^4\binom{5}{1}(1)^3 + (-1)^5\binom{5}{0}(0)^3 = 125 - 320 + 70 - 80 + 5 = 0$

For $n = 5$, $m = 4$, $\sum_{k=0}^{5}(-1)^k\binom{5}{5-k}(5-k)^4$
$= (-1)^0\binom{5}{5}(5)^4 + (-1)^1\binom{5}{4}(4)^4 + (-1)^2\binom{5}{3}(3)^4 + (-1)^3\binom{5}{2}(2)^4 + (-1)^4\binom{5}{1}(1)^4 + (-1)^5\binom{5}{0}(0)^4$
$= 625 - (5)(256) + (10)(81) - (10)(16) + (5)(1)$
$= 625 - 1280 + 810 - 160 + 5 = 0$

7. (a) (i) Here we want the number of onto functions for a seven-element domain and a two-element codomain. From the result following Example 5.26 this number is $2!S(7,2)$.
(ii) There are $\binom{5}{2}$ ways to select two of the elements from B. So from part (i) it follows that there are $\binom{5}{2}[2!S(7,2)]$ functions $f : A \to B$ where the range contains two elements.

(iii) $3!S(7,3)$ (iv) $\binom{5}{3}[3!S(7,3)]$

(v) $4!S(7,4)$ (vi) $\binom{5}{4}[4!S(7,4)]$

(b) $\binom{n}{k}[k!S(m,k)]$

9. For each $r \in \mathbf{R}$ there is at least one $a \in \mathbf{R}$ such that $a^5 - 2a^2 + a - r = 0$ because the polynomial $x^5 - 2x^2 + x - r$ has odd degree and real coefficients. Consequently, f is onto. However, $f(0) = 0 = f(1)$, so f is not one-to-one.

11.

m \ n	1	2	3	4	5	6	7	8	9	10
9	1	255	3025	7770	6951	2646	462	36	1	
10	1	511	9330	34105	42525	22827	5880	750	45	1

13. (a) Since $156,009 = 3 \times 7 \times 17 \times 19 \times 23$, it follows that there are $S(5,2) = 15$ two-factor unordered factorizations of 156,009, where each factor is greater than 1.

(b) $\sum_{i=2}^{5} S(5,i) = S(5,2) + S(5,3) + S(5,4) + S(5,5) = 15 + 25 + 10 + 1 = 51.$

(c) $\sum_{i=2}^{n} S(n,i).$

15. a) $n = 4$: $\sum_{i=1}^{4} i!S(4,i)$; $n = 5$: $\sum_{i=1}^{5} i!S(5,i)$

In general, the answer is $\sum_{i=1}^{n} i!S(n,i)$.

b) $\binom{15}{12}\sum_{i=1}^{12} i!S(12,i)$.

17. Let $a_1, a_2, \ldots, a_m, x$ denote the $m+1$ distinct objects. Then $S_r(m+1,n)$ counts the number of ways these objects can be distributed among n identical containers so that each container receives at least r of the objects.

Each of these distributions falls into exactly one of two categories:

1) The element x is in a container with r or more other objects: Here we start with $S_r(m,n)$ distributions of $a_1, a_2, \ldots, a_m$ into n identical containers – each container receiving at least r of the objects. Now we have n *distinct* containers – distinguished by their contents. Consequently, there are n choices for locating the object x. As a result, this category provides $nS_r(m,n)$ of the distributions.

2) The element x is in a container with $r-1$ of the other objects: These other $r-1$ objects can be chosen in $\binom{m}{r-1}$ ways, and then these objects – along with x – can be placed in one of the n containers. The remaining $m+1-r$ distinct objects can then be distributed among the $n-1$ identical containers – where each container receives at least r of the objects – in $S_r(m+1-r, n-1)$ ways. Hence this category provides the remaining $\binom{m}{r-1}S_r(m+1-r,n-1)$ distributions.

19. (a) We know that $s(m,n)$ counts the number of ways we can place m people — call them $p_1, p_2, \ldots, p_m$ — around n circular tables, with at least one occupant at each table. These arrangements fall into two disjoint sets: (1) The arrangements where p_1 is alone: There are $s(m-1,n-1)$ such arrangements; and, (2) The arrangements where p_1 shares a table with at least one of the other $m-1$ people: There are $s(m-1,n)$ ways where $p_2, p_3, \ldots, p_m$ can be seated around the n tables so that every table is occupied. Each such arrangement determines a total of $m-1$ locations (at all the n tables) where p_1 can now be seated — this for a total of $(m-1)s(m-1,n)$ arrangements. Consequently, $s(m,n) = (m-1)s(m-1,n) + s(m-1,n-1)$, for $m \geq n > 1$.

(b) For $m = 2$, we have $s(m,2) = 1 = 1!(1/1) = (m-1)!\sum_{i=1}^{m-1}\frac{1}{i}$. So the result is true in this case; this establishes the basis step for a proof by mathematical induction. Assuming the result for $m = k(\geq 2)$ we have $s(k,2) = (k-1)!\sum_{i=1}^{k-1}\frac{1}{i}$. Using the result from part (a) we now find that $s(k+1,2) = ks(k,2)+s(k,1) = k(k-1)!\sum_{i=1}^{k-1}\frac{1}{i}+(k-1)! = k!\sum_{i=1}^{k-1}\frac{1}{i}+(1/k)k! = k!\sum_{i=1}^{k}\frac{1}{i}$. The result now follows for all $m \geq 2$ by the Principle of Mathematical Induction.

Section 5.4

1. Here we find, for example, that
$f(f(a,b),c) = f(a,c) = c$, while
$f(a, f(b,c)) = f(a,b) = a$, so f is *not* associative.

3. (a) $f(x,y) = x + y - xy = y + x - yx = f(y,x)$, so the binary operation is commutative.
$f(f(w,x),y) = f(w,x) + y - f(w,x)y = (w + x - wx) + y - (w + x - wx)y = w + x + y - wx - wy - xy + wxy$.
$f(w, f(x,y)) = w + f(x,y) - w \cdot f(x,y) = w + (x + y - xy) - w(x + y - xy) = w + x + y - wx - wy - xy + wxy$.
Since $f(f(w,x),y) = f(w, f(x,y))$, the (closed) binary operation is associative.
(b), (d) Commutative and associative
(c) Neither commutative nor associative. For example, $f(2,3) = 2^3 = 8 \neq 9 = 3^2 = f(3,2)$. Also, $f(2, f(2,3)) = f(2, 2^3) = f(2,8) = 2^8 = 256$, while $f(f(2,2),3) = f(2^2,3) = f(4,3) = 4^3 = 64$.

5. (a) 25
(b) There are five choices (from A) for each of the 25 ordered pairs in $A \times A$. Consequently, the number of functions $f : A \times A \to A$ is 5^{25}.
(c) 5^{25}
(d) There are five ordered pairs in $A \times A$ of the form (x,x), where $x \in A$. There are five choices (from A) for each of these ordered pairs. This leaves $5^2 - 5 = 20$ ordered pairs of the form (x,y), where $x, y \in A$ and $x \neq y$. These 20 ordered pairs give rise to $20/2 = 10$ sets of two ordered pairs of the form (x,y), (y,x) – where $x, y \in A$, $x \neq y$. Here there are five choices (from A) for each set of two ordered pairs. Hence there are $5^5 \cdot 5^{10} = 5^5 \cdot 5^{(5^2-5)/2} = 5^{15}$ commutative closed binary operations on A.

7. (a) Yes (b) Yes (c) No

9. (a) $|A| = (32)(38) = 1216$. (b) The identity element for f is $p^{31}q^{37}$.

11. By the Well-Ordering Principle A has a least element and this same element is the identity for g. If A is finite then A will have a largest element and this same element will be the identity for f. If A is infinite then f cannot have an identity.

13. (a) 5 (b) {(25,25,6), (25,2,4), (60,40,20), (25,40,10)} (c) A_1, A_2

Section 5.5

1. Here the socks are the pigeons and the colors are the pigeonholes.

3. $26^2 + 1 = 677$

5. (a) For each $x \in \{1, 2, 3, \ldots, 300\}$ wrote $x = 2^n \cdot m$, where $n \geq 0$ and $\gcd(2, m) = 1$.

There are 150 possibilities for m: namely, $1, 3, 5, \ldots, 299$. In selecting 151 numbers from $\{1, 2, 3, \ldots, 300\}$ there must be two numbers of the form $x = 2^s \cdot m$, $y = 2^t \cdot m$. If $x < y$ then $x|y$; otherwise $y < x$ and $y|x$.
(b) If $n+1$ integers are selected from the set $\{1, 2, 3, \ldots, 2n\}$, then there must be two integers x, y in the selection where $x|y$ or $y|x$.

7. (a) Here the pigeons are the integers $1, 2, 3, \ldots, 25$ and the pigeonholes are the 13 sets: $\{1, 25\}, \{2, 24\}, \ldots, \{11, 15\}, \{12, 14\}, \{13\}$. In selecting 14 integers we get the elements in at least one two-element subset, and these sum to 26.
(b) If $S = \{1, 2, 3, \ldots, 2n+1\}$, for n a positive integer, then any subset of size $n+2$ from S must contain two elements that sum to $2n+2$.

9. (a) For any $t \in \{1, 2, 3, \ldots, 100\}, 1 \leq \sqrt{t} \leq 10$. Selecting 11 elements from $\{1, 2, 3, \ldots, 100\}$ there must be two, say x and y, where $\lfloor \sqrt{x} \rfloor = \lfloor \sqrt{y} \rfloor$, so that $0 < |\sqrt{x} - \sqrt{y}| < 1$.
(b) Let $n \in \mathbf{Z}^+$. If $n+1$ elements are selected from $\{1, 2, 3, \ldots, n^2\}$, then there exist two, say x and y, where $0 < |\sqrt{x} - \sqrt{y}| < 1$.

11.

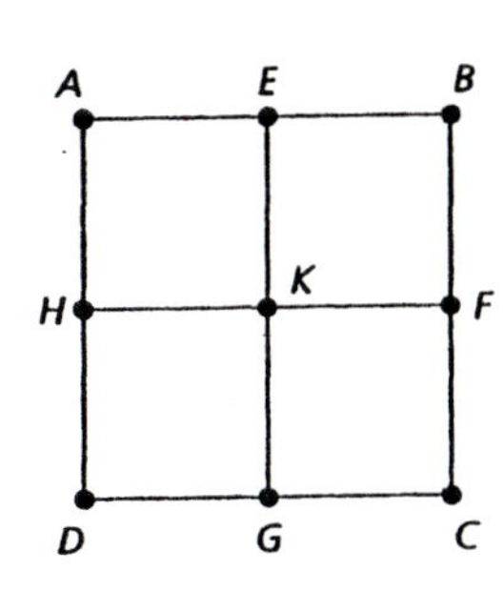

Divide the interior of the square into four smaller congruent squares as shown in the figure. Each smaller square has diagonal length $1/\sqrt{2}$. Let region R_1 be the interior of square AEKH together with the points on segment EK, excluding point E. Region R_2 is the interior of square EBFK together with the points on segment FK, excluding points F,K. Regions R_3, R_4 are defined in a similar way. Then if five points are chosen in the interior of square ABCD, at least two are in R_i for some $1 \leq i \leq 4$ and these points are within $1/\sqrt{2}$ (units) of each other.

13. Consider the subsets A of S where $1 \leq |A| \leq 3$. Since $|S| = 5$, there are $\binom{5}{1} + \binom{5}{2} + \binom{5}{3} = 25$ such subsets A. Let s_A denote the sum of the elements in A. Then $1 \leq s_A \leq 7 + 8 + 9 = 24$. So by the Pigeonhole Principle, there are two subsets of S whose elements yield the same sum.

15. For $(\emptyset \neq) T \subseteq S$, we have $1 \leq s_T \leq m + (m-1) + \cdots + (m-6) = 7m - 21$. The set S has $2^7 - 1 = 128 - 1 = 127$ nonempty subsets. So by the Pigeonhole Principle we need to have $127 > 7m - 21$ or $148 > 7m$. Hence $7 \leq m \leq 21$.

17. (a) 2,4,1,3
(b) 3,6,9,2,5,8,1,4,7
(c) For $n \geq 2$, there exists a sequence of n^2 distinct real numbers with no decreasing or increasing subsequence of length $n+1$. For example, consider $n, 2n, 3n, \ldots, (n-1)n, n^2, (n-1), (2n-1), \ldots, (n^2-1), (n-2), (2n-2), \ldots, (n^2-2), \ldots, 1, (n+1), (2n+1), \ldots, (n-1)n+1$.
(d) The result in Example 5.49 (for $n \geq 2$) is best possible – in the sense that we cannot reduce the length of the sequence from n^2+1 to n^2 and still obtain the desired subsequence of length $n+1$.

19. Proof: If not each pigeonhole contains at most k pigeons – for a total of at most kn pigeons. But we have $kn+1$ pigeons. So we have a contradiction and the result then follows.

21. (a) 1001 (b) 2001
(c) Let $k, n \in \mathbf{Z}^+$. The smallest value for $|S|$ (where $S \subset \mathbf{Z}^+$) so that there exist n elements $x_1, x_2, \ldots, x_n \in S$ where all n of these integers have the same remainder upon division by k is $k(n-1)+1$.

23. Proof: If not, then the number of pigeons roosting in the first pigeonhole is $x_1 \leq p_1 - 1$, the number of pigeons roosting in the second pigeonhole is $x_2 \leq p_2 - 1, \ldots$, and the number roosting in the n-th pigeonhole is $x_n \leq p_n - 1$. Hence the total number of pigeons is $x_1 + x_2 + \cdots + x_n = (p_1 - 1) + (p_2 - 1) + \cdots + (p_n - 1) = p_1 + p_2 + \cdots + p_n - n < p_1 + p_2 + \cdots + p_n - n + 1$, the number of pigeons we started with. The result now follows because of this contradiction.

Section 5.6

1. (a) There are 7! bijective functions on A – of these, 6! satisfy $f(1) = 1$. Hence there are $7! - 6! = 6(6!)$ bijective functions $f : A \to A$ where $f(1) \neq 1$.
(b) $n! - (n-1)! = (n-1)(n-1)!$

3. $9x^2 - 9x + 3 = g(f(x)) = 1 - (ax+b) + (ax+b)^2 = a^2x^2 + (2ab - a)x + (b^2 - b + 1)$. By comparing coefficients on like powers of x, $a = 3, b = -1$ or $a = -3, b = 2$.

5. $g^2(A) = g(T \cap (S \cup A)) = T \cap (S \cup [T \cap (S \cup A)]) =$
$T \cap [(S \cup T) \cap (S \cup (S \cup A))] = T \cap [(S \cup T) \cap (S \cup A)] =$
$[T \cap (S \cup T)] \cap (S \cup A) = T \cap (S \cup A) = g(A)$.

7. (a) $(f \circ g)(x) = 3x - 1$; $(g \circ f)(x) = 3(x-1)$;

$$(g \circ h)(x) = \begin{cases} 0, & x \text{ even}; \\ 3, & x \text{ odd} \end{cases} \qquad (h \circ g)(x) = \begin{cases} 0, & x \text{ even}; \\ 1, & x \text{ odd} \end{cases}$$

$$(f \circ (g \circ h))(x) = f((g \circ h)(x)) = \begin{cases} -1, & x \text{ even}; \\ 2, & x \text{ odd} \end{cases}$$

$$((f \circ g) \circ h)(x) = \begin{cases} (f \circ g)(0), & x \text{ even} \\ (f \circ g)(1), & x \text{ odd} \end{cases} = \begin{cases} -1, & x \text{ even} \\ 2, & x \text{ odd} \end{cases}$$

(b) $f^2(x) = f(f(x)) = x - 2$; $f^3(x) = x - 3$; $g^2(x) = 9x$; $g^3(x) = 27x$; $h^2 = h^3 = h^{500} = h$.

9. (a) We start with $y = e^{2x+5}$. Upon replacing x by y and y by x we get the equation $x = e^{2y+5}$. Now $x = e^{2y+5} \Rightarrow \ln x = \ln e^{2y+5} \Rightarrow \ln x = (2y+5)\ln e \Rightarrow \ln x = (2y+5) \cdot 1 \Rightarrow \ln x = 2y + 5 \Rightarrow y = \frac{1}{2}(\ln x - 5)$. So $f^{-1}(x) = \frac{1}{2}(\ell n x - 5)$.

(b) For $x \in \mathbf{R}^+, (f \circ f^{-1})(x) = f(\frac{1}{2}(\ell n x - 5)) = e^{2((1/2)(\ell n x - 5))+5} = e^{\ell n x - 5 + 5} = e^{\ell n x} = x$; for $x \in \mathbf{R}$, $(f^{-1} \circ f)(x) = f^{-1}(e^{2x+5}) = \frac{1}{2}[\ell n(e^{2x+5}) - 5] = \frac{1}{2}[2x + 5 - 5] = x$.

11. f, g invertible $\Longrightarrow$ each of f, g is both one-to-one and onto $\Longrightarrow g \circ f$ is one-to-one and onto $\Longrightarrow g \circ f$ invertible. Since $(g \circ f) \circ (f^{-1} \circ g^{-1}) = 1_C$ and $(f^{-1} \circ g^{-1}) \circ (g \circ f) = 1_A$, $f^{-1} \circ g^{-1}$ is an inverse of $g \circ f$. By uniqueness of inverses $f^{-1} \circ g^{-1} = (g \circ f)^{-1}$.

13.

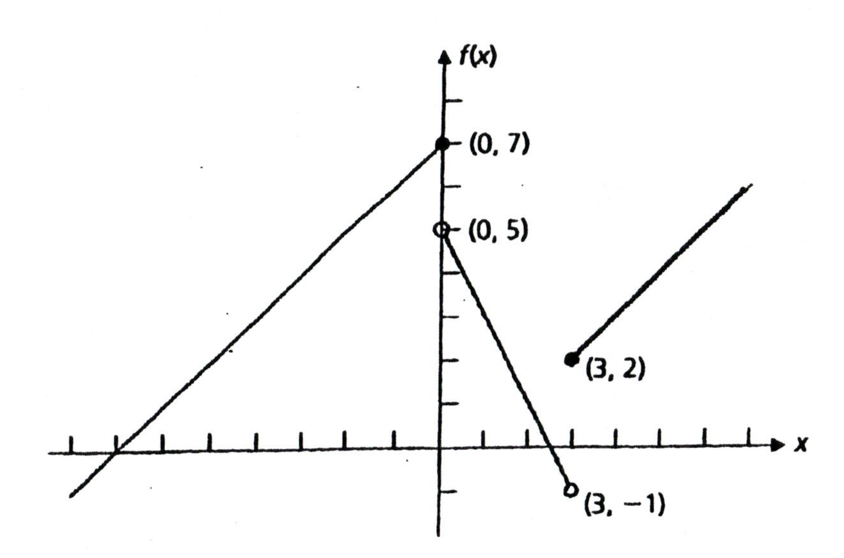

(a) $f^{-1}(-10) = \{x \in \mathbf{R} \mid x \leq 10 \text{ and } x + 7 = -10\} = \{-17\}$
$f^{-1}(0) = \{-7, 5/2\}$
$f^{-1}(4) = \{-3, 1/2, 5\}$
$f^{-1}(6) = \{-1, 7\}$
$f^{-1}(7) = \{0, 8\}$
$f^{-1}(8) = \{9\}$

(b) (i) $f^{-1}([-5,-1]) = \{x \in \mathbf{R} \mid x \leq 0 \text{ and } -5 \leq x + 7 \leq -1\} \cup \{x \in \mathbf{R} \mid 0 < x < 3 \text{ and } -5 \leq -2x + 5 \leq -1\} \cup \{x \in \mathbf{R} \mid 3 \leq x \text{ and } -5 \leq x - 1 \leq -1\} = \{x \in \mathbf{R} \mid x \leq 0 \text{ and } -12 \leq x \leq -8\} \cup \{x \in \mathbf{R} \mid 0 < x < 3 \text{ and } 3 \leq x \leq 5\} \cup \{x \in \mathbf{R} \mid 3 \leq x \text{ and } -4 \leq x \leq 0\} = [-12,-8] \cup \emptyset \cup \emptyset = [-12,-8]$
(ii) $f^{-1}([-5,0]) = [-12,-7] \cup [5/2,3)$
(iii) $f^{-1}([-2,4]) = \{x \in \mathbf{R} \mid x \leq 0 \text{ and } -2 \leq x + 7 \leq 4\} \cup \{x \in \mathbf{R} \mid 0 < x < 3 \text{ and } -2 \leq -2x + 5 \leq 4\} \cup \{x \in \mathbf{R} \mid 3 \leq x \text{ and } -2 \leq x - 1 \leq 4\} = \{x \in \mathbf{R} \mid x \leq 0 \text{ and } -9 \leq x \leq -3\} \cup \{x \in \mathbf{R} \mid 0 < x < 3 \text{ and } 1/2 \leq x \leq 7/2\} \cup \{x \in \mathbf{R} \mid 3 \leq x \text{ and } -1 \leq x \leq 5\} = [-9,-3] \cup [1/2,3) \cup [3,5] = [-9,-3] \cup [1/2,5]$
(iv) $f^{-1}((5,10)) = (-2,0] \cup (6,11)$
(v) $f^{-1}([11,17)) = \{x \in \mathbf{R} \mid x \leq 0 \text{ and } 11 \leq x + 7 < 17\} \cup \{x \in \mathbf{R} \mid 0 < x < 3 \text{ and } 11 \leq -2x + 5 < 17\} \cup \{x \in \mathbf{R} \mid 3 \leq x \text{ and } 11 \leq x - 1 < 17\} = \{x \in \mathbf{R} \mid x \leq 0 \text{ and } 4 \leq x < 10\} \cup \{x \in \mathbf{R} \mid 0 < x < 3 \text{ and } -6 < x \leq -3\} \cup \{x \in \mathbf{R} \mid 3 \leq x \text{ and } 12 \leq x < 18\} = \emptyset \cup \emptyset \cup [12,18) = [12,18)$

15. Since $f^{-1}(\{6,7,8\}) = \{1,2\}$ there are three choices for each of $f(1)$ and $f(2)$ – namely, 6, 7 or 8. Furthermore $3,4,5 \notin f^{-1}(\{6,7,8\})$ so $3,4,5 \in f^{-1}(\{9,10,11,12\})$ and we have four choices for each of $f(3)$, $f(4)$, and $f(5)$. Therefore, it follows by the rule of product that there are $3^2 \cdot 4^3 = 576$ functions $f : A \to B$ where $f^{-1}(\{6,7,8\}) = \{1,2\}$.

17. (a) The range of $f = \{2,3,4,\ldots\} = \mathbf{Z}^+ - \{1\}$.
(b) Since 1 is not in the range of f the function is not onto.
(c) For all $x, y \in \mathbf{Z}^+$, $f(x) = f(y) \Rightarrow x+1 = y+1 \Rightarrow x = y$, so f is one-to-one.
(d) The range of g is $\mathbf{Z}^+$.
(e) Since $g(\mathbf{Z}^+) = \mathbf{Z}^+$, the codomain of g, this function is onto.
(f) Here $g(1) = 1 = g(2)$, and $1 \neq 2$, so g is not one-to-one.
(g) For all $x \in \mathbf{Z}^+$, $(g \circ f)(x) = g(f(x)) = g(x+1) = \max\{1, (x+1)-1\} = \max\{1, x\} = x$, since $x \in \mathbf{Z}^+$. Hence $g \circ f = 1_{\mathbf{Z}^+}$.
(h) $(f \circ g)(2) = f(\max\{1,1\}) = f(1) = 1+1 = 2$
$(f \circ g)(3) = f(\max\{1,2\}) = f(2) = 2+1 = 3$
$(f \circ g)(4) = f(\max\{1,3\}) = f(3) = 3+1 = 4$
$(f \circ g)(7) = f(\max\{1,6\}) = f(6) = 6+1 = 7$
$(f \circ g)(12) = f(\max\{1,11\}) = f(11) = 11+1 = 12$
$(f \circ g)(25) = f(\max\{1,24\}) = f(24) = 24+1 = 25$
(i) No, because the functions f, g are *not* inverses of each other. The calculations in part (h) may suggest that $f \circ g = 1_{\mathbf{Z}^+}$ since $(f \circ g)(x) = x$ for $x \geq 2$. But we also find that $(f \circ g)(1) = f(\max\{1,0\}) = f(1) = 2$, so $(f \circ g)(1) \neq 1$, and, consequently, $f \circ g \neq 1_{\mathbf{Z}^+}$.

19. (a) $a \in f^{-1}(B_1 \cap B_2) \iff f(a) \in B_1 \cap B_2 \iff f(a) \in B_1$ and $f(a) \in B_2 \iff a \in f^{-1}(B_1)$ and $a \in f^{-1}(B_2) \iff a \in f^{-1}(B_1) \cap f^{-1}(B_2)$
(c) $a \in f^{-1}(\overline{B_1}) \iff f(a) \in \overline{B_1} \iff f(a) \notin B_1 \iff a \notin f^{-1}(B_1) \iff a \in \overline{f^{-1}(B_1)}$

21. (a) Suppose that $x_1, x_2 \in \mathbf{Z}$ and $f(x_1) = f(x_2)$. Then either $f(x_1)$, $f(x_2)$ are both even or they are both odd. If they are both even, then $f(x_1) = f(x_2) \Rightarrow -2x_1 = -2x_2 \Rightarrow x_1 = x_2$. Otherwise, $f(x_1)$, $f(x_2)$ are both odd and $f(x_1) = f(x_2) \Rightarrow 2x_1 - 1 = 2x_2 - 1 \Rightarrow 2x_1 = 2x_2 \Rightarrow x_1 = x_2$. Consequently, the function f is one-to-one.

In order to prove that f is an onto function let $n \in \mathbf{N}$. If n is even, then $(-n/2) \in \mathbf{Z}$ and $(-n/2) < 0$, and $f(-n/2) = -2(-n/2) = n$. For the case where n is odd we find that $(n+1)/2 \in \mathbf{Z}$ and $(n+1)/2 > 0$, and $f((n+1)/2) = 2[(n+1)/2] - 1 = (n+1) - 1 = n$. Hence f is onto.

(b) $f^{-1} : \mathbf{N} \to \mathbf{Z}$, where

$$f^{-1}(x) = \begin{cases} (\frac{1}{2})(x+1), & x = 1,3,5,7\ldots \\ -x/2\ , & x = 0,2,4,6,\ldots \end{cases}$$

23. (a) For all $n \in \mathbf{N}$, $(g \circ f)(n) = (h \circ f)(n) = (k \circ f)(n) = n$.
(b) The results in part (a) do not contradict Theorem 5.7. For although $g \circ f = h \circ f = k \circ f = 1_{\mathbf{N}}$, we note that

(i) $(f \circ g)(1) = f(\lfloor 1/3 \rfloor) = f(0) = 3 \cdot 0 = 0 \neq 1$, so $f \circ g \neq 1_{\mathbf{N}}$;
(ii) $(f \circ h)(1) = f(\lfloor 2/3 \rfloor) = f(0) = 3 \cdot 0 = 0 \neq 1$, so $f \circ h \neq 1_{\mathbf{N}}$; and
(iii) $(f \circ k)(1) = f(\lfloor 3/3 \rfloor) = f(1) = 3 \cdot 1 = 3 \neq 1$, so $f \circ k \neq 1_{\mathbf{N}}$.
Consequently, none of g, h, and k, is the inverse of f. (After all, since f is *not* onto it is *not* invertible.)

Section 5.7

1.

(a) $f \in O(n)$ (b) $f \in O(1)$ (c) $f \in O(n^3)$
(d) $f \in O(n^2)$ (e) $f \in O(n^3)$ (f) $f \in O(n^2)$
(g) $f \in O(n^2)$

3. (a) For all $n \in \mathbf{Z}^+, 0 \leq \log_2 n < n$. So let $k = 1$ and $m = 200$ in Definition 5.23. Then $|f(n)| = 100 \log_2 n = 100((1/2) \log_2 n) < 200((1/2)n) = 200|g(n)|$, so $f \in O(g)$.
(b) For $n = 6$, $2^n = 64 < 3096 = 4096 - 1000 = 2^{12} - 1000 = 2^{2n} - 1000$. Assuming that $2^k < 2^{2k} - 1000$ for $n = k \geq 6$, we find that $2 < 2^2 \Longrightarrow 2(2^k) < 2^2(2^{2k} - 1000) < 2^2 2^{2k} - 1000$, or $2^{k+1} < 2^{2(k+1)} - 1000$, so $f(n) < g(n)$ for all $n \geq 6$. Therefore, with $k = 6$ and $m = 1$ in Definition 5.23 we find that for $n \geq k$ $|f(n)| \leq m|g(n)|$ and $f \in O(g)$.
(c) For all $n \geq 4$, $n^2 \leq 2^n$ (A formal proof of this can be given by mathematical induction.) So let $k = 4$ and $m = 3$ in Definition 5.23. Then for $n \geq k$, $|f(n)| = 3n^2 \leq 3(2^n) < 3(2^n + 2n) = m|g(n)|$ and $f \in O(g)$.

5. To show that $f \in O(g)$, let $k = 1$ and $m = 4$ in Definition 5.23. Then for all $n \geq k$, $|f(n)| = n^2 + n \leq n^2 + n^2 = 2n^2 \leq 2n^3 = 4((1/2)(n^3)) = 4|g(n)|$, and f is dominated by g.
To show that $g \notin O(f)$, we follow the idea given in Example 5.66 – namely that

$$\forall m \in \mathbf{R}^+ \ \forall k \in \mathbf{Z}^+ \ \exists n \in \mathbf{Z}^+ \ [(n \geq k) \wedge (|g(n)| > m|f(n)|)].$$

So not matter what the values of m and k are, choose $n > max\{4m, k\}$. Then $|g(n)| = (1/2)n^3 > (1/2)(4m)n^2 = m(2n^2) \geq m(n^2 + n) = m|f(n)|$, so $g \notin O(f)$.

7. For all $n \geq 1, \log_2 n \leq n$, so with $k = 1$ and $m = 1$ in Definition 5.23 we have $|g(n)| = \log_2 n \leq n = m \cdot n = m|f(n)|$. Hence $g \in O(f)$.
To show that $f \in O(g)$ we first observe that $\lim_{n\to\infty} \frac{n}{\log_2 n} = +\infty$. (This can be established by using L'Hospital's Rule from the Calculus.) Since $\lim_{n\to\infty} \frac{n}{\log_2 n} = +\infty$ we find that for every $m \in \mathbf{R}^+$ and $k \in \mathbf{Z}^+$ there is an $n \in \mathbf{Z}^+$ such that $\frac{n}{\log_2 n} > m$, or $|f(n)| = n > m \log_2 n = m|g(n)|$. Hence $f \notin O(g)$.

9. Since $f \in O(g)$, there exists $m \in \mathbf{R}^+, k \in \mathbf{Z}^+$ so that $|f(n)| \leq m|g(n)|$ for all $n \geq k$. But then $|f(n)| \leq [m/|c|]|cg(n)|$ for all $n \geq k$, so $f \in O(cg)$.

11. (a) For all $n \geq 1$, $f(n) = 5n^2 + 3n > n^2 = g(n)$. So with $M = 1$ and $k = 1$, we have $|f(n)| \geq M|g(n)|$ for all $n \geq k$ and it follows that $f \in \Omega(g)$.

(b) For all $n \geq 1$, $g(n) = n^2 = (1/10)(5n^2 + 5n^2) > (1/10)(5n^2 + 3n) = (1/10)f(n)$. So with $M = (1/10)$ and $k = 1$, we find that $|g(n)| \geq M|f(n)|$ for all $n \geq k$ and it follows that $g \in \Omega(f)$.

(c) For all $n \geq 1$, $f(n) = 5n^2 + 3n > n = h(n)$. With $M = 1$ and $k = 1$, we have $|f(n)| \geq M|h(n)|$ for all $n \geq k$ and so $f \in \Omega(h)$.

(d) Suppose that $h \in \Omega(f)$. If so, there exist $M \in \mathbf{R}^+$ and $k \in \mathbf{Z}^+$ with $n = |h(n)| \geq M|f(n)| = M(5n^2+3n)$ for all $n \geq k$. Then $0 < M \leq n/(5n^2+3n) = 1/(5n+3)$. But how can M be a positive *constant* while $1/(5n+3)$ approaches 0 as n (a *variable*) gets larger? From this contradiction it follows that $h \notin \Omega(f)$.

13. (a) For $n \geq 1$, $f(n) = \sum_{i=1}^{n} i = n(n+1)/2 = (n^2/2) + (n/2) > (n^2/2)$. With $k = 1$ and $M = 1/2$, we have $|f(n)| \geq M|n^2|$ for all $n \geq k$. Hence $f \in \Omega(n^2)$.
(b) $\sum_{i=1}^{n} i^2 = 1^2 + 2^2 + \cdots + n^2 > \lceil n/2 \rceil^2 + \cdots + n^2 > \lceil n/2 \rceil^2 + \cdots + \lceil n/2 \rceil^2 = \lceil (n+1)/2 \rceil \lceil n/2 \rceil^2 > n^3/8$. With $k = 1$ and $M = 1/8$, we have $|g(n)| \geq M|n^3|$ for all $n \geq k$. Hence $g \in \Omega(n^3)$.

Alternately, for $n \geq 1$, $g(n) = \sum_{i=1}^{n} i^2 = n(n+1)(2n+1)/6 = (2n^3 + 3n^2 + n)/6 > n^3/6$. With $k = 1$ and $M = 1/6$, we find that $|g(n)| \geq M|n^3|$ for all $n \geq k$ – so $g \in \Omega(n^3)$.
(c) $\sum_{i=1}^{n} i^t = 1^t+2^t+\cdots+n^t > \lceil n/2 \rceil^t+\cdots+n^t > \lceil n/2 \rceil^t+\cdots+\lceil n/2 \rceil^t = \lceil (n+1)/2 \rceil \lceil n/2 \rceil^t > (n/2)^{t+1}$. With $k = 1$ and $M = (1/2)^{t+1}$, we have $|h(n)| \geq M|n^{t+1}|$ for all $n \geq k$. Hence $h \in \Omega(n^{t+1})$.

15. Proof: $f \in \Theta(g) \Rightarrow f \in \Omega(g)$ and $f \in O(g)$ (from Exercise 14 of this section) $\Rightarrow g \in O(f)$ and $g \in \Omega(f)$ (from Exercise 12 of this section) $\Rightarrow g \in \Theta(f)$.

Section 5.8

1. (a) As i, j each vary from 1 to n, the statement

$$sum := sum + 1$$

is executed n^2 times, so $f \in O(n^2)$.
(b) $f \in O(n^3)$
(c) Here the statement is executed $n + (n-1) + \cdots + 2 + 1 = n(n+1)/2$ times. Since $n(n+1)/2 = (1/2)n^2 + (1/2)n$ we have $f \in O(n^2)$.
(d) Let $2^k \leq n < 2^{k+1}$ for $k \in \mathbf{Z}^+$. The number of times the statement is executed is $k+1$. With $2^k \leq n$ we have $k \leq \log_2 n$, so $f(n) = k + 1 \leq \log_2 n + 1$ and $f \in O(\log_2 n)$.
(e) $f \in O(n \log_2 n)$

3. (a) For the following program segment the value of the integer n, and the values of the array entries $A[1], A[2], A[3], \ldots, A[n]$ are supplied beforehand. Also, the variables i, Max, and Location that are used here are integer variables.

```
Begin
    Max := A[1];
    Location := 1;
    If n = 1 then
        Begin
            Writeln ('The first occurrence of the maximum ');
            Write ('entry in the array is at position 1.')
        End;
    If n > 1 then
        Begin
            For i := 2 to n do
                If Max < A[i] then
                    Begin
                        Max := A[i];
                        Location := i
                    End;
                Writeln (' The first occurrence of the maximum ');
                Write (' entry in the array is at position ', i:0, '.')
        End
End;
```

(b) If, as in Exercise 2, we define the worst-case complexity function $f(n)$ as the number of times the comparison Max $< A[i]$ is executed, then $f(n) = n - 1$ for all $n \in \mathbf{Z}^+$, and $f \in O(n)$.

5. (a) Here there are five additions and ten multiplications.
(b) For the general case there are n additions and $2n$ multiplications.

7. Proof: For $n = 1$, we find that $a_1 = 0 = \lfloor 0 \rfloor = \lfloor \log_2 1 \rfloor$, so the result is true in this first case.
Now assume the result true for all $n = 1, 2, 3, \ldots, k$, where $k \geq 1$, and consider the cases for $n = k + 1$.
(i) $n = k + 1 = 2^m$, where $m \in \mathbf{Z}^+$: Here $a_n = 1 + a_{\lfloor n/2 \rfloor} = 1 + a_{2^{m-1}} = 1 + \lfloor \log_2 2^{m-1} \rfloor = 1 + (m - 1) = m = \lfloor \log_2 2^m \rfloor = \lfloor \log_2 n \rfloor$; and
(ii) $n = k + 1 = 2^m + r$, where $m \in \mathbf{Z}^+$ and $0 < r < 2^m$: Here $2^m < n < 2^{m+1}$, so we have
(1) $2^{m-1} < (n/2) < 2^m$;
(2) $2^{m-1} = \lfloor 2^{m-1} \rfloor \leq \lfloor n/2 \rfloor < \lfloor 2^m \rfloor = 2^m$; and
(3) $m - 1 = \log_2 2^{m-1} \leq \log_2 \lfloor n/2 \rfloor < \log_2 2^m = m$.
Consequently, $\lfloor \log_2 \lfloor n/2 \rfloor \rfloor = m - 1$ and $a_n = 1 + a_{\lfloor n/2 \rfloor} = 1 + \lfloor \log_2 \lfloor n/2 \rfloor \rfloor = 1 + (m - 1) = m = \lfloor \log_2 n \rfloor$. Therefore it follows from the Alternative Form of the Principle of Mathematical Induction that $a_n = \lfloor \log_2 n \rfloor$ for all $n \in \mathbf{Z}^+$.

9. Here $np = 3/4$ and $q = 1 - np = 1/4$, so $E(X) = np(n + 1)/2 + nq = (3/4)[(n + 1)/2] + (1/4)n = (3/8)n + (3/8) + (1/4)n = (5/8)n + (3/8)$.

11.

a)
```
procedure LocateRepeat (n: positive integer; a_1, a_2, a_3, ..., a_n: integers)
begin
    location := 0
    i := 2
    while i ≤ n and location = 0 do
        begin
            j := 1
            while j < i and location = 0 do
                if a_j = a_i then location := i
                else j := j + 1
            i := i + 1
        end
end {location is the subscript of the first array entry that repeats a previous
    array entry; location is 0 if the array contains n distinct integers}
```

b) For $n \geq 2$, let $f(n)$ count the maximum number of times the second **while** loop is executed. The second **while** loop is executed at most $n-1$ times for each value of i, where $2 \leq i \leq n$. Consequently, $f(n) = 1 + 2 + 3 + \cdots + (n-1) = (n-1)(n)/2$, which occurs when the array consists of n distinct integers or when the only repeat is a_{n-1} and a_n. Since $(n-1)(n)/2 = (1/2)(n^2 - n)$ we have $f \in O(n^2)$.

Supplementary Exercises

1. (a) If either A or B is $\emptyset$ then $A \times B = \emptyset = A \cap B$ and the result is true.
For A, B nonempty we find that:

$(x,y) \in (A \times B) \cap (B \times A) \Rightarrow (x,y) \in A \times B$ and $(x,y) \in B \times A \Rightarrow (x \in A$ and $y \in B)$ and $(x \in B$ and $y \in A) \Rightarrow x \in A \cap B$ and $y \in A \cap B \Rightarrow (x,y) \in (A \cap B) \times (A \cap B)$; and

$(x,y) \in (A \cap B) \times (A \cap B) \Rightarrow (x \in A$ and $x \in B)$ and $(y \in A$ and $y \in B) \Rightarrow (x,y) \in A \times B$ and $(x,y) \in B \times A \Rightarrow (x,y) \in (A \times B) \cap (B \times A)$.

Consequently, $(A \times B) \cap (B \times A) = (A \cap B) \times (A \cap B)$.

(b) If either A or B is $\emptyset$ then $A \times B = \emptyset = B \times A$ and the result follows.
If not, let $(x,y) \in (A \times B) \cup (B \times A)$. Then

$(x,y) \in (A \times B) \cup (B \times A) \Rightarrow (x,y) \in A \times B$ or $(x,y) \in (B \times A) \Rightarrow (x \in A$ and $y \in B)$ or $(x \in B$ and $y \in A) \Rightarrow (x \in A$ or $x \in B)$ and $(y \in A$ or $y \in B) \Rightarrow x, y \in A \cup B \Rightarrow (x,y) \in (A \cup B) \times (A \cup B)$.

3. (a) $f(1) = f(1 \cdot 1) = 1 \cdot f(1) + 1 \cdot f(1)$, so $f(1) = 0$.
(b) $f(0) = 0$
(c) Proof (by Mathematical Induction): When $a = 0$ the result is true, so consider $a \neq 0$. For $n = 1$, $f(a^n) = f(a) = 1 \cdot a^0 \cdot f(a) = na^{n-1} f(a)$, so the result follows in this first

case, and this establishes our basis step. Assume the result true for $n = k(\geq 1)$ – that is, $f(a^k) = ka^{k-1}f(a)$. For $n = k+1$ we have $f(a^{k+1}) = f(a \cdot a^k) = af(a^k) + a^k f(a) = aka^{k-1}f(a) + a^k f(a) = ka^k f(a) + a^k f(a) = (k+1)a^k f(a)$. Consequently, the truth of the result for $n = k+1$ follows from the truth of the result for $n = k$. So by the Principle of Mathematical Induction the result is true for all $n \in \mathbf{Z}^+$.

5. $(x, y) \in (A \cap B) \times (C \cap D) \iff x \in A \cap B, y \in C \cap D \iff (x \in A, y \in C)$ and $(x \in B, y \in D) \Longrightarrow (x, y) \in A \times C$ and $(x, y) \in B \times D \iff (x, y) \in (A \times C) \cap (B \times D)$

7. If $0 \leq x < 1$, then $\lfloor x \rfloor = 0$ and $x^2 = 1/2$. So $x = 1/\sqrt{2}$.
If $1 \leq x < 2$, then $\lfloor x \rfloor = 1$ and $x^2 = 3/2$. So $x = \sqrt{3/2}$.
For $k \in \mathbf{Z}^+$ and $k \geq 2$, if $k \leq x < k+1$, then $\lfloor x \rfloor = k$ and if x satisfies the given equation we have $x^2 = k + (1/2)$. But for $k \geq 2$ we find that $k(k-1) > 0$, so $k(k-1) \geq 1 > 1/2$, and $k^2 - k > 1/2$. Now $k^2 > k + (1/2) \Rightarrow k > \sqrt{k + (1/2)} = x$ and we do *not* have $k \leq x < k+1$.
Finally, let $k \in \mathbf{Z}^+$ and consider $-k \leq x < -k+1$. Then $x^2 - \lfloor x \rfloor = x^2 - (-k) = x^2 + k$, and $x^2 - \lfloor x \rfloor = 1/2 \Rightarrow x^2 = -k + 1/2 < 0$, so x cannot be a real number.
Consequently, there are only two real numbers that satisfy the equation $x^2 - \lfloor x \rfloor = 1/2$ — namely, $x = 1/\sqrt{2}$ and $x = \sqrt{3/2}$.

9. (a) $f^2(x) = f(f(x)) = a(f(x) + b) - b = a[(a(x+b) - b) + b] - b = a^2(x+b) - b$
$f^3(x) = f(f^2(x)) = f(a^2(x+b) - b) = a[(a^2(x+b) - b) + b] - b = a^3(x+b) - b$

(b) Conjecture: For $n \in \mathbf{Z}^+$, $f^n(x) = a^n(x+b) - b$. Proof (by Mathematical Induction): The formula is true for $n = 1$ – by the definition of $f(x)$. Hence we have our basis step. Assume the formula true for $n = k(\geq 1)$ – that is, $f^k(x) = a^k(x+b) - b$. Now consider $n = k+1$. We find that $f^{k+1}(x) = f(f^k(x)) = f(a^k(x+b) - b) = a[(a^k(x+b) - b) + b] - b = a^{k+1}(x+b) - b$. Since the truth of the formula at $n = k$ implies the truth of the formula at $n = k+1$, it follows that the formula is valid for all $n \in \mathbf{Z}^+$ – by the Principle of Mathematical Induction.

11. (a) $(7 \times 6 \times 5 \times 4 \times 3)/(7^5) \doteq 0.15$.
(b) For the computer program the elements of B are replaced by {1,2,3,4,5,6,7}.

```
10   Random
20   Dim F(5)
30   For I = 1 To 5
40        F(I) = Int(Rnd*7 + 1)
50   Next I
60   For J = 2 To 5
70        For K = 1 To J - 1
80             If F(J) = F(K) then GOTO 120
90        Next K
100  Next J
```

```
110  GOTO 140
120  C = C + 1
130  GOTO 10
140  C = C + 1
150  Print "After "; C; " generations the resulting"
160  Print "function is one-to-one."
170  Print "The one-to-one function is given as:"
180  For I = 1 To 5
190       Print "("; I; ","; F(I); ")"
200  Next I
210  End
```

13. For $1 \leq i \leq 10$, let x_i be the number of letters typed on day i. Then $x_1 + x_2 + x_3 + \ldots + x_8 + x_9 + x_{10} = 84$, or $x_3 + \ldots + x_8 = 54$. Suppose that $x_1 + x_2 + x_3 < 25, x_2 + x_3 + x_4 < 25, \ldots, x_8 + x_9 + x_{10} < 25$. Then $x_1 + 2x_2 + 3(x_3 + \ldots + x_8) + 2x_9 + x_{10} < 8(25) = 200$, or $3(x_3 + \ldots + x_8) < 160$. Consequently, $54 = x_3 + \ldots + x_8 < (160)/3 = 53\ 1/3$.

15. For $\prod_{k=1}^{n}(k - i_k)$ to be odd, $(k - i_k)$ must be odd for all $1 \leq k \leq n$, i.e., one of k, i_k must be even and the other odd. Since n is odd, $n = 2m + 1$ and in the list $1, 2, \ldots, n$, there are m even integers and $m + 1$ odd integers. Let $1, 3, 5, \ldots, n$ be the pigeons and $i_1, i_3, i_5, \ldots, i_n$ the pigeonholes. At most m of the pigeonholes can be even integers, so $(k - i_k)$ must be even for at least one $k = 1, 3, 5, \ldots, n$. Consequently, $\prod_{k=1}^{n}(k - i_k)$ is even.

17. Let the n distinct objects be $x_1, x_2, \ldots, x_n$. Place x_n in a container. Now there are two *distinct* containers. For each of $x_1, x_2, \ldots, x_{n-1}$ there are two choices and this gives 2^{n-1} distributions. Among these there is one where $x_1, x_2, \ldots, x_{n-1}$ are in the container with x_n, so we remove this distribution and find $S(n, 2) = 2^{n-1} - 1$.

19. (a) and (b) $m!S(n, m)$

21. Fix $m = 1$. For $n = 1$ the result is true. Assume $f \circ f^k = f^k \circ f$ and consider $f \circ f^{k+1}$. $f \circ f^{k+1} = f \circ (f \circ f^k) = f \circ (f^k \circ f) = (f \circ f^k) \circ f = f^{k+1} \circ f$. Hence $f \circ f^n = f^n \circ f$ for all $n \in \mathbf{Z}^+$. Now assume that for $t \geq 1, f^t \circ f^n = f^n \circ f^t$. Then $f^{t+1} \circ f^n = (f \circ f^t) \circ f^n = f \circ (f^t \circ f^n) = f \circ (f^n \circ f^t) = (f \circ f^n) \circ f^t = (f^n \circ f) \circ f^t = f^n \circ (f \circ f^t) = f^n \circ f^{t+1}$, so $f^m \circ f^n = f^n \circ f^m$ for all $m, n \in \mathbf{Z}^+$.

23. Proof: Let $a \in A$. Then

$$f(a) = g(f(f(a))) = f(g(f(f(f(a))))) = f(g \circ f^3(a)).$$

From $f(a) = g(f(f(a)))$ we have $f^2(a) = (f \circ f)(a) = f(g(f(f(a))))$. So $f(a) = f(g \circ f^3(a)) = f(g(f(f(f(a))))) = f^2(f(a)) = f^2(g(f^2(a))) = f(f(g(f(f(a))))) = f(g(f(a))) = g(a)$.

Consequently, $f = g$.

25. a) Note that $2 = 2^1$, $16 = 2^4$, $128 = 2^7$, $1024 = 2^{10}$, $8192 = 2^{13}$, and $65536 = 2^{16}$. Consider the exponents on 2. If four numbers are selected from $\{1, 4, 7, 10, 13, 16\}$, there is at least one pair whose sum is 17. Hence if four numbers are selected from S, there are two numbers whose product is $2^{17} = 131072$.

b) Let $a, b, c, d, n \in \mathbf{Z}^+$. Let $S = \{b^a, b^{a+d}, b^{a+2d}, \ldots, b^{a+nd}\}$. If $\lceil \frac{n}{2} \rceil + 1$ numbers are selected from S then there are at least two of them whose product is b^{2a+nd}.

27. $f \circ g = \{(x,z),(y,y),(z,x)\}; g \circ f = \{(x,x),(y,z),(z,y)\};$
$f^{-1} = \{(x,z),(y,x),(z,y)\}; g^{-1} = \{(x,y),(y,x),(z,z)\};$
$(g \circ f)^{-1} = \{(x,x),(y,z),(z,y)\} = f^{-1} \circ g^{-1}; g^{-1} \circ f^{-1} = \{(x,z),(y,y),(z,x)\}.$

29. Under these conditions we know that $f^{-1}(\{6,7,9\}) = \{2,4,5,6,9\}$. Consequently we have
(i) two choices for each of $f(1)$, $f(3)$, and $f(7)$ – namely, 4 or 5;
(ii) two choices for each of $f(8)$ and $f(10)$ – namely, 8 or 10; and
(iii) three choices for each of $f(2)$, $f(4)$, $f(5)$, $f(6)$, and $f(9)$ – namely, 6, 7, or 9.
Therefore, by the rule of product, it follows that the number of functions satisfying these conditions is $2^3 \cdot 2^2 \cdot 3^5 = 7776$.

31. (a) $(\pi \circ \sigma)(x) = (\sigma \circ \pi)(x) = x$

(b) $\pi^n(x) = x - n; \sigma^n(x) = x + n (n \geq 2)$.

(c) $\pi^{-n}(x) = x + n; \sigma^{-n}(x) = x - n (n \geq 2)$.

33. (a) Here there are eight distinct primes and each subset A satisfying the stated property determines a distribution of the eight distinct objects in $X = \{2,3,5,7,11,13,17,19\}$ into four identical containers with no container left empty. There are $S(8,4)$ such distributions.

(b) $S(n,m)$

35. (a) Let $m = 1$ and $k = 1$. Then for all $n \geq k, |f(n)| \leq 2 < 3 \leq |g(n)| = m|g(n)|$, so $f \in O(g)$.

(b) Let $m = 4$ and $k = 1$. Then for all $n \geq k, |g(n)| \leq 4 = 4 \cdot 1 \leq 4|f(n)| = m|f(n)|$, so $g \in O(f)$.

37. First note that if $\log_a n = r$, then $n = a^r$ and $\log_b n = \log_b(a^r) = r \log_b a = (\log_b a)(\log_a n)$. Now let $m = (\log_b a)$ and $k = 1$. Then for all $n \geq k, |g(n)| = \log_b n = (\log_b a)(\log_a n) = m|f(n)|$, so $g \in O(f)$.
Finally, with $m = (\log_b a)^{-1} = \log_a b$ and $k = 1$, we find that for all $n \geq k, |f(n)| = \log_a n = (\log_a b)(\log_b n) = m|g(n)|$. Hence $f \in O(g)$.

CHAPTER 6
LANGUAGES: FINITE STATE MACHINES

Section 6.1

1. (a) $|\Sigma^2| = |\Sigma|^2 = 5^2 = 25$; 125 (b) $\sum_{i=0}^{5} 5^i = 3906$

3. Since $36 = \| x^3 \| = 3 \| x \|$, it follows that $\| x \| = 12$.

5. Consider the strings of length 3 where xy is a proper prefix. These strings look like xya where a can be any of v, w, x, y, z. So there are 5 such strings. Such strings of length 4 look like $xyab$ where each of a, b can be any of v, w, x, y, z. So there are 5^2 such strings. Continuing, there are 5^3 strings of length 5 where xy is a proper prefix and 5^4 such strings of length 6. Consequently, the number of strings in A where xy is a proper prefix is $\sum_{i=1}^{4} 5^i$.

7. (a) { 00,11, 000,111, 0000,1111}
(b) {0,1}
(c) $\Sigma^* - \{\lambda, 00, 11, 000, 111, 0000, 1111\}$
(d) {0,1, 00,11}
(e) Σ^*
(f) $\Sigma^* - \{0, 1, 00, 11\} = \{\lambda, 01, 10\} \cup \{w | \; \| w \| \geq 3\}$

9. (a) $x \in AC \Longrightarrow x = ac$, for some $a \in A, c \in C \Longrightarrow x \in BD$, since $A \subseteq B, C \subseteq D$.

(b) If $A\emptyset \neq \emptyset$, let $x \in A\emptyset$. $x \in A\emptyset \Longrightarrow x = yz$, for some $y \in A, z \in \emptyset$. But $z \in \emptyset$ is impossible. Hence $A\emptyset = \emptyset$. [In like manner $\emptyset A = \emptyset$].

11. For any alphabet Σ, let $B \subseteq \Sigma$. Then, if $A = B^*$, it follows from part (f) of Theorem 6.2 that $A^* = (B^*)^* = B^* = A$.

13. (a) Here A^* consists of all strings x of even length where if $x \neq \lambda$, then x starts with 0 and ends with 1, and the symbols (0 and 1) alternate.
(b) In this case A^* contains precisely those strings made up of $3n$ 0's, for $n \in \mathbf{N}$.
(c) Here a string $x \in A^*$ if (and only if)
(i) x is a string of n 0's, for $n \in \mathbf{N}$; or
(ii) x is a string that starts and ends with 0, and has at least one 1 – but no consecutive 1's.
(d) For this last case A^* consists of the following:
(i) Any string of n 1's, for $n \in \mathbf{N}$; and

(ii) Any string that starts with 1 and contains at least one 0, but no consecutive 0's.

15. Let $\sum$ be an alphabet with $\emptyset \neq A \subseteq \sum^*$. If $|A| = 1$ and $x \in A$, then $xx = x$ since $A^2 = A$. But $\| xx \| = 2 \| x \| = \| x \| \Longrightarrow \| x \| = 0 \Longrightarrow x = \lambda$. If $|A| > 1$, let $x \in A$ where $\| x \| > 0$ but $\| x \|$ is minimal. Then $x \in A^2 \Longrightarrow x = yz$, for some $y, z \in A$. Since $\| x \| = \| y \| + \| z \|$, if $\| y \|, \| z \| > 0$, then one of y, z is in A with length smaller than $\| x \|$. Consequently, one of $\| y \|$ or $\| z \|$ is 0, so $\lambda \in A$.

17. If $A = A^2$ then it follows by mathematical induction that $A = A^n$ for all $n \in \mathbf{Z}^+$. Hence $A = A^+$. From Exercise 15 we know that $A = A^2 \Longrightarrow \lambda \in A$, so $A = A^*$.

19. By Definition 6.11 $AB = \{ab|a \in A, b \in B\}$, and since it is possible to have $a_1b_1 = a_2b_2$ with $a_1, a_2 \in A, a_1 \neq a_2$, and $b_1, b_2 \in B, b_1 \neq b_2$, it follows that $|AB| \leq |A \times B| = |A||B|$.

21. (a) The words 001 and 011 have length 3 and are in A. The words 00011 and 00111 have length 5 and they are also in A.
(b) From step (1) we know that $1 \in A$. Then by applying step (2) three times we get
(i) $1 \in A \Longrightarrow 011 \in A$;
(ii) $011 \in A \Longrightarrow 00111 \in A$; and
(iii) $00111 \in A \Longrightarrow 0001111 \in A$.
(c) If 00001111 were in A, then from step (2) we see that this word would have to be generated from 000111 (in A). Likewise, 000111 in $A \Longrightarrow$ 0011 is in $A \Longrightarrow$ 01 is in A. However, there are no words in A of length 2 — in fact, there are no words of even length in A.

23.

(a) **Steps**	**Reasons**
1. ( ) is in A.	Part (1) of the recursive definition
2. (( )) is in A.	Step 1 and part (2(ii)) of the definition
3. (( ))( ) is in A.	Steps 1, 2, and part (2(i)) of the definition

(b) **Steps**	**Reasons**
1. ( ) is in A.	Part (1) of the recursive definition
2. (( )) is in A.	Step 1 and part (2(ii)) of the definition
3. (( ))( ) is in A.	Steps 1, 2, and part (2(i)) of the definition
4. (( ))( )( ) is in A.	Steps 1, 3, and part (2(i)) of the definition

(c) **Steps**	**Reasons**
1. ( ) is in A.	Part (1) of the recursive definition
2. ( )( ) is in A.	Step 1 and part (2(i)) of the definition
3. (( )( )) is in A.	Step 2 and part (2(ii)) of the definition
4. ( )(( )( )) is in A.	Steps 1, 3, and part (2(i)) of the definition

25. Length 3: $\binom{3}{0} + \binom{2}{1} = 3$
Length 4: $\binom{4}{0} + \binom{3}{1} + \binom{2}{2} = 5$

Length 5: $\binom{5}{0} + \binom{4}{1} + \binom{3}{2} = 8$
Length 6: $\binom{6}{0}+\binom{5}{1}+\binom{4}{2}+\binom{3}{3} = 13$ [Here the summand $\binom{6}{0}$ counts the strings where there are no 0s; the summand $\binom{5}{1}$ counts the strings where we arrange the symbols $1, 1, 1, 1, 00$; the summand $\binom{4}{2}$ is for the arrangements of $1, 1, 00, 00$; and the summand $\binom{3}{3}$ counts the arrangements of $00, 00, 00$.]

27.

A: (1) $\lambda \in A$
(2) If $a \in A$, then $0a0, 0a1, 1a0, 1a1 \in A$.

B: (1) $0, 1 \in A$.
(2) If $a \in A$, then $0a0, 0a1, 1a0, 1a1 \in A$.

Section 6.2

1. (a)

State	s_0	s_1	s_2	s_1	s_2	s_1	s_2	s_1
Input	1	0	1	0	1	0	1	
Output	0	0	1	0	1	0	1	

(b)

State	s_0	s_1	s_2	s_0	s_1	s_2	s_0	s_1
Input	1	0	0	1	0	0	1	
Output	0	0	0	0	0	0	0	

(c)

State	s_0	s_1	s_2	s_1	s_2	s_0	s_1	s_2	s_0	s_0
Input	1	0	1	0	0	1	0	0	0	
Output	0	0	1	0	0	0	0	0	0	

3. (a)

State	s_0	s_0	s_3	s_3	s_0	s_2	s_3
Input	a	b	b	c	c	c	
Output	0	1	0	1	1	0	

(b)

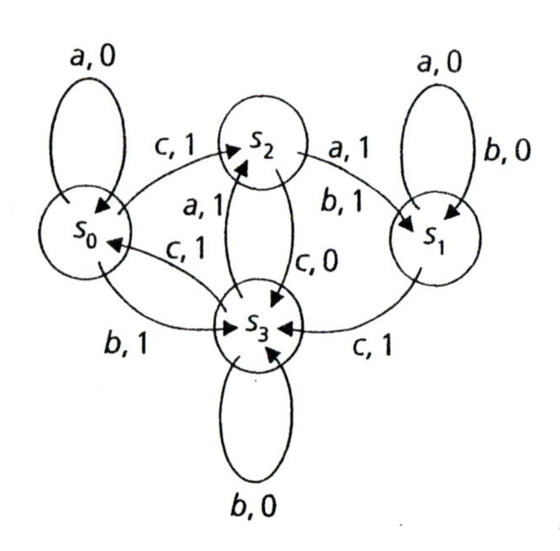

5. (a)

State	s_0	s_1	s_2	s_2	s_2	s_2	s_2
Input	1	1	0	1	1	1	
Output	0	1	0	0	0	0	

(b)

State	s_1	s_2	s_2	s_2	s_2	s_2	s_2
Input	1	1	0	1	1	1	
Output	1	0	0	0	0	0	

State	s_2	s_2	s_2	s_2	s_2	s_2	s_2
Input	1	1	0	1	1	1	
Output	0	0	0	0	0	0	

State	s_3	s_3	s_3	s_0	s_1	s_2	s_2
Input	1	1	0	1	1	1	
Output	1	1	0	0	1	0	

(c)

	ν		ω	
	0	1	0	1
s_0	s_0	s_1	0	0
s_1	s_1	s_2	1	1
s_2	s_2	s_2	0	0
s_3	s_0	s_3	0	1
s_4	s_2	s_3	0	1

(d) s_1

(e) $x = 101$ (unique)

7. (a) (i) $|S \times I| = |S||I| = 3 \cdot 5 = 15$

(ii) Since $|S \times I| = 15$ and $|S| = 3$, there are 3^{15} such functions.

(iii) 2^{15}

(b) Each finite state machine is determind by our choice for ν and our choice for ω. In total we have $3^{15} \cdot 2^{15} = 6^{15}$ choices.

9.

(a)

	ν		ω	
	0	1	0	1
s_0	s_4	s_1	0	0
s_1	s_3	s_2	0	0
s_2	s_3	s_2	0	1
s_3	s_3	s_3	0	0
s_4	s_5	s_3	0	0
s_5	s_5	s_3	1	0

(b) There are only two possibilities: $x = 1111$ or $x = 0000$.

(c) $A = \{111\}\{1\}^* \cup \{000\}\{0\}^*$
(d) Here $A = \{11111\}\{1\}^* \cup \{00000\}\{0\}^*$.

Section 6.3

1. (a)

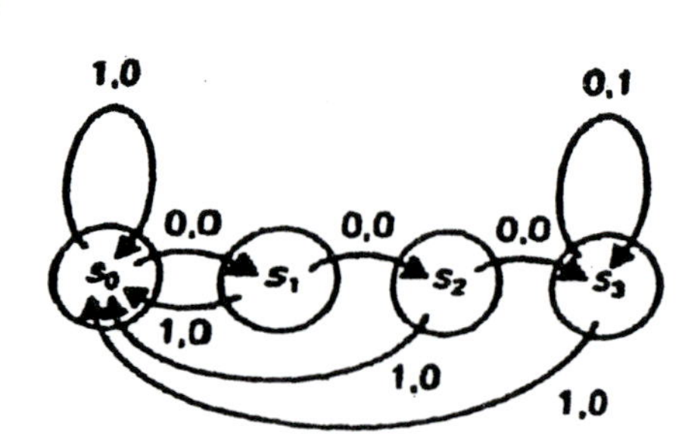

(b)

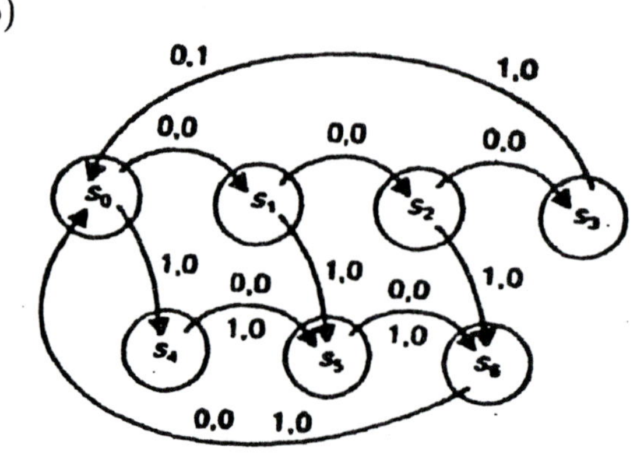

3.

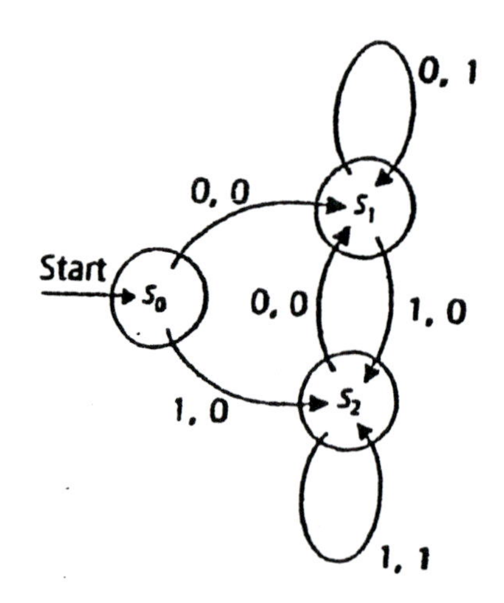

5.

(a)

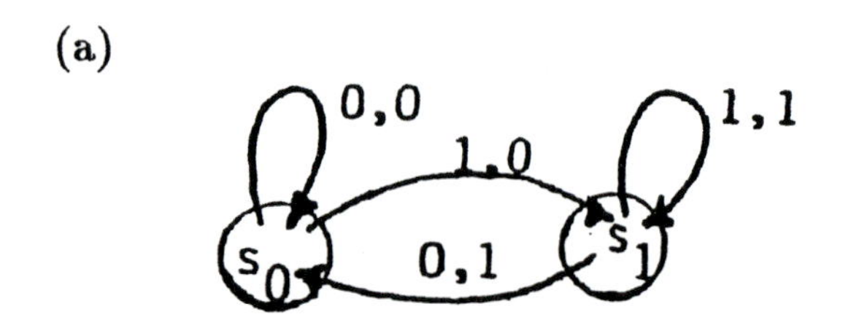

(b) (i) Input 111
Output 011
(ii) Input 1010
Output 0101
(iii) Input 00011
Output 00001

(c) The machine outputs a 0 followed by the first $n - 1$ symbols of the n symbol input string x. Hence the machine is a unit delay.

(d) The machine here performs the same tasks as the one in Fig. 6.13 (and has only two states.)

7. (a) The transient states are s_0, s_1. State s_4 is a sink state. $\{s_1, s_2, s_3, s_4, s_5\}$, $\{s_4\}$, $\{s_2, s_3, s_5\}$ (with the corresponding restrictions on the given function ν) constitute submachines. The strongly connected submachines are $\{s_4\}$ and $\{s_2, s_3, s_5\}$.

(b) States s_2, s_3 are transient. The only sink state is s_4. The set $\{s_0, s_1, s_3, s_4\}$ provides the states for a submachine; $\{s_0, s_1\}$, $\{s_4\}$ provide strongly connected submachines.

(c) Here there are no transient states. State s_6 is a sink state. There are three submachines: $\{s_2, s_3, s_4, s_5, s_6\}$, $\{s_3, s_4, s_5, s_6\}$, and $\{s_6\}$. The only strongly connected submachine is $\{s_6\}$.

Supplementary Exercises

1. (a) True (b) False, because $w \notin \Sigma_2$ (c) True
(d) True (e) True (f) True

3. Let $x \in \Sigma$ and $A = \{x\}$. Then $A^2 = \{x^2\}$ and $(A^2)^* = \{\lambda, x^2, x^4, \ldots\}$. However $A^* = (\lambda, x, x^2, \ldots\}$ and $(A^*)^2 = A^*$, so $(A^*)^2 \neq (A^2)^*$.

5. O_{02} : Starting at s_0 we can return to s_0 for any input from $\{1, 00\}^*$. To finish at state s_2 requires an input of 0. Hence $O_{02} = \{1, 00\}^*\{0\}$
$O_{22} : \{0\}\{1, 00\}^*\{0\}$
$O_{11} : \emptyset$
$O_{00} : \{1, 00\}^* - \{\lambda\}$
$O_{10} : \{1\}\{1, 00\}^* \cup \{10\}\{1, 00\}^*$

7. (a) By the Pigeonhole Principle there is a first state s that is encountered twice. Let y be the output string that resulted since s was first encountered until we reach this state a second time. Then from that point on the output is $yyy\ldots$.
(b) n (c) n

9.

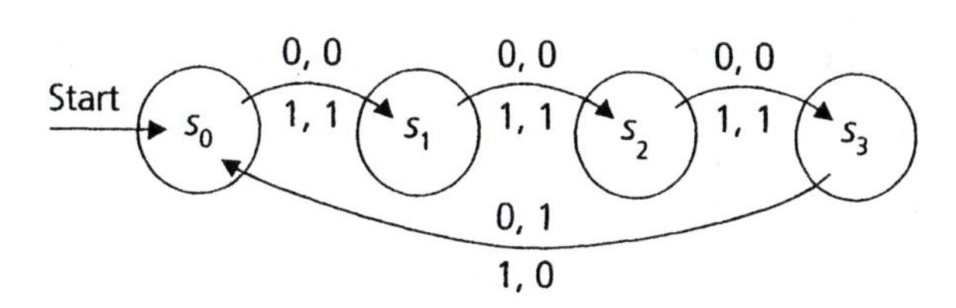

11.

(a)

	ν		ω	
	0	1	0	1
(s_0, s_3)	(s_0, s_4)	(s_1, s_3)	1	1
(s_0, s_4)	(s_0, s_3)	(s_1, s_4)	0	1
(s_1, s_3)	(s_1, s_3)	(s_2, s_4)	1	1
(s_1, s_4)	(s_1, s_4)	(s_2, s_3)	1	1
(s_2, s_3)	(s_2, s_3)	(s_0, s_4)	1	1
(s_2, s_4)	(s_2, s_4)	(s_0, s_3)	1	0

(b) $\omega((s_0, s_3), 1101) = 1111; M_1$ is in state s_0 and M_2 is in state s_4.

CHAPTER 7
RELATIONS: THE SECOND TIME AROUND

Section 7.1

1. (a) {(1,1),(2,2),(3,3),(4,4),(1,2),(2,1),(2,3),(3,2)}
(b) {(1,1),(2,2),(3,3),(4,4),(1,2)}
(c) {(1,1),(2,2),(1,2),(2,1)}

3. (a) Let $f_1, f_2, f_3 \in F$ with $f_1(n) = n+1$, $f_2(n) = 5n$, and $f_3(n) = 4n + 1/n$.
(b) Let $g_1, g_2, g_3 \in F$ with $g_1(n) = 3$, $g_2(n) = 1/n$, and $f_3(n) = \sin n$.

5. (a) For all $a \in \mathbf{Z}^+$, $a|a$, so $\mathcal{R}$ is reflexive. Now $2|6$ but $6 \not| 2$, so the relation is not symmetric. For $a, b \in \mathbf{Z}^+$, $a|b$ and $b|a \Rightarrow a = b$, so $\mathcal{R}$ is antisymmetric. (Here we need a, b to be positive integers – for if a, b were merely integers, then with $a = 2$ and $b = -2$ we could have $a|b$ and $b|a$, but $a \neq b$.) Finally, let $a, b, c \in \mathbf{Z}^+$ where $a|b$ and $b|c$. $a|b \Rightarrow b = am$ for some $m \in \mathbf{Z}^+$. Likewise, $b|c \Rightarrow c = bn$ for some $n \in \mathbf{Z}^+$. Consequently, $c = bn = a(mn)$, with $mn \in \mathbf{Z}^+$, so $a|c$ and $\mathcal{R}$ is transitive.
(b) Here the relation is not reflexive because $0 \not| 0$, but it is transitive.
(c) reflexive, symmetric, transitive
(d) symmetric
(e) For any $x \in \mathbf{Z}$, $x + x = 2x$ is even, so no integer is related to itself and $\mathcal{R}$ is not reflexive. If $x, y \in \mathbf{Z}$, then $x\mathcal{R}y \Rightarrow x + y$ is odd $\Rightarrow y + x$ is odd $\Rightarrow y\mathcal{R}x$, so $\mathcal{R}$ is symmetric. Also, $2\mathcal{R}3$ and $3\mathcal{R}2$, but $2 \neq 3$, so $\mathcal{R}$ is not antisymmetric. Finally, $2\mathcal{R}3$ and $3\mathcal{R}4$, but $2\not\mathcal{R}4$, so $\mathcal{R}$ is not transitive.
(f) reflexive, symmetric, transitive
(g) reflexive, symmetric
(h) For all $(a, b) \in \mathbf{Z} \times \mathbf{Z}$, $a \leq a$, so $(a, b)\mathcal{R}(a, b)$ and $\mathcal{R}$ is reflexive. Here $(2,3)\mathcal{R}(3,4)$ but $(3,4)\not\mathcal{R}(2,3)$, so the relation is not symmetric. Also, $(2,3)\mathcal{R}(2,4)$ and $(2,4)\mathcal{R}(2,3)$, but $(2,3) \neq (2,4)$ – so $\mathcal{R}$ is not antisymmetric. Finally, let $(a,b), (c,d), (e,f) \in \mathbf{Z} \times \mathbf{Z}$ with $(a,b)\mathcal{R}(c,d)$ and $(c,d)\mathcal{R}(e,f)$. Then $a \leq c$ and $c \leq e$, so $a \leq e$, and it follows that $(a,b)\mathcal{R}(e,f)$, making $\mathcal{R}$ transitive.

7. (a) For all $x \in A$, $(x,x) \in \mathcal{R}_1, \mathcal{R}_2$, so $(x,x) \in \mathcal{R}_1 \cap \mathcal{R}_2$ and $\mathcal{R}_1 \cap \mathcal{R}_2$ is reflexive.

(b) All of these results are true. For example if $\mathcal{R}_1, \mathcal{R}_2$ are both transitive and $(x,y), (y,z) \in \mathcal{R}_1 \cap \mathcal{R}_2$ then $(x,y), (y,z) \in \mathcal{R}_1, \mathcal{R}_2$, so $(x,z) \in \mathcal{R}_1, \mathcal{R}_2$ (transitive property) and $(x,z) \in \mathcal{R}_1 \cap \mathcal{R}_2$.

If $\mathcal{R}_1$, $\mathcal{R}_2$ are both symmetric, let $(x, y) \in \mathcal{R}_1 \cap \mathcal{R}_2$. Then $(x, y) \in \mathcal{R}_1 \cap \mathcal{R}_2 \Rightarrow (x, y) \in \mathcal{R}_1$ and $(x, y) \in \mathcal{R}_2 \Rightarrow (y, x) \in \mathcal{R}_1$ and $(y, x) \in \mathcal{R}_2$ [because $\mathcal{R}_1, \mathcal{R}_2$ are symmetric] $\Rightarrow (y, x) \in \mathcal{R}_1 \cap \mathcal{R}_2$, so $\mathcal{R}_1 \cap \mathcal{R}_2$ is symmetric.

Finally, suppose that each of $\mathcal{R}_1, \mathcal{R}_2$ is antisymmetric and that $(x, y), (y, x) \in \mathcal{R}_1 \cap \mathcal{R}_2$. Then $(x, y), (y, x) \in \mathcal{R}_1 \cap \mathcal{R}_2 \Rightarrow (x, y), (y, x) \in \mathcal{R}_1$ and $(x, y), (y, x) \in \mathcal{R}_2 \Rightarrow x = y$ [because $\mathcal{R}_1, \mathcal{R}_2$ are antisymmetric], so $\mathcal{R}_1 \cap \mathcal{R}_2$ is antisymmetric. [Note: This result remains true if only one of $\mathcal{R}_1, \mathcal{R}_2$ is antisymmetric.]

9.

(a) False: Let $A = \{1, 2\}$ and $\mathcal{R} = \{(1, 2), (2, 1)\}$.
(b) (i) If $\mathcal{R}_1$ is reflexive, then $(a, a) \in \mathcal{R}_1$ for all $a \in A$. With $\mathcal{R}_2 \supseteq \mathcal{R}_1$ it follows that $(a, a) \in \mathcal{R}_2$ for all $a \in A$. Consequently, $\mathcal{R}_2$ is reflexive.
(ii) Symmetric: False. Let $A = \{1, 2\}, \mathcal{R}_1 = \{(1, 1)\}, \mathcal{R}_2 = \{(1, 1), (1, 2)\}$.
(iii) Antisymmetric & Transitive: False. Let $A = \{1, 2\}, \mathcal{R}_1 = \{(1, 2)\}, \mathcal{R}_2 = \{(1, 2), (2, 1)\}$.
(c) (i) Reflexive: False. Let $A = \{1, 2\}, \mathcal{R}_1 = \{(1, 1)\}, \mathcal{R}_2 = \{(1, 1), (2, 2)\}$.
(ii) Symmetric: False. Let $A = \{1, 2\}, \mathcal{R}_1 = \{(1, 2)\}, \mathcal{R}_2 = \{(1, 2), (2, 1)\}$.
(iii) Suppose that $(x, y), (y, x) \in \mathcal{R}_1$, for $x, y \in A$. Then $(x, y), (y, x) \in \mathcal{R}_2$, since $\mathcal{R}_1 \subseteq \mathcal{R}_2$. Since $\mathcal{R}_2$ is antisymmetric it follows that $x = y$. Consequently, $\mathcal{R}_1$ is antisymmetric.
(iv) Transitive: False. Let $A = \{1, 2\}, \mathcal{R}_1 = \{(1, 2), (2, 1)\}, \mathcal{R}_2 = \{(1, 1), (1, 2), (2, 1), (2, 2)\}$.
(d) For any nonempty finite set A the smallest equivalence relation on A is the "equality" relation. If $|A| = n$, then this relation consists of the n ordered pairs $(a, a), a \in A$. The largest equivalence relation A is $A \times A$, where $|A \times A| = n^2$ when $|A| = n$. Hence for any equivalence relation $\mathcal{R}$ on A, we have $n \leq |\mathcal{R}| \leq n^2$, and the given result is true.

11. (a) $\binom{2+2-1}{2}\binom{2+2-1}{2} = \binom{3}{2}\binom{3}{2} = 9$
(b) $\binom{3+2-1}{2}\binom{2+2-1}{2} = \binom{4}{2}\binom{3}{2} = 18$
(c) $\binom{4+2-1}{2}\binom{2+2-1}{2} = \binom{5}{2}\binom{3}{2} = 30$
(d) $\binom{4+2-1}{2}\binom{3+2-1}{2} = \binom{5}{2}\binom{4}{2} = 60$
(e) $\binom{2+2-1}{2}^4 = \binom{3}{2}^4 = 3^4 = 81$
(f) Since $13,860 = 2^2 \cdot 3^2 \cdot 5 \cdot 7 \cdot 11$, it follows that $\mathcal{R}$ contains $\binom{3+2-1}{2}^2\binom{2+2-1}{2}^3 = \binom{4}{2}^2\binom{3}{2}^3 = (36)(27) = 972$ ordered pairs.

13. There may exist an element $a \in A$ such that for all $b \in B$ neither (a, b) nor $(b, a) \in \mathcal{R}$.

15. $r - n$ counts the elements in $\mathcal{R}$ of the form $(a, b), a \neq b$. Since $\mathcal{R}$ is symmetric, $r - n$ is even.

17. (a) Write $A \times A = A_1 \cup A_2$ where $A_1 = \{(x, x)\}|x \in A\}$ and $A_2 = \{(x, y)|x, y \in A, x \neq y\}$. Now $|A_1| = 7$ and $|A_2| = 7^2 - 7 = 42$. Consider the 42 ordered pairs in A_2 as 21 sets of two ordered pairs of the form $(x, y), (y, x)$ – where $x, y \in A$ and $x \neq y$. For a symmetric relation on A to contain four ordered pairs three situations are possible:
(1) We have four ordered pairs from A_1 – this can happen in $\binom{7}{4} = \binom{7}{4}\binom{21}{0}$ ways.

(2) We have two ordered pairs from A_1 and one of the 21 sets of two ordered pairs from A_2. This occurs in $\binom{7}{2}\binom{21}{1}$ ways.

(3) We have two of the 21 sets of two ordered pairs from A_2. This occurs in $\binom{21}{2} = \binom{7}{0}\binom{21}{2}$ ways.

Consequently, the answer is $\binom{7}{4}\binom{21}{0} + \binom{7}{2}\binom{21}{1} + \binom{7}{0}\binom{21}{2}$.

(b) $\binom{7}{5}\binom{21}{0} + \binom{7}{3}\binom{21}{1} + \binom{7}{1}\binom{21}{2}$

(c) $\binom{7}{7}\binom{21}{0} + \binom{7}{5}\binom{21}{1} + \binom{7}{3}\binom{21}{2} + \binom{7}{1}\binom{21}{3}$

(d) $\binom{7}{6}\binom{21}{1} + \binom{7}{4}\binom{21}{2} + \binom{7}{2}\binom{21}{3} + \binom{7}{0}\binom{21}{4}$

Section 7.2

1. $\mathcal{R} \circ \mathcal{S} = \{(1,3),(1,4)\}$; $\mathcal{S} \circ \mathcal{R} = \{(1,2),(1,3),(1,4),(2,4)\}$;
$\mathcal{R}^2 = \mathcal{R}^3 = \{(1,4),(2,4),(4,4)\}$;
$\mathcal{S}^2 = \mathcal{S}^3 = \{(1,1),(1,2),(1,3),(1,4)\}$.

3. $(a,d) \in (\mathcal{R}_1 \circ \mathcal{R}_2) \circ \mathcal{R}_3 \Longrightarrow (a,c) \in \mathcal{R}_1 \circ \mathcal{R}_2, (c,d) \in \mathcal{R}_3$ for some $c \in C \Longrightarrow (a,b) \in \mathcal{R}_1, (b,c) \in \mathcal{R}_2, (c,d) \in \mathcal{R}_3$ for some $b \in B, c \in C \Longrightarrow (a,b) \in \mathcal{R}_1, (b,d) \in \mathcal{R}_2 \circ \mathcal{R}_3 \Longrightarrow (a,d) \in \mathcal{R}_1 \circ (\mathcal{R}_2 \circ \mathcal{R}_3)$, and $(\mathcal{R}_1 \circ \mathcal{R}_2) \circ \mathcal{R}_3 \subseteq \mathcal{R}_1 \circ (\mathcal{R}_2 \circ \mathcal{R}_3)$.

5. $\mathcal{R}_1 \circ (\mathcal{R}_2 \cap \mathcal{R}_3) = \mathcal{R}_2 \circ \{(m,3),(m,4)\} = \{(1,3),(1,4)\}$
$(\mathcal{R}_1 \circ \mathcal{R}_2) \cap (\mathcal{R}_1 \circ \mathcal{R}_3) = \{(1,3),(1,4)\} \cap \{(1,3),(1,4)\} = \{(1,3),(1,4)\}$.

7. This follows by the Pigeonhole Principle. Here the pigeons are the $2^{n^2}+1$ integers between 0 and 2^{n^2}, inclusive, and the pigeonholes are the 2^{n^2} relations on A.

9. Here there are two choices for each $a_{ii}, 1 \leq i < 6$. For each pair $a_{ij}, a_{ji}, 1 \leq i < j \leq 6$, there are two choices, and there are $(36-6)/2 = 15$ such pairs. Consequently there are $(2^6)(2^{15}) = 2^{21}$ such matrices.

11. Consider the entry in the i-th row and j-th column of $M(\mathcal{R}_1 \circ \mathcal{R}_2)$. If this entry is a 1 then there exists $b_k \in B$ where $1 \leq k \leq n$ and $(a_i, b_k) \in \mathcal{R}_1, (b_k, c_j) \in \mathcal{R}_2$. Consequently, the entry in the i-th row and k-th column of $M(\mathcal{R}_1)$ is 1 and the entry in the k-th row and j-th column of $M(\mathcal{R}_2)$ is 1. This results in a 1 in the i-th row and j-th column in the product $M(\mathcal{R}_1) \cdot M(\mathcal{R}_2)$.

Should the entry in row i and column j of $M(\mathcal{R}_1 \circ \mathcal{R}_2)$ be 0, then for each $b_k, 1 \leq k \leq n$, either $(a_i, b_k) \notin \mathcal{R}_1$ or $(b_k, c_j) \notin \mathcal{R}_2$. This means that in the matrices $M(\mathcal{R}_1), M(\mathcal{R}_2)$, if the entry in the i-th row and k-th column of $M(\mathcal{R}_1)$ is 1 then the entry in the k-th row and j-th column of $M(\mathcal{R}_2)$ is 0. Hence the entry in the i-th row and j-th column of $M(\mathcal{R}_1) \cdot M(\mathcal{R}_2)$ is 0.

13. (a) $\mathcal{R}$ reflexive $\Longleftrightarrow (x,x) \in \mathcal{R}$, for all $x \in A \Longleftrightarrow m_{xx} = 1$ in $M = (m_{ij})_{n \times n}$, for all

$x \in A \iff I_n \leq M.$

(b) $\mathcal{R}$ symmetric $\iff [\forall x, y \in A \ \ (x,y) \in \mathcal{R} \implies (y,x) \in \mathcal{R}] \iff [\forall x, y \in A \ \ m_{xy} = 1$ in $M \implies m_{yx} = 1$ in $M] \iff M = M^{tr}$.

15. (a)

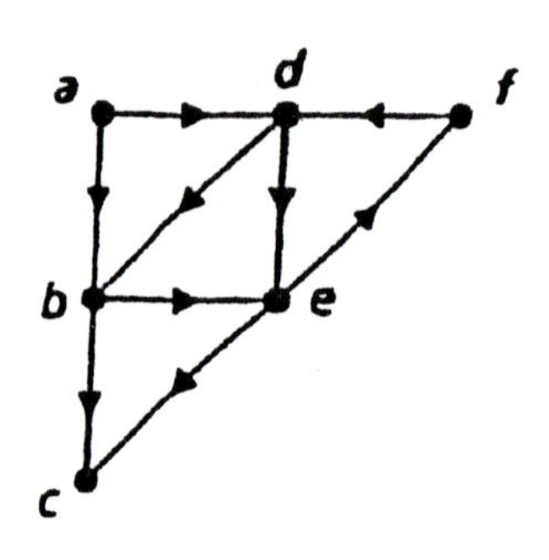

(b)

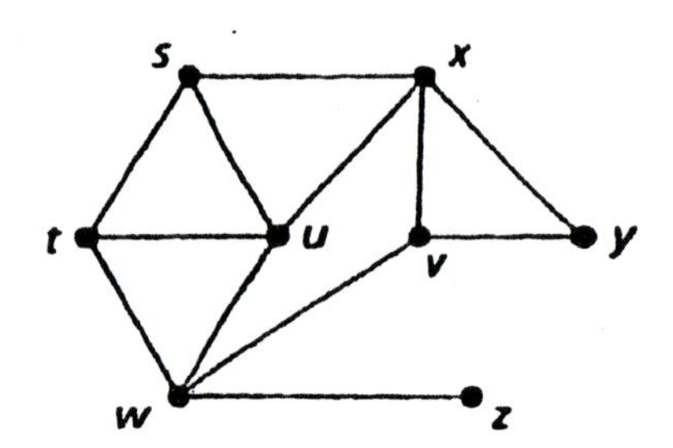

17. (i) $\mathcal{R} = \{(a,b),(b,a),(a,e),(e,a),(b,c),(c,b),(b,d),(d,b),(b,e),(e,b),(d,e),(e,d),(d,f),(f,d)\}$

$$M(\mathcal{R}) = \begin{array}{c} \\ (a) \\ (b) \\ (c) \\ (d) \\ (e) \\ (f) \end{array} \begin{array}{c} (a)\,(b)\,(c)\,(d)\,(e)\,(f) \\ \begin{bmatrix} 0 & 1 & 0 & 0 & 1 & 0 \\ 1 & 0 & 1 & 1 & 1 & 0 \\ 0 & 1 & 0 & 0 & 0 & 0 \\ 0 & 1 & 0 & 0 & 1 & 1 \\ 1 & 1 & 0 & 1 & 0 & 0 \\ 0 & 0 & 0 & 1 & 0 & 0 \end{bmatrix} \end{array}$$

For parts (ii), (iii), and (iv), the rows and columns of the relation matrix are indexed as in part (i).

(ii) $\mathcal{R} = \{(a,b),(b,e),(d,b),(d,c),(e,f)\}$

$$M(\mathcal{R}) = \begin{bmatrix} 0 & 1 & 0 & 0 & 0 & 0 \\ 0 & 0 & 0 & 0 & 1 & 0 \\ 0 & 0 & 0 & 0 & 0 & 0 \\ 0 & 1 & 1 & 0 & 0 & 0 \\ 0 & 0 & 0 & 0 & 0 & 0 \\ 0 & 0 & 0 & 0 & 0 & 0 \end{bmatrix}$$

(iii) $\mathcal{R} = \{(a,a),(a,b),(b,a),(c,d),(d,c),(d,e),(e,d),(d,f),(f,d),(e,f),(f,e)\}$

$$M(\mathcal{R}) = \begin{bmatrix} 1 & 1 & 0 & 0 & 0 & 0 \\ 1 & 0 & 0 & 0 & 0 & 0 \\ 0 & 0 & 0 & 1 & 0 & 0 \\ 0 & 0 & 1 & 0 & 1 & 1 \\ 0 & 0 & 0 & 1 & 0 & 1 \\ 0 & 0 & 0 & 1 & 1 & 0 \end{bmatrix}$$

(iv) $\mathcal{R} = \{(b,a),(b,c),(c,b),(b,e),(c,d),(e,d)\}$

$$M(\mathcal{R}) = \begin{bmatrix} 0 & 0 & 0 & 0 & 0 & 0 \\ 1 & 0 & 1 & 0 & 1 & 0 \\ 0 & 1 & 0 & 1 & 0 & 0 \\ 0 & 0 & 0 & 0 & 0 & 0 \\ 0 & 0 & 0 & 1 & 0 & 0 \\ 0 & 0 & 0 & 0 & 0 & 0 \end{bmatrix}$$

19. $\mathcal{R}$: $\mathcal{R}^2$: $\mathcal{R}^3$ and $\mathcal{R}^4$:

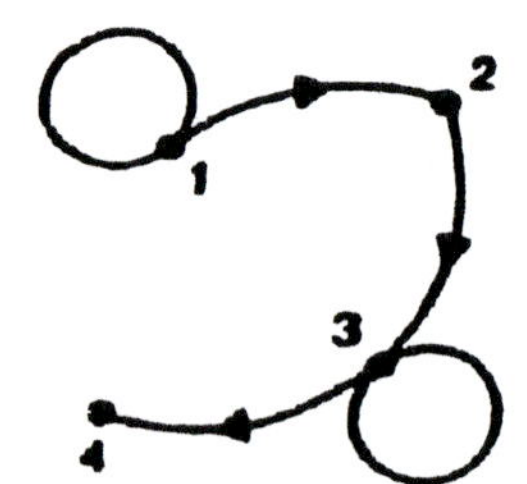

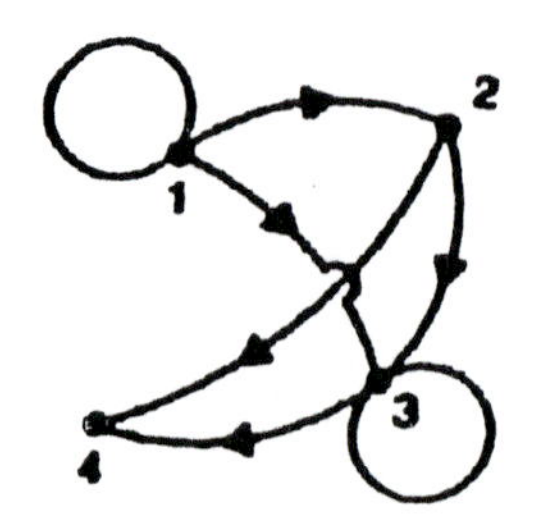

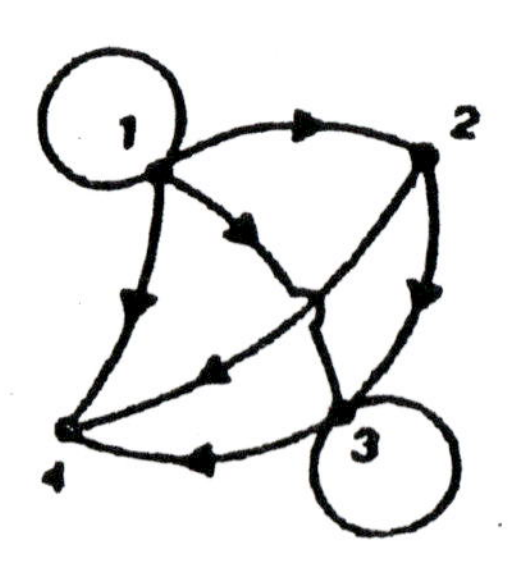

21. Since $|A| = 5$, we have $|A \times A| = 25$ – so there are 25 arcs (or, directed edges). Each such edge can be included or excluded from the directed graph, so there are 2^{25} possible directed graphs.

These 25 directed edges fall into two categories: 5 (directed) loops (a, a), for $a \in A$ and 20 directed edges of the form (a, b), where $a, b \in A$ and $a \neq b$. Each loop can be included or excluded. This gives us 2^5 possibilities. The other 20 directed edges can be considered in 10 sets of two – of the form $(x, y), (y, x)$, for $x, y \in A$ with $x \neq y$. Each set of two can be included or excluded and this provides 2^{10} possibilities. So, in total, the number of undirected graphs is $2^5 \cdot 2^{10} = 2^{15}$.

23. (a) $\mathcal{R}_1$: $\mathcal{R}_2$:

$$\begin{bmatrix} 1 & 1 & 0 & 0 & 0 \\ 1 & 1 & 0 & 0 & 0 \\ 0 & 0 & 1 & 1 & 0 \\ 0 & 0 & 1 & 1 & 0 \\ 0 & 0 & 0 & 0 & 1 \end{bmatrix} \qquad \begin{bmatrix} 1 & 1 & 1 & 0 & 0 \\ 1 & 1 & 1 & 0 & 0 \\ 1 & 1 & 1 & 0 & 0 \\ 0 & 0 & 0 & 1 & 1 \\ 0 & 0 & 0 & 1 & 1 \end{bmatrix}$$

(b) Given an equivalence relation $\mathcal{R}$ on a finite set A, list the elements of A so that elements in the same cell of the partition (See Section 7.4.) are adjacent. The resulting relation matrix will then have square blocks of 1's along the diagonal (from upper left to lower right).

25.

(s_1) a := 1
(s_2) b := 2
(s_3) a := a + 3
(s_4) c := b
(s_5) a := 2 * a − 1
(s_6) b := a * c
(s_7) c := 7
(s_8) d := c + 2

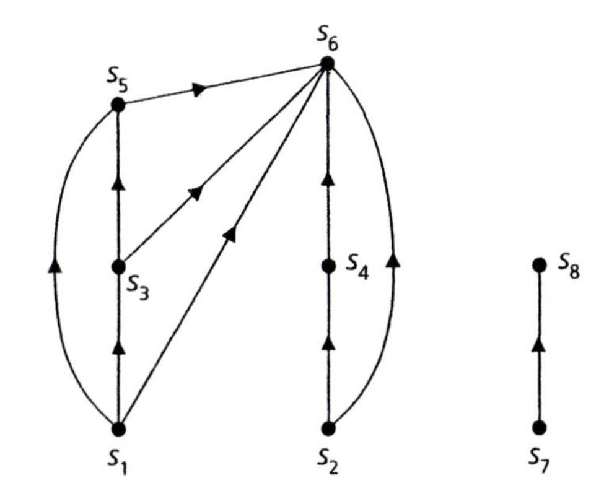

27. $\binom{n}{2} = 703 \Rightarrow n(n-1)/2 = 703 \Rightarrow n(n-1) = 1406 \Rightarrow n^2 - n - 1406 = 0 \Rightarrow (n-38)(n+37) = 0 \Rightarrow n = 38$

Section 7.3

1.

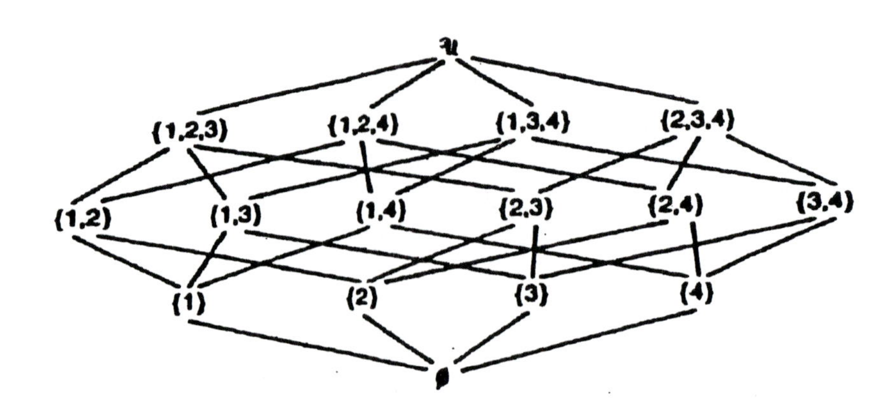

3. For all $a \in A, b \in B, a\mathcal{R}_1 a$ and $b\mathcal{R}_2 b$ so $(a,b)\mathcal{R}(a,b)$, and $\mathcal{R}$ is reflexive. Next $(a,b)\mathcal{R}(c,d), (c,d)\mathcal{R}(a,b) \Longrightarrow a\mathcal{R}_1 c, c\mathcal{R}_1 a$ and $b\mathcal{R}_2 d, d\mathcal{R}_2 b \Longrightarrow a = c, b = d \Longrightarrow (a,b) = (c,d)$, so $\mathcal{R}$ is antisymmetric. Finally, $(a,b)\mathcal{R}(c,d), (c,d)\mathcal{R}(e,f) \Longrightarrow a\mathcal{R}_1 c, c\mathcal{R}_1 e$ and $b\mathcal{R}_2 d, d\mathcal{R}_2 f \Longrightarrow a\mathcal{R}_1 e, b\mathcal{R}_2 f \Longrightarrow (a,b)\mathcal{R}(e,f)$, and $\mathcal{R}$ is transitive. Consequently, $\mathcal{R}$ is a partial order.

5. $\emptyset < \{1\} < \{2\} < \{3\} < \{1,2\} < \{1,3\} < \{2,3\} < \{1,2,3\}$. (There are other possibilities.)

7. (a)

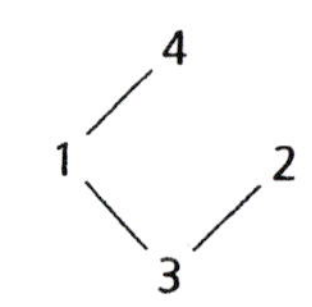

(b) $3 < 2 < 1 < 4$ or $3 < 1 < 2 < 4$.

(c) 2

9. Let x, y both be greatest lower bounds. Then $x\mathcal{R}y$ since x is a lower bound and y is a greatest lower bound. By similar reasoning $y\mathcal{R}x$. Since $\mathcal{R}$ is antisymmetric, $x = y$. [The proof for the *lub* is similar.]

11. Let $\mathcal{U} = \{1,2\}, A = \mathcal{P}(\mathcal{U})$, and $\mathcal{R}$ the inclusion relation. Then $(A, \mathcal{R})$ is a poset but not a total order. Let $B = \{\emptyset, \{1\}\}$. Then $(B \times B) \cap \mathcal{R}$ is a total order.

13. Since a total order is reflexive we have n loops – one at each vertex. Further, for all vertices x, y, where $x \neq y$, since $\mathcal{R}$ is a total order we have exactly one of the edges (x,y) or (y,x) in G. This provides the remaining $\binom{n}{2}$ non-loop edges in G. Consequently, G contains $n + \binom{n}{2}$ edges in total.

15. (a) The n elements of A are arranged along a vertical line. For if $A = \{a_1, a_2, \ldots a_n\}$, where $a_1\mathcal{R}a_2\mathcal{R}a_3\mathcal{R}\ldots\mathcal{R}a_n$, then the diagram can be drawn as

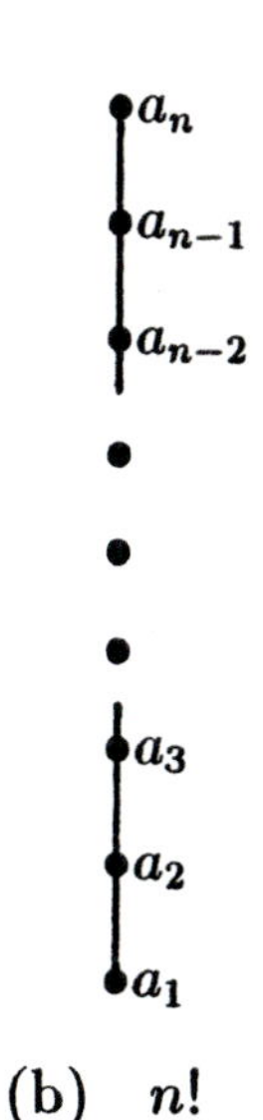

(b) $n!$

17.

	lub	*glb*
(a)	{1,2}	$\emptyset$
(b)	{1,2,3}	$\emptyset$
(c)	{1,2}	$\emptyset$
(d)	{1,2,3}	{1}
(e)	{1,2,3}	$\emptyset$

19. For each $a \in \mathbf{Z}$ it follows that $a\mathcal{R}a$ because $a - a = 0$, an even nonnegative integer. Hence $\mathcal{R}$ is *reflexive*. If $a, b, c \in \mathbf{Z}$ with $a\mathcal{R}b$ and $b\mathcal{R}c$ then

$$a - b = 2m, \text{ for some } m \in \mathbf{N}$$
$$b - c = 2n, \text{ for some } n \in \mathbf{N},$$

and $a - c = (a - b) + (b - c) = 2(m + n)$, where $m + n \in \mathbf{N}$. Therefore, $a\mathcal{R}c$ and $\mathcal{R}$ is *transitive*. Finally, suppose that $a\mathcal{R}b$ and $b\mathcal{R}a$ for some $a, b \in \mathbf{Z}$. Then $a - b$ and $b - a$ are both nonnegative integers. Since this can only occur for $a - b = b - a$, we find that $[a\mathcal{R}b \wedge b\mathcal{R}a] \Rightarrow a = b$, so $\mathcal{R}$ is *antisymmetric*.

Consequently, the relation $\mathcal{R}$ is a partial order for $\mathbf{Z}$. But it is *not* a total order. For example, $2, 3 \in \mathbf{Z}$ and we have neither $2\mathcal{R}3$ nor $3\mathcal{R}2$, because neither -1 nor 1, respectively, is a nonnegative even integer.

21. (a) For all $(a, b) \in A$, $a = a$ and $b \leq b$, so $(a, b)\mathcal{R}(a, b)$ and the relation is reflexive. If $(a, b), (c, d) \in A$ with $(a, b)\mathcal{R}(c, d)$ and $(c, d)\mathcal{R}(a, b)$, then if $a \neq c$ we find that

$(a, b)\mathcal{R}(c, d) \Rightarrow a < c$, and
$(c, d)\mathcal{R}(a, b) \Rightarrow c < a$,

and we obtain $a < a$. Hence we have $a = c$.
And now we find that

$(a, b)\mathcal{R}(c, d) \Rightarrow b \leq d$, and
$(c, d)\mathcal{R}(a, b) \Rightarrow d \leq b$,

so $b = d$. Therefore, $(a, b)\mathcal{R}(c, d)$ and $(c, d)\mathcal{R}(a, b) \Rightarrow (a, b) = (c, d)$, so the relation is antisymmetric. Finally, consider $(a, b), (c, d), (e, f) \in A$ with $(a, b)\mathcal{R}(c, d)$ and $(c, d)\mathcal{R}(e, f)$. Then

(i) $a < c$, or (ii) $a = c$ and $b \leq d$; and
(i)′ $c < e$, or (ii)′ $c = e$ and $d \leq f$.
Consequently,
(i)″ $a < e$ or (ii)″ $a = e$ and $b \leq f$ — so, $(a, b)\mathcal{R}(e, f)$ and the relation is transitive.
The preceding shows that $\mathcal{R}$ is a partial order on A.

(b) & (c) Here the least element (and only minimal element) is (0,0). The element (2,2) is the greatest element (and the only maximal element).
(d) We obtain a total order, for

$$(0,0)\mathcal{R}(0,1)\mathcal{R}(0,2)\mathcal{R}(1,0)\mathcal{R}(1,1)\mathcal{R}(1,2)\mathcal{R}(2,0)\mathcal{R}(2,1)\mathcal{R}(2,2).$$

23. (a) False. Let $\mathcal{U} = \{1,2\}, A = \mathcal{P}(\mathcal{U})$, and $\mathcal{R}$ be the inclusion relation. Then $(A, \mathcal{R})$ is a lattice where for all $S, T \in A$, $lub\{S,T\} = S \cup T$ and $glb\{S,T\} = S \cap T$. However, $\{1\}$ and $\{2\}$ are not related, so $(A, \mathcal{R})$ is not a total order.

(b) If $(A, \mathcal{R})$ is a total order, then for all $x, y \in A, x\mathcal{R}y$ or $y\mathcal{R}x$. For $x\mathcal{R}y, lub\{x,y\} = y$ and $glb\{x,y\} = x$. Consequently, $(A, \mathcal{R})$ is a lattice.

25. (a) a (b) a (c) c (d) e (e) z (f) e (g) v
$(A, \mathcal{R})$ is a lattice with z the greatest (and only maximal) element and a the least (and only minimal) element.

27. Consider the vertex $p^a q^b r^c$, $0 \leq a < m$, $0 \leq b < n$, $0 \leq c < k$. There are mnk such vertices; each determines three edges — going to the vertices $p^{a+1}q^b r^c, p^a q^{b+1} r^c, p^a q^b r^{c+1}$. This accounts for $3mnk$ edges.

Now consider the vertex $p^m q^b r^c$, $0 \leq b < n$, $0 \leq c < k$. There are nk of these vertices; each detemines two edges — going to the vertices $p^m q^{b+1} r^c$, $p^m q^b r^{c+1}$. This accounts for $2nk$ edges. And similar arguments for the vertices $p^a q^n r^c (0 \leq a < m, 0 \leq c < k)$ and $p^a q^b r^k (0 \leq a < m, 0 \leq b < n)$ account for $2mk$ and $2mn$ edges, respectively.

Finally, each of the k vertices $p^m q^n r^c$, $0 \leq c < k$, determines one edge (going to $p^m q^n r^{c+1}$) and so these vertices account for k new edges. Likewise, each of the n vertices $p^m q^b r^k$, $0 \leq b < n$, determines one edge (going to $p^m q^{b+1} r^k$), and so these vertices account for n new edges. Lastly, each of the m vertices $p^a q^n r^k$, $0 \leq a < m$, determines one edge (going to $p^{a+1} q^n r^k$) and these vertices account for m new edges.

The preceding results give the total number of edges as $(m+n+k)+2(mn+mk+nk)+3mnk$.

29. $429 = (\frac{1}{8})\binom{14}{7}$ so $k = 6$, and there are $2 \cdot 7 = 14$ positive integer divisors of $p^6 q$.

Section 7.4

1. (a) Here the collection A_1, A_2, A_3 provides a partition of A.
(b) Although $A = A_1 \cup A_2 \cup A_3 \cup A_4$, we have $A_1 \cap A_2 \neq \emptyset$, so the collection A_1, A_2, A_3, A_4 does *not* provide a partition for A.

3. $\mathcal{R} = \{(1,1),(1,2),(2,1),(2,2),(3,3),(3,4),(4,3),(4,4),(5,5)\}$
$= (\{1,2\} \times \{1,2\}) \cup (\{3,4\} \times \{3,4\}) \cup (\{5\} \times \{5\})$.

5. $\mathcal{R}$ is not transitive since $1\mathcal{R}2$, $2\mathcal{R}3$ but $1 \not\mathcal{R} 3$.

7. (a) For all $(x,y) \in A, x+y = x+y \Longrightarrow (x,y)\mathcal{R}(x,y)$.
$(x_1,y_1)\mathcal{R}(x_2,y_2) \Longrightarrow x_1+y_1 = x_2+y_2 \Longrightarrow x_2+y_2 = x_1+y_1 \Longrightarrow$ $(x_2,y_2)\mathcal{R}(x_1,y_1)$. $(x_1,y_1)\mathcal{R}(x_2,y_2), (x_2,y_2)\mathcal{R}(x_3,y_3) \Longrightarrow$ $x_1+y_1 = x_2+y_2, x_2+y_2 = x_3+y_3$, so $x_1+y_1 = x_3+y_3$ and $(x_1,y_1)\mathcal{R}(x_3,y_3)$. Since $\mathcal{R}$ is reflexive, symmetric and transitive, it is an equivalence relation.

(b) $[(1,3)] = \{(1,3),(2,2),(3,1)\}$;
$[(2,4)] = \{(1,5),(2,4),(3,3),(4,2),(5,1)\}$; $[(1,1)] = \{(1,1)\}$.

(c) $A = \{(1,1)\} \cup \{(1,2),(2,1)\} \cup \{(1,3),(2,2),(3,1)\} \cup$
$\{(1,4),(2,3),(3,2),(4,1)\} \cup \{(1,5),(2,4),(3,3),(4,2),(5,1)\} \cup$
$\{(2,5),(3,4),(4,3),(5,2)\} \cup \{(3,5),(4,4),(5,3)\} \cup \{(4,5),(5,4)\} \cup \{(5,5)\}$.

9. (a) For all $(a,b) \in A$ we have $ab = ab$, so $(a,b)\mathcal{R}(a,b)$ and $\mathcal{R}$ is reflexive. To see that $\mathcal{R}$ is symmetric, suppose that $(a,b),(c,d) \in A$ and that $(a,b)\mathcal{R}(c,d)$. Then $(a,b)\mathcal{R}(c,d) \Rightarrow ad = bc \Rightarrow cb = da \Rightarrow (c,d)\mathcal{R}(a,b)$, so $\mathcal{R}$ is symmetric. Finally, let $(a,b),(c,d),(e,f) \in A$ with $(a,b)\mathcal{R}(c,d)$ and $(c,d)\mathcal{R}(e,f)$. Then $(a,b)\mathcal{R}(c,d) \Rightarrow ad = bc$ and $(c,d)\mathcal{R}(e,f) \Rightarrow cf = de$, so $adf = bcf = bde$ and since $d \neq 0$, we have $af = be$. But $af = be \Rightarrow (a,b)\mathcal{R}(e,f)$, and consequently $\mathcal{R}$ is transitive.
It follows from the above that $\mathcal{R}$ is an equivalence relation on A.

(b) $[(2,14)] = \{(2,14)\}$
$[(-3,-9)] = \{(-3,-9),(-1,-3),(4,12)\}$
$[(4,8)] = \{(-2,-4),(1,2),(3,6),(4,8)\}$
(c) There are five cells in the partition — in fact,

$$A = [(-4,-20)] \cup [(-3,-9)] \cup [(-2,-4)] \cup [(-1,-11)] \cup [(2,14)].$$

11. (a) $(\frac{1}{2})\binom{6}{3}$ – The factor $(\frac{1}{2})$ is needed because each selection of size 3 should account for only one such equivalence relation, not two. For example, if $\{a,b,c\}$ is selected we get the partition $\{a,b,c\} \cup \{d,e,f\}$ that corresponds with an equivalence relation. But the selection $\{d,e,f\}$ gives us the same partition and corresponding equivalence relation.
(b) $\binom{6}{3}[1+3] = 4\binom{6}{3}$ – After selecting 3 of the elements we can partition the remaining 3 in

(i) 1 way into three equivalence classes of size 1; or
(ii) 3 ways into one equivalence class of size 1 and one of size 2.
(c) $\binom{6}{4}[1+1] = 2\binom{6}{4}$
(d) $(\frac{1}{2})\binom{6}{3} + 4\binom{6}{3} + 2\binom{6}{4} + \binom{6}{5} + \binom{6}{6}$

13. Here $A = A_1 \cup A_2 \cup A_3$ is a partition with $|A| = 30$ and $|A_1| = |A_2| = |A_3| = 10$. Also, $\mathcal{R} = (A_1 \times A_1) \cup (A_2 \times A_2) \cup (A_3 \times A_3)$, so $\mathcal{R}$ contains $10^2 + 10^2 + 10^2 = 300$ ordered pairs.

15. Let $\{A_i\}_{i \in I}$ be a partition of a set A. Define $\mathcal{R}$ on A by $x\mathcal{R}y$ if for some $i \in I, x, y \in A_i$. For each $x \in A, x, x \in A_i$ for some $i \in I$, so $x\mathcal{R}x$ and $\mathcal{R}$ is reflexive. $x\mathcal{R}y \Longrightarrow x, y \in A_i$, for some $i \in I \Longrightarrow y, x \in A_i$, for some $i \in I \Longrightarrow y\mathcal{R}x$, so $\mathcal{R}$ is symmetric. If $x\mathcal{R}y$ and $y\mathcal{R}z$, then $x, y \in A_i$ and $y, z \in A_j$ for some $i, j \in I$. Since $A_i \cap A_j$ contains y and $\{A_i\}_{i \in I}$ is a partition, from $A_i \cap A_j = \emptyset$ it follows that $A_i = A_j$, so $i = j$. Hence $x, z \in A_i$, so $x\mathcal{R}z$ and $\mathcal{R}$ is transitive.

17. Proof: Since $\{B_1, B_2, B_3, \ldots, B_n\}$ is a partition of B, we have $B = B_1 \cup B_2 \cup B_3 \cup \ldots \cup B_n$. Therefore $A = f^{-1}(B) = f^{-1}(B_1 \cup \ldots \cup B_n) = f^{-1}(B_1) \cup \ldots \cup f^{-1}(B_n)$ [by generalizing part (b) of Theorem 5.10]. For $1 \leq i < j \leq n$, $f^{-1}(B_i) \cap f^{-1}(B_j) = f^{-1}(B_i \cap B_j) = f^{-1}(\emptyset) = \emptyset$. Consequently, $\{f^{-1}(B_i) | 1 \leq i \leq n, f^{-1}(B_i) \neq \emptyset\}$ is a partition of A.

Note: Part (b) of Example 7.55 is a special case of this result.

Section 7.5

1. (a) $P_1 : \{s_1, s_4\}, \{s_2, s_3, s_5\}$
$(\nu(s_1, 0) = s_4)E_1(\nu(s_4, 0) = s_1)$ but $(\nu(s_1, 1) = s_1)\not{E}_1(\nu(s_4, 1) = s_3)$, so $s_1 \not{E}_2 s_4$.
$(\nu(s_2, 1) = s_3)\not{E}_1(\nu(s_3, 1) = s_4)$ so $s_2 \not{E}_2 s_3$.
$(\nu(s_2, 0) = s_3)E_1(\nu(s_5, 0) = s_3)$ and $(\nu(s_2, 1) = s_3)E_1(\nu(s_5, 1) = s_3)$ so $s_2 E_2 s_5$.
Since $s_2 \not{E}_2 S_3$ and $s_2 E_2 s_5$, it follows that $s_3 \not{E}_2 s_5$.
Hence P_2 is given by $P_2 : \{s_1\}, \{s_2, s_5\}, \{s_3, \}, \{s_4\}$. $(\nu(s_2, x) = s_3)E_2(\nu(s_5, x) = s_3)$ for $x = 0, 1$. Hence $s_2 E_3 s_5$ and $P_2 = P_3$.

Consequently, states s_2 and s_5 are equivalent.

(b) States s_2 and s_5 are equivalent.

(c) States s_2 and s_7 are equivalent; s_3 and s_4 are equivalent.

3. (a)

	ν		ω	
	0	1	0	1
s_1	s_5	s_7	1	0
s_2	s_7	s_2	1	0
s_3	s_6	s_1	1	0
s_4	s_3	s_4	0	0
s_5	s_3	s_5	0	0
s_6	s_2	s_7	1	0
s_7	s_4	s_1	1	0

$P_1 : \{s_1, s_2, s_3, s_6, s_7\}, \{s_4, s_5\}$
$(\nu(s_1, 0) = s_5)\not E_1(\nu(s_2, 0) = s_7)$
$P_2 : \{s_1, s_7\}, \{s_2, s_3, s_6\}, \{s_4, s_5\}$
$(\nu(s_2, 0) = s_7)\not E_2(\nu(s_3, 0) = s_6)$
$P_3 : \{s_1, s_7\}, \{s_2\}, \{s_3, s_6\}, \{s_4, s_5\}$
$(\nu(s_3, 0) = s_6)\not E_3(\nu(s_6, 0) = s_2)$
$P_5 = P_4 : \{s_1, s_7\}, \{s_2\}, \{s_3\}, \{s_6\}, \{s_4, s_5\}$
Consequently, s_1 and s_7 are equivalent; s_4 and s_5 are equivalent.
(b) (i)

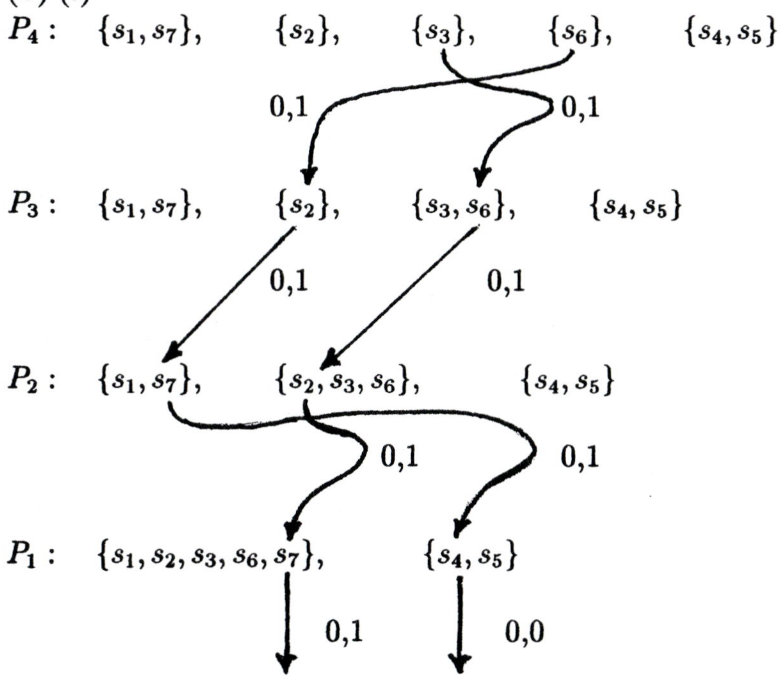

Here $\omega(s_3, 0000) = 1111 \neq 1110 = \omega(s_6, 0000)$.

(ii) 0 (iii) 00

M:	ν 0	ν 1	ω 0	ω 1
s_1	s_4	s_1	1	0
s_2	s_1	s_2	1	0
s_3	s_6	s_1	1	0
s_4	s_3	s_4	0	0
s_6	s_2	s_1	1	0

Supplementary Exercises

1. (a) False. Let $A = \{1,2\}, I = \{1,2\}, \mathcal{R}_1 = \{(1,1)\}, \mathcal{R}_2 = \{(2,2)\}$. Then $\cup_{i \in I} \mathcal{R}_i$ is reflexive but neither $\mathcal{R}_1$ nor $\mathcal{R}_2$ is reflexive. Conversely, however, if $\mathcal{R}_i$ is reflexive for all (actually at least one) $i \in I$, then $\cup_{i \in I} \mathcal{R}_i$ is reflexive.

(b) True. $\cap_{i \in I} \mathcal{R}_i$ reflexive $\iff (a,a) \in \cap_{i \in I} \mathcal{R}_i$ for all $a \in A \iff (a,a) \in \mathcal{R}_i$ for all $a \in A$ and all $i \in I \iff \mathcal{R}_i$ is reflexive for all $i \in I$.

3. $(a,c) \in \mathcal{R}_2 \circ \mathcal{R}_1 \implies$ for some $b \in A, (a,b) \in \mathcal{R}_2, (b,c) \in \mathcal{R}_1$. With $\mathcal{R}_1, \mathcal{R}_2$ symmetric, $(b,a) \in \mathcal{R}_2, (c,b) \in \mathcal{R}_1$, so $(c,a) \in \mathcal{R}_1 \circ \mathcal{R}_2 \subseteq \mathcal{R}_2 \circ \mathcal{R}_1$. $(c,a) \in \mathcal{R}_2 \circ \mathcal{R}_1 \implies (c,d) \in \mathcal{R}_2, (d,a) \in \mathcal{R}_1$, for some $d \in A$. Then $(d,c) \in \mathcal{R}_2, (a,d) \in \mathcal{R}_1$ by symmetry, and $(a,c) \in \mathcal{R}_1 \circ \mathcal{R}_2$, so $\mathcal{R}_2 \circ \mathcal{R}_1 \subseteq \mathcal{R}_1 \circ \mathcal{R}_2$ and the result follows.

5. $(c,a) \in (\mathcal{R}_1 \circ \mathcal{R}_2)^c \iff (a,c) \in \mathcal{R}_1 \circ \mathcal{R}_2 \iff (a,b) \in \mathcal{R}_1, (b,c) \in \mathcal{R}_2$, for some $b \in B \iff (c,b) \in \mathcal{R}_2^c, (b,a) \in \mathcal{R}_1^c$, for some $b \in B \iff (c,a) \in \mathcal{R}_2^c \circ \mathcal{R}_1^c$.

7. Let $\mathcal{U} = \{1,2,3,4,5\}, A = \mathcal{P}(\mathcal{U}) - \{\mathcal{U}, \emptyset\}$. Under the inclusion relation A is a poset with the five minimal elements $\{x\}, 1 \leq x \leq 5$, but no least element. Also, A has five maximal elements – the five subsets of $\mathcal{U}$ of size 4 – but no greatest element.

9. $\binom{n}{2} = 45 \Rightarrow \frac{n(n-1)}{2} = 45 \Rightarrow n(n-1) = 90 \Rightarrow n^2 - n - 90 = 0 \Rightarrow (n-10)(n+9) = 0 \Rightarrow n = 10$

11.

(a)

Adjacency List		Index List	
1	2	1	1
2	3	2	2
3	1	3	3
4	4	4	5
5	5	5	6
6	3	6	8
7	5		

(b)

Adjacency List		Index List	
1	2	1	1
2	3	2	2
3	1	3	3
4	5	4	4
5	4	5	5
		6	6

(c)

Adjacency List		Index List	
1	2	1	1
2	3	2	2
3	1	3	3
4	4	4	6
5	5	5	7
6	1	6	8
7	4		

13. (a) For each $v \in V, v = v$ so $v\mathcal{R}v$. If $v\mathcal{R}w$ then there is a path from v to w. Since the graph G is undirected, the path from v to w is also a path from w to v, so $w\mathcal{R}v$ and $\mathcal{R}$ is symmetric. Finally, if $v\mathcal{R}w$ and $w\mathcal{R}x$, then a subset of the edges in the paths from v to w and w to x provide a path from v to x. Hence $\mathcal{R}$ is transitive and $\mathcal{R}$ is an equivalence relation.

(b) The cells of the partition are the (connected) components of G.

15. One possible order is 10, 3, 8, 6, 7, 9, 1, 4, 5, 2, where program 10 is run first and program 2 last.

17. (b) $[(0.3, 0.7)] = \{(0.3, 0.7)\}$ $\quad [(0.5, 0)] = \{(0.5, 0)\}$ $\quad [(0.4, 1)] = \{(0.4, 1)\}$
$[(0, 0.6)] = \{(0, 0.6), (1, 0.6)\}$ $\quad [(1, 0.2)] = \{(0, 0.2), (1, 0.2)\}$
In general, if $0 < a < 1$, then $[(a, b)] = \{(a, b)\}$; otherwise, $[(0, b)] = \{(0, b), (1, b)\} = [(1, b)]$.
(c) The lateral surface of a cylinder of height 1 and base radius $1/2\pi$.

19. Since $|\mathcal{U}| = n$, $|\mathcal{P}(\mathcal{U}| = 2^n$ and so there are $(2^n)(2^n) = 4^n$ ordered pairs of the form (A, B) where $A, B \subseteq \mathcal{U}$. From Exercise 18 (above) there are 3^n order pairs of the form (A, B) where $A \subseteq B$. [Note: If $(A, B) \in \mathcal{R}$, then so is (B, A).] Hence there are $3^n + 3^n - 2^n$ ordered pairs (A, B) where either $A \subseteq B$ or $B \subseteq A$, or both. We subtract 2^n because we have counted the 2^n ordered pairs (A, B), where $A = B$, twice. Therefore the number of ordered pairs in this relation is $4^n - (2 \cdot 3^n - 2^n) = 4^n - 2 \cdot 3^n + 2^n$.

21. (a) (i) $B\mathcal{R}A\mathcal{R}C$; (ii) $B\mathcal{R}C\mathcal{R}F$

$B\mathcal{R}A\mathcal{R}C\mathcal{R}F$ is a maximal chain. There are six such maximal chains.

(b) Here 11 $\mathcal{R}$ 385 is a maximal chain of length 2, while 2 $\mathcal{R}$ 6 $\mathcal{R}$ 12 is one of length 3. The length of a longest chain for this poset is 3.
(c) (i) $\emptyset \subseteq \{1\} \subseteq \{1, 2\} \subseteq \{1, 2, 3\} \subseteq \mathcal{U}$;
(ii) $\emptyset \subseteq \{2\} \subseteq \{2, 3\} \subseteq \{1, 2, 3\} \subseteq \mathcal{U}$.
There are $4! = 24$ such maximal chains.
(d) $n!$

23. Let $a_1\mathcal{R}a_2\mathcal{R} \ldots \mathcal{R}a_{n-1}\mathcal{R}a_n$ be a longest (maximal) chain in $(A, \mathcal{R})$. Then a_n is a maximal element in $(A, \mathcal{R})$ and $a_1\mathcal{R}a_2\mathcal{R} \ldots \mathcal{R}a_{n-1}$ is a maximal chain in $(B, \mathcal{R}')$. Hence the length of a longest chain in $(B, \mathcal{R}')$ is at least $n-1$. If there is a chain $b_1\mathcal{R}'b_2\mathcal{R}' \ldots \mathcal{R}'b_n$ in $(B, \mathcal{R}')$ of length n, then this is also a chain of length n in $(A, \mathcal{R})$. But then b_n must be a maximal element of $(A, \mathcal{R})$, and this contradicts $b_n \in B$.

25. If $n = 1$, then for all $x, y \in A$, if $x \neq y$ then $x\not\mathcal{R}y$ and $y\not\mathcal{R}x$. Hence $(A, \mathcal{R})$ is an antichain, and the result follows.

Now assume the result true for $n = k \geq 1$, and let $(A, \mathcal{R})$ be a poset where the length of a longest chain is $k + 1$. If M is the set of all maximal elements in $(A, \mathcal{R})$, then $M \neq \emptyset$ and M is an antichain in $(A, \mathcal{R})$. Also, by virtue of Exercise 23 above, $(A - M, \mathcal{R}')$, for $\mathcal{R}' = ((A - M) \times (A - M)) \cap \mathcal{R}$, is a poset with k the length of a longest chain. So by the induction hypothesis $A - M = C_1 \cup C_2 \cup \ldots \cup C_k$, a partition into k antichains.

Consequently, $A = C_1 \cup C_2 \cup \ldots \cup C_k \cup M$, a partition into $k+1$ antichains.

27. (a) There are n edges – namely, $(0,1), (1,2), (2,3), \ldots, (n-1, n)$.
(b) The number of partitions, as described here, equals the number of compositions of n. So the answer is 2^{n-1}.
(c) The number of such partitions is $2^{3-1} \cdot 2^{5-1} = 64$, for there are 2^{3-1} compositions of 3 and 2^{5-1} compositions of 5 ($= 12 - 7$).

PART 2

FURTHER TOPICS

IN

ENUMERATION

CHAPTER 8
THE PRINCIPLE OF INCLUSION AND EXCLUSION

Section 8.1

1. Let $x \in S$ and let n be the number of conditions (from among c_1, c_2, c_3, c_4) satisfied by x:
($n = 0$): Here x is counted once in $N(\overline{c}_2\overline{c}_3\overline{c}_4)$ and once in $N(\overline{c}_1\overline{c}_2\overline{c}_3\overline{c}_4)$.
($n = 1$): If x satisfies c_1 (and not c_2, c_3, c_4), then x is counted once in $N(\overline{c}_2\overline{c}_3\overline{c}_4)$ and once in $N(c_1\overline{c}_2\overline{c}_3\overline{c}_4)$.

If x satisfies c_i, for $i \neq 1$, then x is not counted in any of the three terms in the equation.
($n = 2, 3, 4$): If x satisfies at least two of the four conditions, then x is not counted in any of the three terms in the equation.

The preceding observations show that the two sides of the given equation count the same elements from S, and this provides a combinatorial proof for the formula $N(\overline{c}_2\overline{c}_3\overline{c}_4) = N(c_1\overline{c}_2\overline{c}_3\overline{c}_4) + N(\overline{c}_1\overline{c}_2\overline{c}_3\overline{c}_4)$.

3. $N = 100$
$N(c_1) = 35$; $N(c_2) = 30$; $N(c_3) = 30$; $N(c_4) = 41$
$N(c_1c_2) = 9$; $N(c_1c_3) = 11$; $N(c_1c_4) = 13$; $N(c_2c_3) = 10$; $N(c_2c_4) = 14$; $N(c_3c_4) = 10$.
$N(c_1c_2c_3) = 5$; $N(c_1c_2c_4) = 6$; $N(c_1c_3c_4) = 6$; $N(c_2c_3c_4) = 6$
$N(c_1c_2c_3c_4) = 4$

(a) $N(\overline{c}_1\overline{c}_2c_3\overline{c}_4) = N(\overline{c}_1\overline{c}_2\overline{c}_4) - N(\overline{c}_1\overline{c}_2\overline{c}_3\overline{c}_4)$
$N(\overline{c}_1\overline{c}_2\overline{c}_4) = N - [N(c_1) + N(c_2) + N(c_4)]$
$+[N(c_1c_2) + N(c_1c_4) + N(c_2c_4)] - N(c_1c_2c_4) = 100 - [35 + 30 + 41] + [9 + 13 + 14] - 6$
$= 100 - 106 + 36 - 6 = 24$

$N(\overline{c}_1\overline{c}_2\overline{c}_3\overline{c}_4) = 12$ (as shown in Example 8.3)

So $N(\overline{c}_1\overline{c}_2c_3\overline{c}_4) = 24 - 12 = 12$

Alternately,
$N(\overline{c}_1\overline{c}_2\overline{c}_4) = N - [N(c_1) + N(c_2) + N(c_4)] + [N(c_1c_2) + N(c_1c_4) + N(c_2c_4)] - N(c_1c_2c_4)$, so
$N(\overline{c}_1\overline{c}_2c_3\overline{c}_4) = N(c_3) - [N(c_1c_3) + N(c_2c_3) + N(c_3c_4)] + [N(c_1c_2c_3) + N(c_1c_3c_4) + N(c_2c_3c_4)]$
$-N(c_1c_2c_3c_4) = 30 - [11 + 10 + 10] + [5 + 6 + 6] - 4 = 30 - 31 + 17 - 4 = 12$.

(b) $N(\overline{c}_1\overline{c}_4) = N - [N(c_1) + N(c_4)] + N(c_1c_4)$, so $N(\overline{c}_1c_2c_3\overline{c}_4) = N(c_2c_3) - [N(c_1c_2c_3) + N(c_2c_3c_4)] + N(c_1c_2c_3c_4) = 10 - [5 + 6] + 4 = 3$.

5. (a) c_1: number n is divisible by 2

c_2: number n is divisible by 3

c_3: number n is divisible by 5

$N(c_1) = \lfloor 2000/2 \rfloor = 1000$, $N(c_2) = \lfloor 2000/3 \rfloor = 666$,
$N(c_3) = \lfloor 2000/5 \rfloor = 400$, $N(c_1c_2) = \lfloor 2000/(2)(3) \rfloor = 333$,
$N(c_2c_3) = \lfloor 2000/(3)(5) \rfloor = 133$, $N(c_1c_3) = \lfloor 2000/(2)(5) \rfloor = 200$,
$N(c_1c_2c_3) = \lfloor 2000/(2)(3)(5) \rfloor = 66$.
$N(\bar{c}_1\bar{c}_2\bar{c}_3) = 2000 - (1000 + 666 + 400) + (333 + 200 + 133) - 66 = 534$

(b) Let c_1, c_2, c_3 be as in part (a). Let c_4 denote the number n is divisible by 7. Then $N(c_4) = 285$, $N(c_1c_4) = 142$, $N(c_2c_4) = 95$, $N(c_3c_4) = 57$, $N(c_1c_2c_4) = 47$, $N(c_1c_3c_4) = 28$, $N(c_2c_3c_4) = 19$, $N(c_1c_2c_3c_4) = 9$. $N(\bar{c}_1\bar{c}_2\bar{c}_3\bar{c}_4) = 2000 - (1000 + 666 + 400 + 285) + (333 + 200 + 133 + 142 + 95 + 57) - (66 + 47 + 28 + 19) + 9 = 458$
(c) $534 - 458 = 76$.

7. Let c_1 denote the condition where an arrangement of these 11 letters contains two occurrences of the consecutive pair IN. Define similar conditions c_2, c_3, c_4, c_5, and c_6, for the consecutive pairs NI, IO, OI, NO, and ON, respectively. Then

$N = S_0 = 11!/(2!)^3$;
$N(c_1) = 9!/(2!)^2$, $S_1 = \binom{6}{1}[9!/(2!)^2]$;
$N(c_1c_2) = N(c_1c_3) = N(c_1c_6) = N(c_2c_4) = N(c_2c_5) = N(c_3c_4) = N(c_3c_5) = N(c_4c_6) = N(c_5c_6) = 0$, $N(c_1c_4) = 7!/2!$, and $S_2 = (6)[7!/2!]$; and
$S_3 = S_4 = S_5 = S_6 = 0$.

Consequently, the number of arrangements under the given restrictions is $N(\bar{c}_1\bar{c}_2\bar{c}_3\bar{c}_4\bar{c}_5\bar{c}_6) = S_0 - S_1 + S_2 = [11!/(2!)^3] - \binom{6}{1}[9!/(2!)^2] + (6)[7!/2!] = 4,989,600 - 544,320 + 15,120 = 4,460,400$.

9. Let x be written (in base 10) as $x_1x_2\ldots x_7$. Then the answer to the problem is the number of nonnegative integer solutions to $x_1 + x_2 + \ldots + x_7 = 31$, $0 \leq x_i \leq 9$ for $1 \leq i \leq 7$.

If $1 \leq j \leq 7$, let c_j denote the condition that $x_1, x_2, \ldots, x_7$ is an integer solution of $x_1 + x_2 + \ldots + x_7 = 31$, $0 \leq x_i$, $1 \leq i \leq 7$, but $x_j > 9$ (or $x_j \geq 10$).
$N(c_1)$ is the number of integer solutions for $y_1 + x_2 + x_3 + \ldots + x_7 = 21$, $0 \leq y_1$, $0 \leq x_i$ for $2 \leq i \leq 7$. Here $N(c_1) = \binom{27}{21}$ and $S_1 = \binom{7}{1}\binom{27}{21}$.
$N(c_1c_2)$ is the number of integer solutions for $y_1 + y_2 + x_3 + \ldots + x_7 = 11$, $0 \leq y_1, y_2$, $0 \leq x_i$ for $3 \leq i \leq 7$. One finds $N(c_1c_2) = \binom{17}{11}$ and $S_2 = \binom{7}{2}\binom{17}{11}$.
In a similar way we obtain $S_3 = \binom{7}{3}\binom{7}{1}$ and $S_4 = S_5 = S_6 = S_7 = 0$. Since $N = S_0 = \binom{37}{31}$, we have $N(\bar{c}_1\bar{c}_2\ldots\bar{c}_7) = \binom{37}{31} - \binom{7}{1}\binom{27}{21} + \binom{7}{2}\binom{17}{11} - \binom{7}{3}\binom{7}{1}$.

11. For each distribution of the 15 plants there are 15! arrangements. Consequently, in order to answer this question we need to know the number of positive integer solutions for $x_1 + x_2 + x_3 + x_4 + x_5 = 15$, where $1 \leq x_i \leq 4$ for all $1 \leq i \leq 5$.
This is equal to the number of nonnegative integer solutions for

$y_1 + y_2 + y_3 + y_4 + y_5 = 10$, where $0 \le y_i \le 3$ for all $1 \le i \le 5$. [Here $y_i + 1 = x_i$ for all $1 \le i \le 5$.]
For $1 \le i \le 5$ let c_i denote the condition that $y_1 + y_2 + y_3 + y_4 + y_5 = 10$ where $y_i \ge 4$ (or $y_i > 3$) and $y_j \ge 0$ for $1 \le j \le 5$ and $j \ne i$. Then $N(c_1)$ is the number of nonnegative integer solutions for
$z_1 + z_2 + z_3 + z_4 + z_5 = 6$. [Here $z_1 + 4 = y_1$, and $z_i = y_i$ for all $2 \le i \le 5$.] This is $\binom{5+6-1}{6} = \binom{10}{6}$, and so $S_1 = \binom{5}{1}\binom{10}{6}$.
If $1 \le i < j \le 5$, $N(c_ic_j)$ is the number of nonnegative integer solutions for
$w_1 + w_2 + w_3 + w_4 + w_5 = 2$. [Here $w_i + 4 = y_i$, $w_j + 4 = y_j$ and $w_k = y_k$ for all $1 \le k \le 5$, $k \ne i, j$.]
This is $\binom{5+2-1}{2} = \binom{6}{2}$, and so $S_2 = \binom{5}{2}\binom{6}{2}$.
Similar calculations show us that $S_3 = S_4 = S_5 = 0$, and so $N(\overline{c}_1\overline{c}_2\overline{c}_3\overline{c}_4\overline{c}_5) = S_0 - S_1 + S_2 = \binom{5+10-1}{10} - \binom{5}{1}\binom{10}{6} + \binom{5}{2}\binom{6}{2} = \binom{14}{10} - \binom{5}{1}\binom{10}{6} + \binom{5}{2}\binom{6}{2}$.
Consequently, Flo can arrange these 25 plants, according to the restrictions given, in $(15!)[\binom{14}{10} - \binom{5}{1}\binom{10}{6} + \binom{5}{2}\binom{6}{2}]$ ways.

13. Let c_1 denote that the arrangement contains the pattern *spin*. Likewise, let c_2, c_3, c_4 denote this for the patterns *game*, *path*, and *net*, respectively. Then $N = 26!$, $N(c_1) = N(c_2) = N(c_3) = 23!$, $N(c_4) = 24!$, $N(c_1c_2) = 20!$, $N(c_1c_4) = 21!$; $S_0 = 26!$, $S_1 = 3(23!) + 24!$, $S_2 = 20! + 21!$, $S_3 = S_4 = 0$, and $N(\overline{c}_1\overline{c}_2\overline{c}_3\overline{c}_4) = 26! - [3(23!) + 24!] - (20! + 21!)$.

15. For $1 \le i \le 6$, let
c_i: the six dice are rolled but there is no occurrence of i.
$N = 6^8 = S_0$
$N(c_i) = 5^8, 1 \le i \le 6$; $S_1 = \binom{6}{1}5^8$
$N(c_ic_j) = 4^8, 1 \le i < j \le 6$; $S_2 = \binom{6}{2}4^8$
Likewise, $S_3 = \binom{6}{3}3^8$; $S_4 = \binom{6}{4}2^8$; $S_5 = \binom{6}{5}1^8$; and $S_6 = 0$.
Therefore, the answer is $N(\overline{c}_1\overline{c}_2\overline{c}_3\overline{c}_4\overline{c}_5\overline{c}_6)/6^8 = [6^8 - \binom{6}{1}5^8 + \binom{6}{2}4^8 - \binom{6}{3}3^8 + \binom{6}{4}2^8 - \binom{6}{5}]/6^8$.

17. Let c_1 : the three x's are together; c_2 : the three y's are together; and c_3 : the three z's are together.
$N = 9!/[(3!)^3]$ $\quad N(c_1) = N(c_2) = N(c_3) = 7!/[(3!)^2]$
$N(c_ic_j) = (5!)/(3!)$, $1 \le i < j \le 3$ $\quad N(c_1c_2c_3) = 3!$
$N(\overline{c}_1\overline{c}_2\overline{c}_3) = 9!/[(3!)^3] - 3[7!/[(3!)^2]] + 3(5!/3!) - 3!$

19. Here we need to know the number of integer solutions for

$$x_1 + x_2 + x_3 + x_4 + x_5 = 20,$$

where $1 \le x_i \le 6$ for $1 \le i \le 5$.
This is equal to the number of integer solutions for

$$y_1 + y_2 + y_3 + y_4 + y_5 = 15,$$

with $0 \le y_i \le 5$ for $1 \le i \le 5$.
If $1 \le i \le 5$ then let c_i denote the condition that y_1, y_2, y_3, y_4, y_5 is a solution for $y_1 + y_2 + y_3 + y_4 + y_5 = 15$, where $0 \le y_j$ for $1 \le j \le 5$ and $j \ne i$, but $y_i \ge 6$. Then the number of integer solutions for

$$y_1 + y_2 + y_3 + y_4 + y_5 = 15,$$

where $0 \le y_i \le 5$ for $1 \le i \le 5$, is $N(\overline{c}_1\overline{c}_2\overline{c}_3\overline{c}_4\overline{c}_5)$.
N: Here N counts the number of nonnegative integer solutions for $y_1+y_2+y_3+y_4+y_5 = 15$. This number is $\binom{5+15-1}{15} = \binom{19}{15}$. [Hence $S_0 = \binom{19}{15}$.]
$N(c_1)$: To determine $N(c_1)$ we need to find the number of nonnegative integer solutions for

$$z_1 + z_2 + z_3 + z_4 + z_5 = 9,$$

where $z_i = y_i$ for $i \ne 1$, and $y_1 = z_1+6$. Consequently, $N(c_1) = \binom{5+9-1}{9} = \binom{13}{9}$, and $S_1 = \binom{5}{1}\binom{13}{9}$.
$N(c_1c_2)$: Now we need to count the number of nonnegative integer solutions for

$$w_1 + w_2 + w_3 + w_4 + w_5 = 3,$$

where $w_i = y_i$, for $i = 3, 4, 5$; $y_1 = w_1 + 6$, and $y_2 = w_2 + 6$. This number is $\binom{5+3-1}{3} = \binom{7}{3}$, and, as a result, we have $S_2 = \binom{5}{2}\binom{7}{3}$.
Since $S_3 = S_4 = S_5 = 0$, it follows that $N(\overline{c}_1\overline{c}_2\overline{c}_3\overline{c}_4\overline{c}_5) = S_0 - S_1 + S_2 = \binom{19}{15} - \binom{5}{1}\binom{13}{9} + \binom{5}{2}\binom{7}{3} = 3876 - (5)(715) + (10)(35) = 3876 - 3575 + 350 = 651$.
The sample space here is $S = \{(x_1, x_2, x_3, x_4, x_5) | 1 \le x_i \le 6$, for $1 \le i \le 5\}$. And since $|S| = 6^5 = 7776$, it follows that the probability that the sum of Zachary's five rolls is 20 equals $651/7776 \doteq 0.08372$.

21. (a) $\phi(51) = \phi(3 \cdot 17) = 51(1 - \frac{1}{3})(1 - \frac{1}{17}) = (51)(\frac{2}{3})(\frac{16}{17}) = (2)(16) = 32$
(b) $\phi(420) = \phi(2^2 \cdot 3 \cdot 5 \cdot 7) = (420)(1 - \frac{1}{2})(1 - \frac{1}{3})(1 - \frac{1}{5})(1 - \frac{1}{7}) = 420(\frac{1}{2})(\frac{2}{3})(\frac{4}{5})(\frac{6}{7}) = (420)(\frac{1}{3})(\frac{4}{5})(\frac{6}{7}) = 96$
(c) $\phi(12300) = \phi(2^2 \cdot 3 \cdot 5^2 \cdot 41) = (12300)(1 - \frac{1}{2})(1 - \frac{1}{3})(1 - \frac{1}{5})(1 - \frac{1}{41}) = 12300(\frac{1}{2})(\frac{2}{3})(\frac{4}{5})(\frac{40}{41}) = (10)(2)(4)(40) = 3200$

23. (a) $\phi(2^n) = 2^n(1 - \frac{1}{2}) = 2^n(\frac{1}{2}) = 2^{n-1}$
(b) $\phi(2^n p) = (2^n p)(1 - \frac{1}{2})(1 - \frac{1}{p}) = (2^n p)(\frac{1}{2})(\frac{p-1}{p}) = 2^{n-1}(p - 1)$

25. (a) $\phi(6000) = \phi(2^4 \cdot 3 \cdot 5^3) = 6000(1 - (1/2))(1 - (1/3))(1 - (1/5)) = 1600$.
(b) $6000 - 1600 - 1$ (for 6000) $= 4399$.

27. $\phi(17) = \phi(32) = \phi(48) = 16$.

29. If 4 divides $\phi(n)$ then one of the following must hold:
(1) n is divisible by 8;
(2) n is divisible by two (or more) distinct odd primes;
(3) n is divisible by an odd prime p (such as 5, 13, and 17) where 4 divides $p - 1$; and

(4) n is divisible by 4 (and not 8) and at least one odd prime.

Section 8.2

1. From Example 8.10 we know that
$N = S_0 = 2^{10}$, $S_1 = \binom{5}{1}2^6$, $S_2 = \binom{5}{2}2^3$, $S_3 = \binom{5}{3}2^1$, $S_4 = \binom{5}{4}2^0$, and $S_5 = \binom{5}{5}2^0$.
Consequently,
$E_0 = S_0 - S_1 + S_2 - S_3 + S_4 - S_5 = 2^{10} - 5(2^6) + 10(2^3) - 10(2) + 5(2^0) - 2^0 = 1024 - 320 + 80 - 20 + 5 - 1 = 768$;
$E_1 = S_1 - \binom{2}{1}S_2 + \binom{3}{2}S_3 - \binom{4}{3}S_4 + \binom{5}{4}S_5 = 5(2^6) - 2(10)(2^3) + 3(10)(2) - 4(5) + 5 = 320 - 160 + 60 - 20 + 5 = 205$;
$E_2 = S_2 - \binom{3}{1}S_3 + \binom{4}{2}S_4 - \binom{5}{3}S_5 = 10(8) - 3(10)(2) + 6(5) - 10 = 80 - 60 + 30 - 10 = 40$;
$E_3 = S_3 - \binom{4}{1}S_4 + \binom{5}{2}S_5 = 10(2) - 4(5) + 10(1) = 10$;
$E_4 = S_4 - \binom{5}{1}S_5 = 5 - 5 = 0$; and $E_5 = S_5 = 1$.
$\sum_{i=0}^{5} E_i = E_0 + E_1 + E_2 + E_3 + E_4 + E_5 = 768 + 205 + 40 + 10 + 0 + 1 = 1024 = N$.

3. Let c_1 denote the presence of consecutive E's in the arrangement. Likewise, $c_2, c_3, c_4,$ and c_5 are defined for consecutive N's, O's, R's, and S's, respectively.
(a) $N = (14!)/(2!)^5$
$N(c_1) = (13!)/(2!)^4$; $S_1 = \binom{5}{1}[(13!)/(2!)^4]$
$N(c_1c_2) = (12!)/(2!)^3$; $S_2 = \binom{5}{2}[(12!)/(2!)^3]$
$N(c_1c_2c_3) = (11!)/(2!)^2$; $S_3 = \binom{5}{3}[(11!)/(2!)^2]$
$N(c_1c_2c_3c_4) = 10!/2!$; $S_4 = \binom{5}{4}(10!/2!)$
$N(c_1c_2c_3c_4c_5) = 9! = S_5$
$N(\bar{c}_1\bar{c}_2\bar{c}_3\bar{c}_4\bar{c}_5) = 1,286,046,720$
(b) $E_2 = S_2 - \binom{3}{1}S_3 + \binom{4}{2}S_4 - \binom{5}{3}S_5 = 350,179,200$
(c) $L_3 = S_3 - \binom{3}{2}S_4 + \binom{4}{2}S_5 = 74,753,280$

5. For $A = \{1,2,3,\ldots,10\}$, $B = \{1,2,3,4\}$ we want to count the number of functions $f : A \to B$ where $|f(A)| = 2$.
For $1 \leq i \leq 4$ let c_i denote the condition $c_i : f : A \to B$ and i is not in the range of f. Then $S_0 = N = 4^{10}$; $N(c_i) = 3^{10}$ for $1 \leq i \leq 4$, $S_1 = \binom{4}{3}3^{10}$; $N(c_ic_j) = 2^{10}$ for $1 \leq i < j \leq 4$, $S_2 = \binom{4}{2}2^{10}$; $N(c_ic_jc_k) = 1^{10}$ for $1 \leq i < j < k \leq 4$, $S_3 = \binom{4}{5}1^{10}$; and $N(c_1c_2c_3c_4) = S_4 = 0$. Consequently, the answer here is $E_2 = S_2 - \binom{3}{1}S_3 + \binom{4}{2}S_4 = 6(2^{10}) - 3(4) = 6132$. Also, $L_2 = S_2 - \binom{2}{1}S_3 + \binom{3}{1}S_4 = 6(2^{10}) - 2(4) = 6136$.

7. For $1 \leq i \leq 4$, let c_i denote a void in ($i = 1$) clubs, ($i = 2$) diamonds, ($i = 3$) hearts, and ($i = 4$) spades.
$N(c_i) = \binom{39}{13}$, $1 \leq i \leq 4$; $N(c_ic_j) = \binom{26}{13}$, $1 \leq i < j \leq 4$; $N(c_ic_jc_k) = \binom{13}{13}$, $1 \leq i < j < k \leq$

4; $N(c_1c_2c_3c_4) = 0$.
$N(\bar{c}_1\bar{c}_2\bar{c}_3\bar{c}_4) = \binom{52}{13} - \binom{4}{1}\binom{39}{13} + \binom{4}{2}\binom{26}{13} - \binom{4}{3}\binom{13}{13}$.

The probability that the 13 cards include at least one card from each suit is $N(\bar{c}_1\bar{c}_2\bar{c}_3\bar{c}_4)/\binom{52}{13}$.
(b) $E_1 = S_1 - \binom{2}{1}S_2 + \binom{3}{2}S_3 - \binom{4}{3}S_4 = \binom{4}{1}\binom{39}{13} - 2\binom{4}{2}\binom{26}{13} + 3\binom{4}{3}\binom{13}{13} - 0$. The probability of exactly one void is $E_1/\binom{52}{13}$.

(c) $E_2 = S_2 - \binom{3}{1}S_3 = \binom{4}{2}\binom{26}{13} - 3\binom{4}{3}\binom{13}{13}$. The probability of exactly two voids is $E_2/\binom{52}{13}$.

Section 8.3

1. For $1 \le i \le 5$ let c_i be the condition that $2i$ is in position $2i$.
$N = 10!$; $N(c_i) = 9!$, $1 \le i \le 5$; $N(c_ic_j) = 8!$, $1 \le i < j \le 5$; ... ; $N(c_1c_2c_3c_4c_5) = 5!$
$N(\bar{c}_1\bar{c}_2\bar{c}_3\bar{c}_4\bar{c}_5) = 10! - \binom{5}{1}9! + \binom{5}{2}8! - \binom{5}{3}7! + \binom{5}{4}6! - \binom{5}{5}5!$

3. The number of derangements for 1,2,3,4,5 is $5![1 - 1 + (1/2!) - (1/3!) + (1/4!) - (1/5!)] = 5![(1/2!) - (1/3!) + (1/4!) - (1/5!)] = (5)(4)(3) - (5)(4) + 5 - 1 = 60 - 20 + 5 - 1 = 44$.

5. (a) $7! - d_7$ $(d_7 \doteq (7!)e^{-1})$; (b) $d_{26} \doteq (26!)e^{-1}$

7. Let $n = 5 + m$. Then $11,660 = d_5 \cdot d_m = 44(d_m)$, and so $d_m = 265 = d_6$. Consequently, $n = 11$.

9. Mrs. Ford can distribute the ten distinct books among her ten children in 10! ways. The number of ways she can then distribute these ten books a second time so that none of her children receives the same book again is the number of derangements of $1, 2, 3, \ldots, 10$ – namely, d_{10}. Consequently, by the rule of product, the answer is $(10!)d_{10} \doteq (10!)^2(e^{-1})$.

11. (a) $(d_{10})^2$
(b) For $1 \le i \le 10$ let c_i denote that woman i gets back both of her possessions.
$N = (10!)^2$; $N(c_i) = (9!)^2$, $1 \le i \le 10$; $N(c_ic_j) = (8!)^2$, $1 \le i < j \le 10$; etc.
$N(\bar{c}_1\bar{c}_2 \ldots \bar{c}_{10}) = (10!)^2 - \binom{10}{1}(9!)^2 + \binom{10}{2}(8!)^2 - \ldots + (-1)^{10}\binom{10}{0}(0!)^2$.

13. For each $n \in \mathbf{Z}^+$, $n!$ counts the total number of permutations of $1, 2, 3, \ldots, n$. Each such permutation will have k elements that are deranged (that is, there are k elements $x_1, x_2, \ldots, x_k$ in $\{1, 2, 3, \ldots, n\}$ where x_1 is *not* in position x_1, x_2 is *not* in position x_2,..., and x_k is *not* in position x_k) and $n - k$ elements are fixed (that is, the $n - k$ elements $y_1, y_2, \ldots, y_{n-k}$ in $\{1, 2, 3, \ldots, n\} - \{x_1, x_2, \ldots, x_k\}$ are such that y_1 is in position y_1, y_2 is in position y_2,..., and y_{n-k} is in position y_{n-k}).
The $n - k$ fixed elements can be chosen in $\binom{n}{n-k}$ ways and the remaining k elements

can then be permuted (that is, deranged) in d_k ways. Hence there are $\binom{n}{n-k}d_k = \binom{n}{k}d_k$ permutations of $1, 2, 3, \ldots, n$ with $n-k$ fixed elements (and k deranged elements). As k varies from 0 to n we count all of the $n!$ permutations of $1, 2, 3, \ldots, n$ according to the number k of deranged elements.
Consequently,

$$n! = \binom{n}{0}d_0 + \binom{n}{1}d_1 + \binom{n}{2}d_2 + \ldots + \binom{n}{n}d_n = \sum_{k=0}^{n}\binom{n}{k}d_k.$$

15. For $1 \le i \le n-1$, let c_i denote the condition c_i : $i(i+1)$ occurs in the clockwise circular arrangement of $1, 2, 3, \ldots, n$. Let c_n denote the condition for when $n1$ occurs in the clockwise circular arrangement of $1, 2, 3, \ldots n$.
Then $N = S_0 = n!/n = (n-1)!$
$N(c_i) = (n-1)!/(n-1) = (n-2)!, 1 \le i \le n$; $S_1 = \binom{n}{1}(n-2)!$
$N(c_i c_j) = (n-2)!/(n-2) = (n-3)!, 1 \le i < j \le n$; $S_2 = \binom{n}{2}(n-3)!$
$N(c_i c_j c_k) = (n-3)!/(n-3) = (n-4)!, 1 \le i < j < k \le n$; $S_3 = \binom{n}{3}(n-4)!, \ldots,$
$N(c_1 c_2 c_3 \ldots c_n) = 1$
Consequently, the answer is $N(\overline{c}_1\overline{c}_2\overline{c}_3 \ldots \overline{c}_n) = \binom{n}{0}(n-1)! - \binom{n}{1}(n-2)! + \binom{n}{2}(n-3)! - \ldots + (-1)^{n-1}\binom{n}{n-1}(0!) + (-1)^n\binom{n}{n}$.

Sections 8.4 and 8.5

1. These results follow by counting the possible locations for the desired numbers of rooks on each chessboard.

3. (a) $\binom{8}{0} + \binom{8}{1}8x + \binom{8}{2}(8 \cdot 7)x^2 + \binom{8}{3}(8 \cdot 7 \cdot 6)x^3 + \binom{8}{4}(8 \cdot 7 \cdot 6 \cdot 5)x^4 + \ldots + \binom{8}{8}(8!)x^8 = \sum_{i=0}^{8}\binom{8}{i}P(8,i)x^i$. (b) $\sum_{i=0}^{n}\binom{n}{i}P(n,i)x^i$

5. (a) (i) $(1+2x)^3$ (ii) $1 + 8x + 14x^2 + 4x^3$
(iii) $1 + 9x + 25^2 + 21x^3$ (iv) $1 + 8x + 16x^2 + 7x^3$
(b) If the board C consists of n steps, and each step has k blocks, then $r(C, x) = (1+kx)^n$.

7.

	Java	C++	VHDL	Perl	SQL
(1) Jeanne					■
(2) Charles					■
(3) Todd				■	■
(4) Paul	■	■			
(5) Sandra		■	■		

$r(C, x) = (1 + 4x + 3x^2)(1 + 4x + 2x^2) = 1 + 8x + 21x^2 + 20x^3 + 6x^4$

For $1 \le i \le 5$ let c_i be the condition that an assignment is made with person (i) assigned to a language he or she wishes to avoid.
$N(\bar{c}_1\bar{c}_2\bar{c}_3\bar{c}_4\bar{c}_5) = 5! - 8(4!) + 21(3!) - 20(2!) + 6(1!) = 20.$

9.

(a)	Calculus I	Calculus III	Calculus II	Combinatorics	Statistics
(1) Violet					
(2) Mary Lou					
(3) Al					
(4) Jack					
(5) Lynn					

For $1 \le i \le 5$, let c_i denote the condition where an assignment is made but teacher i is assigned a course she, or he, does not want to teach.

The rook polynomial for the shaded board is $(1+4x+2x^2)(1+3x+x^2)(1+x) = 1+8x+22x^2+25x^3+12x^4+2x^5$.

$N(\bar{c}_1\bar{c}_2\bar{c}_3\bar{c}_4\bar{c}_5) = 5! - 8(4!) + 22(3!) - 25(2!) + 12(1!) - 2(0!) = 120 - 192 + 132 - 50 + 12 - 2 = 264 - 244 = 20.$

(b)	Calculus I	Calculus III	Calculus II	Statistics
(1) Mary Lou				
(2) Al				
(3) Jack				
(4) Lynn				

Here the rook polynpomial is $(1+2x)(1+2x)(1+x) = 1+5x+8x^2+4x^3$ and the number of correct assignments is $4! - 5(3!) + 8(2!) - 4(1!) = 6$.

So the given probability is $6/20 = 3/10$.

11.

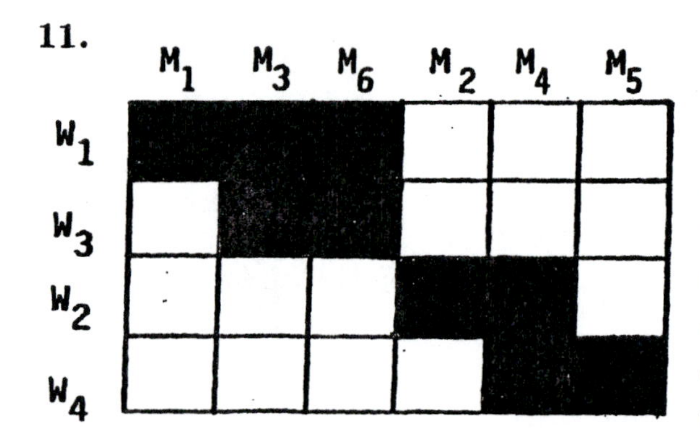

$r(C, x) = (1 + 5x + 4x^2)(1 + 4x + 3x^2) = 1 + 9x + 27x^2 + 31x^3 + 12x^4$.
For $1 \le i \le 4$, let c_i denote the condition where each of the four women has been matched with one of the six men but woman i is paired with an incompatible partner. Then

$N(\bar{c}_1\bar{c}_2\bar{c}_3\bar{c}_4) = (6 \cdot 5 \cdot 4 \cdot 3) - 9(5 \cdot 4 \cdot 3) + 27(4 \cdot 3) - 31(3) + 12 = 63.$

Supplementary Exercises

1. We need only consider the divisors 2,3, and 5. Let c_1 denote divisibility by 2, c_2

divisibility by 3, and c_3 divisibility by 5.

$N = 500$; $N(c_1) = \lfloor 500/2 \rfloor = 250$; $N(c_2) = \lfloor 500/3 \rfloor = 166$; $N(c_3) = \lfloor 500/5 \rfloor = 100$; $N(c_1c_2) = \lfloor 500/6 \rfloor = 83$; $N(c_1c_3) = \lfloor 500/10 \rfloor = 50$; $N(c_2c_3) = \lfloor 500/15 \rfloor = 33$; $N(c_1c_2c_3) = \lfloor 500/30 \rfloor = 16$.

$N(\bar{c}_1\bar{c}_2\bar{c}_3) = 500 - (250 + 166 + 100) + (83 + 50 + 33) - 16 = 134$.

3. For each distribution of the 24 balls (among the four shelves) there are $(24!)/(6!)^4$ possible arrangements. Hence we need to know in how many ways the boys can distribute the balls for the given restrictions. This is the number of integer solutions for

$$x_1 + x_2 + x_3 + x_4 = 24,$$

where $2 \le x_i \le 7$ for all $1 \le i \le 4$.
This equals the number of integer solutions for

$$y_1 + y_2 + y_3 + y_4 = 16,$$

where $0 \le y_i \le 5$ for all $1 \le i \le 4$. [Here $y_i + 2 = x_i$ for each $1 \le i \le 4$.]
For $1 \le i \le 4$ define c_i to be the condition that y_1, y_2, y_3, y_4 is a solution of

$$y_1 + y_2 + y_3 + y_4 = 16,$$

where $y_i > 5$ (or $y_i \ge 6$) and $y_j \ge 0$ for all $1 \le j \le 4$, $j \ne i$. Then, for example, $N(c_1)$ is the number of nonnegative integer solutions for

$$w_1 + w_2 + w_3 + w_4 = 10.$$

[Here $w_1 + 6 = y_1$ and $w_i = y_i$ for $i = 2, 3, 4$.] So $N(c_1) = \binom{4+10-1}{10} = \binom{13}{10}$ and $S_1 = \binom{4}{1}\binom{13}{10}$.

Similar arguments show us that $N(c_1c_2) = \binom{4+4-1}{4} = \binom{7}{4}$ and $S_2 = \binom{4}{2}\binom{7}{4}$; and $S_3 = S_4 = 0$.
Therefore the number of distributions for the given restrictions is

$$N(\bar{c}_1\bar{c}_2\bar{c}_3\bar{c}_4) = \binom{19}{16} - \binom{4}{1}\binom{13}{10} + \binom{4}{2}\binom{7}{4},$$

and Joseph and Jeffrey can arrange the 24 balls in

$$[(24!)/(6!)^4]\left[\binom{19}{16} - \binom{4}{1}\binom{13}{10} + \binom{4}{2}\binom{7}{4}\right]$$

ways.

5. Let c_i denote the occurrence of the pattern $i(i+1)$ for $1 \le i \le 7$.

The occurrence of the pattern 81 is denoted by c_8.

For $1 \le i \le 8$, $N(c_i) = 7!$; $N(c_ic_j) = 6!$, $1 \le i < j \le 8$; etc.

$$N(\overline{c}_1\overline{c}_2\ldots\overline{c}_8) = 8! - \binom{8}{1}7! + \binom{8}{2}6! - \binom{8}{3}5! + \ldots + (-1)^7\binom{8}{7}1! = 14832.$$

7. For $1 \leq i \leq 10$, let c_i denote the condition where student i occupies the same chair before and after the break. Then the answer to this exercise is $N(\overline{c}_1\overline{c}_2\overline{c}_3\ldots\overline{c}_{10}) =$ $S_0 - S_1 + S_2 - S_3 + \cdots + S_{10}$.

Here $S_0 = \binom{14}{10}10! = (14)(13)(12)\cdots(5)$.

$N(c_1) = \binom{13}{9}9! = (13)(12)\cdots(5)$, and by symmetry $N(c_i) = N(c_1)$ for $2 \leq i \leq 10$. So $S_1 = \binom{10}{1}\binom{13}{9}9!$

$N(c_1c_2) = \binom{12}{8}8!$ and $S_2 = \binom{10}{2}\binom{12}{8}8!$

In general for $0 \leq k \leq 10$,

$$S_k = \binom{10}{k}\binom{14-k}{10-k}(10-k)!$$

and $N(\overline{c}_1\overline{c}_2\overline{c}_3\ldots\overline{c}_{10}) = \sum_{k=0}^{10}(-1)^k S_k = \sum_{k=0}^{10}(-1)^k\binom{10}{k}\binom{14-k}{10-k}(10-k)! = 1,764,651,461.$

9. The total number of arrangements is $T = (13!)/[(2!)^5]$.

(a) $S_3 = \binom{5}{3}[(10!)/(2!)^2]$

$S_4 = \binom{5}{4}[(9!)/(2!)]$

$S_5 = \binom{5}{5}(8!)$

$E_3 = [S_3 - \binom{4}{1}S_4 + \binom{5}{2}S_5]/T$

(b) $E_4 = [S_4 - \binom{5}{1}S_5]$ $\quad E_5 = S_5$

The answer is $[T - (E_4 + E_5)]/T$.

1. (a) $\binom{n-m}{r-m} = \binom{n-m}{n-r}$

(b) Let $A = \{x_1, x_2, \ldots, x_m, y_{m+1}, \ldots, y_n\}$. For $1 \leq i \leq m$ let c_i denote that r elements are selected from A with $r \geq m$ and x_i is not in the selection.

$N = \binom{n}{r}$; $\quad N(c_i) = \binom{n-1}{r}, 1 \leq i \leq m$; $S_1 = \binom{m}{1}\binom{n-1}{r}$

$N(c_ic_j) = \binom{n-2}{r}, 1 \leq i < j \leq m$; $S_2 = \binom{m}{2}\binom{n-2}{r}$, etc.

$\binom{n-m}{n-r} = N(\overline{c}_1\overline{c}_2\ldots\overline{c}_m) = \sum_{i=0}^{m}(-1)^i\binom{m}{i}\binom{n-i}{r}$.

Consider the derangements of the symbols L,A,P_1,T,O,P_2. There are d_6 such arrangements. Of these there are

(i) d_4 arrangements where P_2 is in position 3 and P_1 is in position 6;

(ii) d_5 arrangements where P_1 is in position 6 and P_2 is not in position 3; and,

(iii) d_5 arrangements where P_2 is in position 3 and P_1 is not in position 6.

There are $d_6 - 2d_5 - d_4$ such arrangements of L,A,P_1,T,O,P_2. Hence there are $(1/2)[d_6 - 2d_5 - d_4] = (1/2)[265 - 2(44) - 9] = 84$ ways to arrange the letters in LAPTOP so that none of L,A,T,O is in its original position and P is not in the third or sixth position. [Why the 1/2? Because we do not distinguish arrangements such as P_1 L A P_2 T O and P_2 L A P_1 T O.]

15.

(a) $S_1 = \{1, 5, 7, 11, 13, 17\}$ $\quad S_2 = \{2, 4, 8, 10, 14, 16\}$
$S_3 = \{3, 15\}$ $\quad S_6 = \{6, 12\}$
$S_9 = \{9\}$ $\quad S_{18} = \{18\}$

(b) $|S_1| = 6 = \phi(18)$ $\quad |S_3| = 2 = \phi(6)$ $\quad |S_9| = 1 = \phi(2)$
$|S_2| = 6 = \phi(9)$ $\quad |S_6| = 2 = \phi(3)$ $\quad |S_{18}| = 1 = \phi(1)$

17. Proof:
(a) If n is even then by the Fundamental Theorem of Arithmetic (Theorem 4.11) we may write $n = 2^k m$, where $k \geq 1$ and m is odd. Then $2n = 2^{k+1}m$ and $\phi(2n) =$ $(2^{k+1})(1 - \frac{1}{2})\phi(m) = 2^k\phi(m) = 2(2^k)(\frac{1}{2})\phi(m) = 2[2^k(1 - \frac{1}{2})\phi(m)] = 2[\phi(2^k m)] = 2\phi(n)$.
(b) When n is odd we find that $\phi(2n) = (2n)(1 - \frac{1}{2})\prod_{p|n}(1 - \frac{1}{p})$, where the product is taken over all (odd) primes dividing n. (If $n = 1$ then $\prod_{p|n}(1 - \frac{1}{p})$ is 1.) But $(2n)(1 - \frac{1}{2})\prod_{p|n}(1 - \frac{1}{p}) =$ $n\prod_{p|n}(1 - \frac{1}{p}) = \phi(n)$.

19. a) There are d_4 ways to derange the four subjects and 12! ways to arrange the 12 books on any one subject on a shelf. Hence the number of arrangements here is $d_4(12!)^4$.
b) We can select the one subject to be on its original shelf in $\binom{4}{1}$ ways and then arrange these 12 books in 12! ways. There are d_3 ways to derange the three other subjects and 12! ways to arrange the 12 books on any one of these subjects on a shelf. Consequently, the number of arrangements in this case is $\binom{4}{1}d_3(12!)^4$.
c) There are d_4 ways to derange the four subjects and d_{12} ways to derange the 12 books on any one subject on a shelf. Hence, in this case, we have $d_4(d_{12})^4$ arrangements.

CHAPTER 9
GENERATING FUNCTIONS

Section 9.1

1. The number of integer solutions for the given equations is the coefficient of
(a) x^{20} in $(1+x+x^2+\ldots+x^7)^4$.
(b) x^{20} in $(1+x+x^2+\ldots+x^{20})^2(1+x^2+x^4+\ldots+x^{20})^2$ or
$(1+x+x^2+\ldots)^2(1+x^2+x^4+\ldots)^2$.
(c) x^{30} in $(x^2+x^3+x^4)(x^3+x^4+\ldots+x^8)^4$.
(d) x^{30} in $(1+x+x^2+\ldots+x^{30})^3(1+x^2+x^4+\ldots+x^{30})\cdot$
$(x+x^3+x^5+\ldots+x^{29})$ or
$(1+x+x^2+\ldots)^3\,(1+x^2+x^4+\ldots)(x+x^3+x^5+\ldots)$.

3. (a) The generating function is either $(1+x+x^2+x^3+\ldots+x^{10})^6$ or $(1+x+x^2+x^3+\ldots)^6$. [The number of ways to select 10 candy bars is the coefficient of x^{10} in either case.]
(b) The generating function is either $(1+x+x^2+x^3+\ldots+x^r)^n$ or $(1+x+x^2+x^3+\ldots)^n$. [The number of selections of r objects is the coefficient of x^r in either case.]

5. $c_1+c_2+c_3+c_4=20,\ -3\le c_1, c_2,\ -5\le c_3\le 5,\ 0\le c_4$
$(3+c_1)+(3+c_2)+(5+c_3)+c_4=31$
Let $x_1=3+c_1$, $x_2=3+c_2$, $x_3=5+c_3$, and $x_4=c_4$. Then
$x_1+x_2+x_3+x_4=31,\ 0\le x_1, x_2, x_4,\ 0\le x_3\le 10$.
Consequently, the answer is the coefficient of x^{31} in the generating function
$(1+x+x^2+\ldots)^3(1+x+x^2+\ldots+x^{10})$.

Section 9.2

1. (a) $(1+x)^8$ (b) $8(1+x)^7$ (c) $(1+x)^{-1}$
(d) $6x^3/(1+x)$ (e) $(1-x^2)^{-1}$ (f) $x^2/(1-ax)$

3. (a) $g(x)=f(x)-a_3x^3+3x^3$
(b) $g(x)=f(x)-a_3x^3+3x^3-a_7x^7+7x^7$
(c) $g(x)=2f(x)-2a_1x+x-2a_3x^3+3x^3$
(d) $g(x)=2f(x)+[5/(1-x)]+(1-2a_1-5)x+(3-2a_3-5)x^3+(7-2a_7-5)x^7$

5. (a) $(1+x+x^2+x^3+\cdots)^{15}=[1/(1-x)]^{15}=(1-x)^{-15}=\binom{-15}{0}(-x)^0+\binom{-15}{1}(-x)^1+\binom{-15}{2}(-x)^2+\cdots$. So the coefficient of x^7 is $\binom{-15}{7}(-1)^7=(-1)^7\binom{15+7-1}{7}(-1)^7=\binom{21}{7}$.

(b) Now we have $(1+x+x^2+x^3+\cdots)^n = [1/(1-x)]^n = (1-x)^{-n} = \binom{-n}{0}(-x)^0+\binom{-n}{1}(-x)^1+\binom{-n}{2}(-x)^2+\cdots$. So the coefficient of x^7 is $\binom{-n}{7}(-1)^7 = (-1)^7\binom{n+7-1}{7}(-1)^7 = \binom{n+6}{7}$.

7. $(x^2+x^3+x^4+x^5+x^6)^5 = x^{10}(1+x+x^2+x^3+x^4)^5$ so the answer is the coefficient of x^{10} in $(1+x+x^2+x^3+x^4)^5$.
Now $(1+x+x^2+x^3+x^4) = 1+x+x^2+x^3+\cdots-(x^5+x^6+x^7+\cdots) = \frac{1}{1-x} - \frac{x^5}{1-x}$. So $(1+x+x^2+x^3+x^4)^5 = [(1-x^5)/(1-x)]^5 = (1-x^5)^5(1-x)^{-5} = [\binom{5}{0} - \binom{5}{1}x^5 + \binom{5}{2}x^{10} - \binom{5}{3}x^{15} + \binom{5}{4}x^{20} - \binom{5}{5}x^{25}]\cdot[\binom{-5}{0} + \binom{-5}{1}(-x) + \binom{-5}{2}(-x)^2 + \cdots]$, and here the coefficient of x^{10} is $\binom{-5}{10}(-1)^{10} - \binom{5}{1}\binom{-5}{5}(-1)^5 + \binom{5}{2}\binom{-5}{0} = \binom{14}{10} - \binom{5}{1}\binom{9}{5} + \binom{5}{2}$.

9. (a) The highest (nonzero) power of x in $x^3(1-2x)^{10}$ is x^{13}. Consequently, the coefficient of x^{15} is 0.
(b) $(x^3-5x)/(1-x)^3 = (x^3-5x)(1-x)^{-3} = x(x^2-5)(1-x)^{-3}$. So the answer is the coefficient of x^{14} in $(x^2-5)(1-x)^{-3} = (x^2-5)[\binom{-3}{0} + \binom{-3}{1}(-x) + \binom{-3}{2}(-x)^2 + \cdots]$. This coefficient is $\binom{-3}{12}(-1)^{12} - 5\binom{-3}{14}(-1)^{14} = \binom{14}{12} - 5\binom{16}{14}$.

(c) $(1+x)^4/(1-x)^4 = [\binom{4}{0}+\binom{4}{1}x+\binom{4}{2}x^2+\binom{4}{3}x^3+\binom{4}{4}x^4][\binom{-4}{0}+\binom{-4}{1}(-x)+\binom{-4}{2}(-x)^2+\cdots]$.
Here the coefficient of x^{15} is $\binom{-4}{15}(-1)^{15}+\binom{4}{1}\binom{-4}{14}(-1)^{14}+\binom{4}{2}\binom{-4}{13}(-1)^{13}+\binom{4}{3}\binom{-4}{12}(-1)^{12}+\binom{4}{4}\binom{-4}{11}(-1)^{11} = \binom{18}{15} + \binom{4}{1}\binom{17}{14} + \binom{4}{2}\binom{16}{13} + \binom{4}{3}\binom{15}{12} + \binom{14}{11}$.

11. Consider each package of 25 envelopes as one unit. Then the answer to the problem is the coefficient of x^{120} in $(x^6+x^7+\ldots+x^{39}+x^{40})^4 = x^{24}(1+x+\ldots+x^{34})^4$. This is the same as the coefficient of x^{96} in $[(1-x^{35})/(1-x)]^4 = (1-x^{35})^4(1-x)^{-4} = [1-4x^{35}+6x^{70}-\ldots+x^{140}][\binom{-4}{0}+\ldots+\binom{-4}{26}(-x)^{26}+\ldots+\binom{-4}{61}(-x)^{61}+\ldots+\binom{-4}{96}(-x)^{96}+\ldots]$.

Consequently the answer is $\binom{-4}{96}(-1)^{96}-4\binom{-4}{61}(-1)^{61}+6\binom{-4}{26}(-1)^{26} = \binom{99}{96}-4\binom{64}{61}+6\binom{29}{26}$.

13. $(x+x^2+x^3+x^4+x^5+x^6)^{12} = x^{12}[(1-x^6)/(1-x)]^{12} = x^{12}((1-x)^6)^{12}[\binom{-12}{0}+\binom{-12}{1}(-x)+\binom{-12}{2}(-x)^2+\ldots]$. The numerator of the answer is the coefficient of x^{18} in $(1-x^6)^{12}[\binom{12}{0}+\binom{-12}{1}(-x)+\ldots] = [1-\binom{12}{1}x^6+\binom{12}{2}x^{12}-\binom{12}{3}x^{18}+\ldots+x^{72}][\binom{-12}{0}+\binom{-12}{1}(-x)+\ldots]$ and this equals $\binom{-12}{18}(-1)^{18}-\binom{12}{1}\binom{-12}{12}(-1)^{12}+\binom{12}{2}\binom{-12}{6}(-1)^6-\binom{12}{3}\binom{-12}{0} = \binom{29}{18}-\binom{12}{1}\binom{23}{12}+\binom{12}{2}\binom{17}{6}-\binom{12}{3}$. The final answer is obtained by dividing the last result by 6^{12}, the size of the sample space.

15. Here we need the coefficient of x^n in $(1+x+x^2+x^3+\ldots)^2(1+x^2+x^4+\ldots) = (1/(1-x))^2(1/(1-x^2)) = (1/(1-x))^3(1/(1+x))$.
Using a partial fraction decomposition, $\frac{1}{1+x}\cdot\frac{1}{(1-x)^3} = \frac{(1/8)}{(1+x)}+\frac{(1/8)}{(1-x)}+\frac{(1/4)}{(1-x)^2}+\frac{(1/2)}{(1-x)^3}$, where the coefficient of x^n is $(-1)^n(1/8)+(1/8)+(1/4)\binom{-2}{n}(-1)^n+(1/2)\binom{-3}{n}(-1)^n = (1/8)[1+(-1)^n]+(1/4)\binom{n+1}{n}+(1/2)\binom{n+2}{n}$.

17. $(1-x-x^2-x^3-x^4-x^5-x^6)^{-1} = [1-(x+x^2+x^3+x^4+x^5+x^6)]^{-1}$

$$= 1 + \underbrace{(x+x^2+\ldots+x^6)}_{\text{one roll}} + \underbrace{(x+x^2+\ldots+x^6)^2}_{\text{two rolls}} + \underbrace{(x+x^2+\ldots+x^6)^3}_{\text{three rolls}} + \ldots,$$

where the 1 takes care of the case where the die is not rolled.

19. (a) There are $2^{8-1} = 2^7$ compositions of 8 and $2^{\lfloor 8/2 \rfloor} = 2^4$ palindromes of 8. Assuming each composition of 8 has the same probability of being generated, the probability a palindrome of 8 is generated is $2^4/2^7 = 1/8$.

(b) Assuming each composition of n has the same probability of being generated, the probability a palindrome of n is generated is $2^{\lfloor n/2 \rfloor}/2^{n-1} = 2^{\lfloor n/2 \rfloor - n+1} = 2^{1-\lceil n/2 \rceil}$.

21. The number of palindromes of n that start (and end) with t is the number of palindromes of $n-2t$. This is $2^{\lfloor (n-2t)/2 \rfloor}$.

23. Let $n = 2k$. The palindromes of n with an even number of summands have a plus sign at the center and their number is the number of compositions of k – namely, $2^{k-1} = 2^{(n/2)-1}$. Since there are $2^{n/2}$ palindromes in total, the number with an odd number of summands is $2^{n/2} - 2^{(n/2)-1} = 2^{n/2}(1-\frac{1}{2}) = 2^{n/2}(\frac{1}{2}) = 2^{(n/2)-1}$.

25. (a) $Pr(Y = y) = (\frac{5}{6})^{y-1}(\frac{1}{6})$, $y = 1, 2, 3, \ldots$.
(b) and (c) Using the general formulas at the end of Example 9.18, with $p = \frac{1}{6}$ and $q = 1-p = \frac{5}{6}$, it follows that
$E(Y) = \frac{1}{p} = \frac{1}{(1/6)} = 6$, and
$\sigma_Y = \sqrt{\text{Var }(Y)} = \sqrt{q/p^2} = \sqrt{(\frac{5}{6})/(\frac{1}{6})^2} = \sqrt{(\frac{5}{6})(36)} = \sqrt{30} \doteq 5.477226$.

27. Let the discrete random variable Y count the number of tosses Leroy makes until he gets the first tail. Then $Pr(Y = y) = (\frac{2}{3})^{y-1}(\frac{1}{3})$, $y = 1, 2, 3, \ldots$.

Here we are interested in $Pr(Y = 1) + Pr(Y = 3) + Pr(Y = 5) + \cdots = (\frac{1}{3}) + (\frac{2}{3})^2(\frac{1}{3}) + (\frac{2}{3})^4(\frac{1}{3}) + \cdots = (\frac{1}{3})[1 + (\frac{2}{3})^2 + (\frac{2}{3})^4 + \cdots] = (\frac{1}{3})\frac{1}{1-(\frac{2}{3})^2} = (\frac{1}{3})\frac{1}{5/9} = (\frac{1}{3})(\frac{9}{5}) = \frac{3}{5}$.

29. (a) The differences are $3-1, 6-3, 8-6, 15-8$, and $15-15$ – that is 2,3,2,7, and 0, where $2+3+2+7+0 = 14$.

(b) $\{3,5,8,15\}$

(c) $\{1+a, 1+a+b, 1+a+b+c, 1+a+b+c+d\}$

31. $c_k = \sum_{i=0}^{k} i(k-i)^2 = \sum_{i=0}^{k} i(k^2-2ki+i^2) = k^2\sum_{i=0}^{k} i - 2k\sum_{i=0}^{k} i^2 + \sum_{i=0}^{k} i^3 = k^2[k(k+1)/2] -$

$2k[k(k+1)(2k+1)/6] + [(k^2)(k+1)^2/4] = (k^4+k^3)/2 - (k^2)(k+1)(2k+1)/3 + (k^2)(k+1)^2/4 =$

$(1/12)[6k^4+6k^3-4k^2(2k^2+3k+1)+3k^2(k^2+2k+1)] = (1/12)[k^4-k^2] = (1/12)(k^2)(k^2-1)$.

33. (a) $(1+x+x^2+x^3+x^4)(0+x+2x^2+3x^3+\ldots) = \sum_{i=0}^{\infty} c_i x^i$ where $c_0 = 0$, $c_1 = 1$, $c_2 = 1+2 = 3$, $c_3 = 1+2+3 = 6$, $c_4 = 1+2+3+4 = 10$, and
$c_n = n+(n-1)+(n-2)+(n-3)+(n-4) = 5n-10$ for all $n \geq 5$.

(b) $(1-x+x^2-x^3+\cdots)(1-x+x^2-x^3+\ldots) = \frac{1}{(1+x)^2} = (1+x)^{-2}$, the generating function for the sequence $\binom{-2}{0}, \binom{-2}{1}, \binom{-2}{2}, \binom{-2}{3}, \ldots$. Hence the convolution of the given pair of sequences is $c_0, c_1, c_2, \ldots$, where
$c_n = \binom{-2}{n} = (-1)^n\binom{2+n-1}{n} = (-1)^n\binom{n+1}{n} = (-1)^n(n+1)$, $n \in \mathbf{N}$.
[This is the alternating sequence $1, -2, 3, -4, 5, -6, 7, \ldots$.]

Section 9.3

1. 7; 6+1; 5+2; 5+1+1; 4+3; 4+2+1; 4+1+1+1; 3+3+1; 3+2+2; 3+2+1+1; 3+1+1+1+1; 2+2+2+1; 2+2+1+1+1; 2+1+1+1+1+1; 1+1+1+1+1+1+1

3. The number of partitions of 6 into 1's, 2's, and 3's is 7.

5. (a) and (b) $(1+x^2+x^4+x^6+\ldots)(1+x^4+x^8+\ldots)(1+x^6+x^{12}+\ldots)\cdots$
$= \prod_{i=1}^{\infty} \frac{1}{1-x^{2i}}$

7. Let $f(x)$ be the generating funcion for the number of partitions of n where no summand appears more than twice. Let $g(x)$ be the generating function for the number of partitions of n where no summand is divisible by 3.

$g(x) = \frac{1}{1-x} \cdot \frac{1}{1-x^2} \cdot \frac{1}{1-x^4} \cdot \frac{1}{1-x^5} \cdot \frac{1}{1-x^7} \cdots$
$f(x) = (1+x+x^2)(1+x^2+x^4)(1+x^3+x^6)(1+x^4+x^8)\cdots$
$= \frac{1-x^3}{1-x} \cdot \frac{1-x^6}{1-x^2} \cdot \frac{1-x^9}{1-x^3} \cdot \frac{1-x^{12}}{1-x^4} \cdots = g(x).$

9. This result follows from the one-to-one correspondence between the Ferrers graphs with summands (rows) not exceeding m and the transpose graphs (also Ferrers graphs) that have m summands (rows).

Section 9.4

1.

(a) e^{-x}	(b) e^{2x}	(c) e^{-ax}
(d) e^{a^2x}	(e) ae^{a^2x}	(f) xe^{2x}

3. (a) $g(x) = f(x) + [3-a_3](x^3/3!)$
(b) $g(x) = f(x) + [-1-a_3](x^3/3!) = e^{5x} - (126x^3)/3!$
(c) $g(x) = 2f(x) + [2-2a_1](x^1/1!) + [4-2a_2](x^2/2!)$

(d) $g(x) = 2f(x) + 3e^x + [2 - 2a_1 - 3](x^1/1!) + [4 - 2a_2 - 3](x^2/2!) + [8 - 2a_3 - 3](x^3/3!)$

5. We find that

$$\begin{aligned}\tfrac{1}{1-x} &= 1 + x + x^2 + x^3 + \cdots \\ &= (0!)\tfrac{x^0}{0!} + (1!)\tfrac{x^1}{1!} + (2!)\tfrac{x^2}{2!} + (3!)\tfrac{x^3}{3!} + \cdots,\end{aligned}$$

so $1/(1-x)$ is the exponential generating function for the sequence $0!, 1!, 2!, 3!, \ldots$.

7. The answer is the coefficient of $\frac{x^{25}}{25!}$ in $\left(\frac{x^3}{3!} + \frac{x^4}{4!} + \ldots + \frac{x^{10}}{10!}\right)^4$.

9. (a) There are 3^{20} ternary sequences of 20 symbols – that is, 0's, 1's, and 2's. This is our sample space. The exponential generating function for the sequences with an even number of 1's is $f(x) = (1 + x + \frac{x^2}{2!} + \frac{x^3}{3!} + \cdots)(1 + \frac{x^2}{2!} + \frac{x^4}{4!} + \cdots)(1 + x + \frac{x^2}{2!} + \frac{x^3}{3!} + \cdots) = e^x[\frac{1}{2}(e^x + e^{-x})]e^x = \frac{1}{2}[e^{3x} + e^x]$. The coefficient of $x^{20}/20!$ in $f(x)$ is $\frac{1}{2}[3^{20} + 1]$, so the probability is $(1/2)[3^{20} + 1]/(3^{20})$.
(b) Here the exponential generating function is $g(x) = e^x[\frac{1}{2}(e^x + e^{-x})]^2 = (\frac{e^{3x}}{4}) + (\frac{e^x}{2}) + (\frac{e^{-x}}{4})$ and the coefficient of $x^{20}/20!$ in $g(x)$ is $\frac{1}{4}[3^{20} + 3]$. Consequently, the probability is $(1/4)[3^{20} + 3]/(3^{20})$.
(c) The exponential generating function is $h(x) = (x + \frac{x^3}{3!} + \frac{x^5}{5!} + \cdots)(1 + x + \frac{x^2}{2!} + \cdots)(1 + x + \frac{x^2}{2!} + \cdots) = (\frac{e^x - e^{-x}}{2})(e^x)(e^x) = \frac{1}{2}[e^{3x} - e^x]$, where the coefficient of $x^{20}/20!$ is $\frac{1}{2}[3^{20} - 1]$. So the resulting probability is $(1/2)[3^{20} - 1]/(3^{20})$.
(d) Here the exponential generating function is $(\frac{e^x + e^{-x}}{2})(\frac{e^x - e^{-x}}{2})e^x + (\frac{e^x - e^{-x}}{2})(\frac{e^x + e^{-x}}{2})e^x = \frac{1}{2}[e^{3x} - e^{-x}]$. The coefficient of $x^{20}/20!$ is $\frac{1}{2}[3^{20} - 1]$ and the resulting probability is $(1/2)[3^{20} - 1]/(3^{20})$.
(e) $(1/2)[3^{20} + 1]/(3^{20})$

Section 9.5

1. (a) $1 + x + x^2$ is the generating function for the sequence $1, 1, 1, 0, 0, 0, \ldots$, so $(1 + x + x^2)/(1 - x)$ is the generating function for the sequence $1, 1 + 1, 1 + 1 + 1, 1 + 1 + 1 + 0, \ldots$ – that is, the sequence $1, 2, 3, 3, \ldots$.

(b) $1 + x + x^2 + x^3$ is the generating function for the sequence $1, 1, 1, 1, 0, 0, 0, \ldots$, so $(1 + x + x^2 + x^3)/(1 - x)$ is the generating function for the sequence $1, 1 + 1, 1 + 1 + 1, 1 + 1 + 1 + 1, 1 + 1 + 1 + 1 + 0, 1 + 1 + 1 + 1 + 0 + 0, \ldots$ – that is, the sequence $1, 2, 3, 4, 4, 4, \ldots$.

(c) $1 + 2x$ is the generating function for the sequence $1, 2, 0, 0, 0, 0, \ldots$, so $(1 + 2x)/(1 - x)$ is the generating function for the sequence $1, 1 + 2, 1 + 2 + 0, 1 + 2 + 0 + 0, \ldots$ – that is, the sequence $1, 3, 3, 3, \ldots$. Consequently, $(1/(1-x))[(1 + 2x)/(1 - x)] = (1 + 2x)/(1 - x)^2$ is the generating function for the sequence $1, 1 + 3, 1 + 3 + 3, 1 + 3 + 3 + 3, \ldots$ – that is, the sequence $1, 4, 7, 10, \ldots$.

3. $f(x) = [x(1+x)]/(1-x)^3$ generates $0^2, 1^2, 2^2, 3^2, \ldots$; $[x(1+x)]/(1-x)^3 = 0^2 + 1^2 x + 2^2 \cdot x^2 + 3^2 \cdot x^3 + \ldots$; $(d/dx)[(x + x^2)/(1-x)^3] = 1^3 + 2^3 \cdot x + 3^3 \cdot x^2 + \cdots$; $x(d/dx)[(x + x^2)/(1-x)^3] =$

$0^3+1^3x+2^3\cdot x^2+3^3\cdot x^3+\ldots$; $(d/dx)[(x+x^2)/(1-x)^3]=(x^2+4x+1)/(1-x)^4$, so $x(x^2+4x+1)/(1-x)^5$ generates $0^3, 0^3+1^3, 0^3+1^3+2^3, \ldots$, and the coefficient of x^n is $\sum_{i=0}^{n} i^3$.

$(x^3+4x^2+x)(1-x)^{-5}=(x^3+4x^2+x)[\binom{-5}{0}+\binom{-5}{1}(-x)+\binom{-5}{2}(-x)^2+\ldots]$. Here the coefficient of x^n is $\binom{-5}{n-3}(-1)^{n-3}+4\binom{-5}{n-2}(-1)^{n-2}+\binom{-5}{n-1}(-1)^{n-1}=\binom{n+1}{n-3}+4\binom{n+2}{n-2}+\binom{n+3}{n-1}=(1/4!)[(n+1)(n)(n-1)(n-2)+4(n+2)(n+1)(n)(n-1)+(n+3)(n+2)(n+1)(n)]=[(n+1)(n)/4!](6n^2+6n)=(1/4)(n+1)(n)(n^2+n)=[(n+1)(n)/2]^2$.

5. $(1-x)f(x)=(1-x)(a_0+a_1x+a_2x^2+a_3x^3+\ldots)=a_0+(a_1-a_0)x+(a_2-a_1)x^2+(a_3-a_2)x^3+\ldots$, so $(1-x)f(x)$ is the generating function for the sequence $a_0, a_1-a_0, a_2-a_1, a_3-a_2, \ldots$

7. Since e^x is the generating function for $1, 1, 1/2!, 1/3!, \ldots$, it follows that $e^x/(1-x)$ generates the sequence $a_0, a_1, a_2, \ldots$, where $a_n=\sum_{i=0}^{n}(1/i!)$.

Supplementary Exercises

1.
(a) $6/(1-x)+1/(1-x)^2$ (b) $1/(1-ax)$
(c) $1/[1-(1+a)x]$ (d) $1/(1-x)+1/(1-ax)$

3. The generating function for each type of bullet is $(x^2+x^3+\ldots+x^7)^4=x^8(1+x+x^2+\ldots+x^5)^4$. The coefficient of x^{12} in $(1-x^6)^4(1-x)^{-4}=[1-\binom{4}{1}x^6+\binom{4}{2}x^{12}-\ldots][\binom{-4}{0}+\binom{-4}{1}(-x)+\binom{-4}{2}(-x)^2+\ldots]$ is $\binom{-4}{12}(-1)^{12}-\binom{4}{1}\binom{-4}{6}(-1)^6+\binom{4}{2}\binom{-4}{0}=\binom{15}{12}-\binom{4}{1}\binom{9}{6}+\binom{4}{2}$. By the rule of product the answer is $[\binom{15}{12}-\binom{4}{1}\binom{9}{6}+\binom{4}{2}]^2$.

5. Let $f(x)$ be the generating function for the number of partitions of n where no even summand is repeated (although an odd summand may be repeated); $g(x)$ is the generating function for the number of partitions of n in which no summand occurs more than three times. Then $g(x)=(1+x+x^2+x^3)(1+x^2+x^4+x^6)(1+x^3+x^6+x^9)\ldots=[(1+x)(1+x^2)][(1+x^2)(1+x^4)][(1+x^3)(1+x^6)]\ldots=[(1-x^2)/(1-x)](1+x^2)[(1-x^4)/(1-x^2)](1+x^4)[(1-x^6)/(1-x^3)](1+x^6)\ldots=[1/(1-x)](1+x^2)[1/(1-x^3)](1+x^4)[1/(1-x^5)](1+x^6)\ldots=(1+x+x^2+x^3+\ldots)(1+x^2)(1+x^3+x^6+x^9+\ldots)(1+x^4)(1+x^5+x^{10}+x^{15}+\ldots)(1+x^6)\ldots=f(x)$.

7. (a) $(1-2x)^{-5/2}=1+\sum_{r=1}^{\infty}\frac{(-5/2)((-5/2)-1)((-5/2)-2)\cdots((-5/2)-r+1)}{r!}(-2x)^r$
$=1+\sum_{r=1}^{\infty}\frac{(5)(7)(9)\cdots(3+2r)}{r!}x^r$, so $g(x)$ is the exponential generating function for $1, 5, 5(7), 5(7)(9), \ldots$

(b) $(1-ax)^b=1+\sum_{r=1}^{\infty}\frac{(b)(b-1)(b-2)\cdots(b-r+1)}{r!}(-ax)^r=1-abx+b(b-1)a^2x^2/2!+\ldots$
Consequently, by comparing coefficients of like powers of x, we have $-ab=7, b(b-1)a^2=7\cdot 11$ and $a=4, b=-7/4$.

9. For each $n \in \mathbf{Z}^+$, $(1+x)^n = \binom{n}{0} + \binom{n}{1}x + \binom{n}{2}x^2 + \binom{n}{3}x^3 + \ldots + \binom{n}{n}x^n$. Taking the derivative of both sides we find that

$$n(1+x)^{n-1} = \binom{n}{1} + 2\binom{n}{2}x + 3\binom{n}{3}x^2 + \ldots + n\binom{n}{n}x^{n-1}.$$

When $x = 1$ we obtain

$$n(1+1)^{n-1} = n(2^{n-1}) = \binom{n}{1} + 2\binom{n}{2} + 3\binom{n}{3} + \ldots + n\binom{n}{n}.$$

11. (a) The coefficient of x^{20} in $(x+x^2+\ldots)^{12} = x^{12}(1+x+x^2+\ldots)^{12}$ is the coefficient of x^8 in $(1+x+x^2+\ldots)^{12} = (1-x)^{-12}$, and this is $\binom{-12}{8}(-1)^8 = (-1)^8\binom{12+8-1}{8}(-1)^8 = \binom{19}{8}$.

(b) The coefficient of x^{10} in $(x+x^2+\ldots)^6 = x^6(1+x+x^2+\ldots)^6$ is $\binom{-6}{4}(-1)^4 = \binom{9}{4}$. The probability for this type of distribution is $\binom{9}{4}\binom{9}{4}/\binom{19}{8}$.

13. (a) We start with $a+(d-a)x$, the generating function for the sequence $a, d-a, 0, 0, 0, \ldots$. Then $[a+(d-a)x]/(1-x)$ is the generating function for the sequence $a, a+(d-a), a+(d-a)+0, a+(d-a)+0+0, \ldots$ – that is, the sequence $a, d, d, d, \ldots$. Consequently, $[a+(d-a)x]/(1-x)^2$ generates the sequence $a, a+d, a+d+d, a+d+d+d, \ldots$ – that is, the sequence $a, a+d, a+2d, a+3d, \ldots$. [Note: Part (c) of Exercise 1 for Section 9.5 is a special case of this result: Let $a = 1, d = 3$.]

(b) Here we need the coefficient of x^{n-1} in $(1/(1-x))[a+(d-a)x]/(1-x)^2 = [a+(d-a)x]/(1-x)^3 = [a+(d-a)x](1-x)^{-3}$. This coefficient is $a\binom{-3}{n-1}(-1)^{n-1} + (d-a)\binom{-3}{n-2}(-1)^{n-2} = a(-1)^{n-1}\binom{3+(n-1)-1}{n-1}(-1)^{n-1} + (d-a)(-1)^{n-2}\binom{3+(n-2)-1}{n-2}(-1)^{n-2} = a\binom{n+1}{n-1} + (d-a)\binom{n}{n-2} = a(\frac{1}{2})(n+1)(n) + (d-a)(\frac{1}{2})(n)(n-1) = a(\frac{1}{2})(n)[(n+1)-(n-1)] + d(\frac{1}{2})(n)(n-1) = na + (\frac{1}{2})(n)(n-1)d$.
[The reader may wish to compare this with the result for the first Supplementary Exercise in Chapter 4.]

15. (a) $x^n f(x)$

(b) $[f(x) - (a_0 + a_1x + a_2x^2 + \cdots + a_{n-1}x^{n-1})]/x^n$

17. For $k \in \mathbf{Z}^+$, k fixed, we find that $Pr(Y \geq k) = \sum_{y=k}^{\infty} q^{y-1}p$ (where $q = 1-p$) $= q^{k-1}p + q^kp + q^{k+1}p + \cdots = q^{k-1}p[1 + q + q^2 + \cdots]$ $= q^{k-1}p\frac{1}{(1-q)} = q^{k-1}p(\frac{1}{p}) = q^{k-1}$. Consequently, $Pr(Y \geq m | Y \geq n) = Pr(Y \geq m \text{ and } Y \geq n)/Pr(Y \geq n) = Pr(Y \geq m)/Pr(Y \geq n) = q^{m-1}/q^{n-1} = q^{m-n}$. [This property is the reason why a geometric random variable is said to be *memoryless*. In fact, the geometric random variable is the only discrete random variable with this property.]

CHAPTER 10
RECURRENCE RELATIONS

Section 10.1

1. (a) For the geometric progression $2, 10, 50, 250, \ldots$, the common ratio $r = \frac{10}{2} = \frac{50}{10} = \frac{250}{50} = \ldots = 5$, so $a_n = 5a_{n-1}$, $n \geq 1$, $a_0 = 2$.
(b) Here the common ratio $r = -18/6 = -3$, so $a_0 = -3a_{n-1}$, $n \geq 1$, $a_0 = 6$.
(c) $a_n = (2/5)a_{n-1}$, $n \geq 1$, $a_0 = 7$

3. $a_{n+1} - da_n = 0$, $n \geq 0$, so $a_n = d^n a_0$. $153/49 = a_3 = d^3 a_0$, $1377/2401 = a_5 = d^5 a_0 \Longrightarrow$ $a_5/a_3 = d^2 = 9/49$ and $d = \pm 3/7$.

5. $P_n = 100(1 + 0.015)^n$, $P_0 = 100$
$200 = 100(1.015)^n \Longrightarrow 2 = (1.015)^n$
$(1.015)^{46} \doteq 1.9835$ and $(1.015)^{47} \doteq 2.0133$.
Hence Laura must wait $(47)(3) = 141$ months for her money to double.

7. (a) $19 + 18 + 17 + \ldots + 10 = 145$
(b) $9 + 8 + 7 + \ldots + 1 = 45$

9. (a) 21345
(b) 52143, 52134, 25134
(c) 25134, 21534, 21354, 21345

Section 10.2

1. (a) $a_n = 5a_{n-1} + 6a_{n-2}$, $n \geq 2$, $a_0 = 1$, $a_1 = 3$.
Let $a_n = cr^n$, $c, r \neq 0$. Then the characteristic equation is $r^2 - 5r - 6 = 0 = (r-6)(r+1)$, so $r = -1, 6$ are the characteristic roots.
$a_n = A(-1)^n + B(6)^n$
$1 = a_0 = A + B$
$3 = a_1 = -A + 6B$, so $B = 4/7$ and $A = 3/7$.
$a_n = (3/7)(-1)^n + (4/7)(6)^n$, $n \geq 0$.

(b) $2a_{n+2} - 11a_{n+1} + 5a_n = 0$, $n \geq 0$, $a_0 = 2, a_1 = -8$.
Let $a_n = cr^n, c, r \neq 0$. Then the characteristic equation is $2r^2 - 11r + 5 = 0 = (2r-1)(r-5)$, so $r = 1/2$, 5 are the characteristic roots.
$a_n = A(1/2)^n + B(5)^n$

$2 = a_0 = A + B$
$-8 = a_1 = A(1/2) + 5B$, so $A = 4$ and $B = -2$.
$a_n = 4(1/2)^n - 2(5)^n,\ n \geq 0$.

(c) $a_{n+2} + a_n = 0,\ n \geq 0,\ a_0 = 0,\ a_1 = 3$.
With $a_n = cr^n,\ c, r \neq 0$, the characteristic equation $r^2 + 1 = 0$ yields the characteristic roots $\pm i$. Hence $a_n = A(i)^n + B(-i)^n = A(\cos(\pi/2) + i\sin(\pi/2))^n + B(\cos(\pi/2) + i\sin(-\pi/2))^n = C\cos(n\pi/2) + D\sin(n\pi/2)$.
$0 = a_0 = C,\ 3 = a_1 = D\sin(\pi/2) = D$, so $a_n = 3\sin(n\pi/2),\ n \geq 0$.

(d) $a_n - 6a_{n-1} + 9a_{n-2} = 0,\ n \geq 2,\ a_0 = 5,\ a_1 = 12$.
Let $a_n = cr^n,\ c, r \neq 0$. Then $r^2 - 6r + 9 = 0 = (r-3)^2$, so the characteristic roots are 3,3 and $a_n = A(3^n) + Bn(3^n)$.
$5 = a_0 = A;\ 12 = a_1 = 3A + 3B = 15 + 3B,\ B = -1$.
$a_n = 5(3^n) - n(3^n) = (5-n)(3^n),\ n \geq 0$.

(e) $a_n + 2a_{n-1} + 2a_{n-2} = 0,\ n \geq 2,\ a_0 = 1,\ a_1 = 3$.
$r^2 + 2r + 2 = 0,\ r = -1 \pm i$
$(-1+i) = \sqrt{2}(\cos(3\pi/4) + i\sin(3\pi/4))$
$(-1-i) = \sqrt{2}(\cos(5\pi/4) + i\sin(5\pi/4)) =$
$\sqrt{2}(\cos(-3\pi/4) + i\sin(-3\pi/4)) = \sqrt{2}(\cos(3\pi/4) - i\sin(3\pi/4))$
$a_n = (\sqrt{2})^n[A\cos(3\pi n/4) + B\sin(3\pi n/4)]$
$1 = a_0 = A$
$3 = a_1 = \sqrt{2}[\cos(3\pi/4) + B\sin(3\pi/4)] =$
$\sqrt{2}[(-1/\sqrt{2}) + B(1/\sqrt{2})]$, so $3 = -1 + B,\ B = 4$
$a_n = (\sqrt{2})^n[\cos(3\pi n/4) + 4\sin(3\pi n/4)],\ n \geq 0$

3. $(n=0):\ a_2 + ba_1 + ca_0 = 0 = 4 + b(1) + c(0)$, so $b = -4$.
$(n=1):\ a_3 - 4a_2 + ca_1 = 0 = 37 - 4(4) + c$, so $c = -21$.
$a_{n+2} - 4a_{n+1} - 21a_n = 0$
$r^2 - 4r - 21 = 0 = (r-7)(r+3),\ r = 7, -3$
$a_n = A(7)^n + B(-3)^n$
$0 = a_0 = A + B \Longrightarrow B = -A$
$1 = a_1 = 7A - 3B = 10A$, so $A = 1/10,\ B = -1/10$ and $a_n = (1/10)[(7)^n - (-3)^n],\ n \geq 0$.

5. For all three parts, let $a_n,\ n \geq 0$, count the number of ways to fill the n spaces under the condition(s) specified.
(a) Here $a_0 = 1$ and $a_1 = 2$. For $n \geq 2$, consider the nth space. If this space is occupied by a motorcycle – in one of two ways, then we have $2a_{n-1}$ of the ways to fill the n spaces. Further, there are a_{n-2} ways to fill the n spaces when a compact car occupies positions $n-1$ and n. These two cases are exhaustive and have nothing in common, so

$$a_n = 2a_{n-1} + a_{n-2},\quad n \geq 2,\quad a_0 = 1,\quad a_1 = 2.$$

Let $a_n = cr^n$, $c \neq 0$, $r \neq 0$. Upon substitution we have $r^2 - 2r - 1 = 0$, so $r = 1 \pm \sqrt{2}$ and $a_n = c_1(1+\sqrt{2})^n + c_2(1-\sqrt{2})^n$, $n \geq 0$. From $1 = a_0 = c_1 + c_2$ and $2 = a_1 = c_1(1+\sqrt{2}) + c_2(1-\sqrt{2})$, we have $c_1 = \frac{2+\sqrt{2}}{4}$ and $c_2 = \frac{2-\sqrt{2}}{4}$. So $a_n = ((\sqrt{2}+2)/4)(1+\sqrt{2})^n + ((2-\sqrt{2})/4)(1-\sqrt{2})^n = (1/2\sqrt{2})[(1+\sqrt{2})^{n+1} - (1-\sqrt{2})^{n+1}]$, $n \geq 0$.
(b) Here $a_0 = 1$ and $a_1 = 1$. For $n \geq 2$, consider the nth space. This space can be occupied by a motorcycle in one way and accounts for a_{n-1} of the a_n ways to fill the n spaces. If a compact car occupies the $(n-1)st$ and nth spaces, then we have the remaining $3a_{n-2}$ ways to fill n spaces. So here $a_n = a_{n-1} + 3a_{n-2}$, $n \geq 2$, $a_0 = 1$, $a_1 = 1$.

Let $a_n = cr^n$, $c \neq 0$, $r \neq 0$. Upon substitution we have $r^2 - r - 3 = 0$, so $r = (1 \pm \sqrt{13})/2$, and $a_n = c_1[(1+\sqrt{13})/2]^n + c_2[(1-\sqrt{13})/2]^n$, $n \geq 0$. From $1 = a_0 = c_1 + c_2$ and $1 = a_1 = c_1[(1+\sqrt{13})/2] + c_2[(1-\sqrt{13})/2]$, we find that $c_1 = [(1+\sqrt{13})/2\sqrt{13}]$ and $c_2 = [(-1+\sqrt{13})/2\sqrt{13}]$. So $a_n = (1/\sqrt{13})[(1+\sqrt{13})/2]^{n+1} - (1/\sqrt{13})[(1-\sqrt{13})/2]^{n+1}$, $n \geq 0$.
(c) Comparable to parts (a) and (b), here we have $a_n = 2a_{n-1} + 3a_{n-2}$, $n \geq 2$, $a_0 = 1$, $a_1 = 2$. Substituting $a_n = cr^n$, $c \neq 0$, $r \neq 0$, into the recurrence relation, we find that $r^2 - 2r - 3 = 0$ so $(r-3)(r+1) = 0$ and $r = 3$, $r = -1$. Consequently, $a_n = c_1(3^n) + c_2(-1)^n$, $n \geq 0$. From $1 = a_0 = c_1 + c_2$ and $2 = a_1 = 3c_1 - c_2$, we learn that $c_1 = 3/4$ and $c_2 = 1/4$. Therefore, $a_n = (3/4)(3^n) + (1/4)(-1)^n$, $n \geq 0$.

7. (a)

$$\begin{aligned} F_1 &= F_2 - F_0 \\ F_3 &= F_4 - F_2 \\ F_5 &= F_6 - F_4 \\ \cdots &\quad \cdots \quad \cdots \\ F_{2n-1} &= F_{2n} - F_{2n-2} \end{aligned}$$

Conjecture: For all $n \in \mathbf{Z}^+$, $F_1 + F_3 + F_5 + \cdots + F_{2n-1} = F_{2n} - F_0 = F_{2n}$.
Proof: (By the Principle of Mathematical Induction).
For $n = 1$ we have $F_1 = F_2$, and this is true since $F_1 = 1 = F_2$. Consequently, the result is true in this first case (and this establishes the basis step for the proof).
Next we assume the result true for $n = k$ (≥ 1) – that is, we assume

$$F_1 + F_3 + F_5 + \cdots + F_{2k-1} = F_{2k}.$$

When $n = k + 1$ we then find that

$$F_1 + F_3 + F_5 + \cdots + F_{2k-1} + F_{2(k+1)-1} =$$

$$(F_1 + F_3 + F_5 + \cdots + F_{2k-1}) + F_{2k+1} = F_{2k} + F_{2k+1} = F_{2k+2} = F_{2(k+1)}.$$

Therefore the truth for $n = k$ implies the truth at $n = k + 1$, so by the Principle of Mathematical Induction it follows that for all $n \in \mathbf{Z}^+$

$$F_1 + F_3 + F_5 + \cdots + F_{2n-1} = F_{2n}.$$

(b)

$$\begin{aligned} F_2 &= F_3 - F_1 \\ F_4 &= F_5 - F_3 \\ F_6 &= F_7 - F_5 \\ \dots & \quad \dots \quad \dots \\ F_{2n} &= F_{2n+1} - F_{2n-1} \end{aligned}$$

Conjecture: For all $n \in \mathbf{N}$, $F_2 + F_4 + \cdots + F_{2n} = F_0 + F_2 + F_4 + \cdots + F_{2n} = F_{2n+1} - F_1 = F_{2n+1} - 1$.
Proof: (By the Principle of Mathematical Induction)
When $n = 0$ we find that $0 = F_0 = F_1 - F_1 = 0$, so the result is true for this initial case, and this provides the basis step for the proof.
Assuming the result true for $n = k\ (\geq 0)$ we have $\sum_{i=0}^{k} F_{2i} = F_{2k+1} - 1$. Then when $n = k+1$ it follows that $\sum_{i=0}^{k+1} F_{2i} = \sum_{i=0}^{k} F_{2i} + F_{2(k+1)} = F_{2k+1} - 1 + F_{2k+2} = (F_{2k+2} + F_{2k+1}) - 1 = F_{2k+3} - 1 = F_{2(k+1)+1} - 1$. Consequently we see how the truth of the result for $n = k$ implies the truth of the result for $n = k+1$. Therefore it follows that for all $n \in \mathbf{N}$.

$$F_0 + F_2 + F_4 + \cdots + F_{2n} = F_{2n+1} - 1,$$

by the Principle of Mathematical Induction.

9. $a_n = a_{n-1} + a_{n-2},\ n \geq 0,\ a_0 = a_1 = 1$
(Append '+1') (Append '+2')
$a_n = A[(1+\sqrt{5})/2]^n + B[(1-\sqrt{5})/2]^n$
$1 = a_0 = A + B$; $1 = a_1 = A(1+\sqrt{5})/2 + B(1-\sqrt{5})/2$ or
$2 = (A+B) + \sqrt{5}(A-B) = 1 + \sqrt{5}(A-B)$ and $A - B = 1/\sqrt{5}$.
$1 = A + B,\ 1/\sqrt{5} = A - B \Longrightarrow A = (1+\sqrt{5})/2\sqrt{5},\ B = (\sqrt{5}-1)/2\sqrt{5}$ and $a_n = (1/\sqrt{5})[((1+\sqrt{5})/2)^{n+1} - ((1-\sqrt{5})/2)^{n+1}],\ n \geq 0$.

11. a) The solution here is similar to that for part (b) of Example 10.16. For $n = 1$, there are two strings – namely, 0 and 1. When $n = 2$, we find three such strings: 00, 10, 01. For $n \geq 3$, we can build the required strings of length n (1) by appending '0' to each of the a_{n-1} strings of length $n-1$; or (2) by appending '01' to each of the a_{n-2} strings of length $n-2$. These two cases have nothing in common and cover all possibilities, so

$$a_n = a_{n-1} + a_{n-2},\quad n \geq 3, a_1 = 2, a_2 = 3.$$

We find that $a_n = F_{n+2} = (\alpha^{n+2} - \beta^{n+2})/(\alpha - \beta)$ where $\alpha = (1+\sqrt{5})/2$ and $\beta = (1-\sqrt{5})/2$.
b) Here $b_1 = 1$ since 0 is the only string of length 1 that satisfies both conditions. For $n = 2$, there are three strings: 00, 10, and 01 – so $b_2 = 3$. For $n \geq 3$, consider the bit in the nth position of such a binary string of length n.
(1) If the nth bit is a 0, then there are a_{n-1} possibilities for the remaining $n-1$ bits.

(2) If the nth bit is a 1, then the $(n-1)$st and 1st bits are 0, and so there are a_{n-3} possibilities for the remaining $n-3$ bits.

Hence $b_n = a_{n-1} + a_{n-3} = F_{n+1} + F_{n-1}$, from part (a). So

$$b_n = (F_n + F_{n-1}) + (F_{n-2} + F_{n-3}) = (F_n + F_{n-2}) + (F_{n-1} + F_{n-3}) = b_{n-1} + b_{n-2}.$$

The characteristic equation $x^2 - x - 1 = 0$ has characteristic roots $\alpha = (1+\sqrt{5})/2$ and $\beta = (1-\sqrt{5})/2$, so $b_n = c_1\alpha^n + c_2\beta^n$. From $1 = b_1 = c_1\alpha + c_2\beta$ and $3 = b_2 = c_1\alpha^2 + c_2\beta^2$ we learn that $c_1 = c_2 = 1$. Hence $b_n = \alpha^n + \beta^n = L_n$, the nth Lucas number. [Recall that in Example 4.20 we showed that $L_n = F_{n+1} + F_{n-1}$.]

13. For $n \geq 0$, let a_n count the number of words of length n in Σ^* where there are no consecutive alphabetic characters. Let $a_n^{(1)}$ count those words that end with a numeric character, while $a_n^{(2)}$ counts those that end with an alphabetic character. Then $a_n = a_n^{(1)} + a_n^{(2)}$.

$$\begin{aligned}\text{For} \quad n \geq 1, \quad a_{n+1} &= 11a_n^{(1)} + 4a_n^{(2)} \\ &= [4a_n^{(1)} + 4a_n^{(2)}] + 7a_n^{(1)} \\ &= 4a_n + 7a_n^{(1)} \\ &= 4a_n + 7(4a_{n-1}) \\ &= 4a_n + 28a_{n-1},\end{aligned}$$

and $a_0 = 1$, $a_1 = 11$.

Now let $a_n = cr^n$, where $c, r \neq 0$ and $n \geq 0$. Then the resulting characteristic equation is

$$r^2 - 4r - 28 = 0,$$

where $r = (4 \pm \sqrt{128})/2 = 2 \pm 4\sqrt{2}$.
Hence $a_n = A[2+4\sqrt{2}]^n + B[2-4\sqrt{2}]^n$, $n \geq 0$.

$$\begin{aligned}1 = a_0 \Rightarrow 1 &= A + B, \quad \text{and} \\ 11 = a_1 \Rightarrow 11 &= A[2+4\sqrt{2}] + B[2-4\sqrt{2}] \\ &= A[2+4\sqrt{2}] + (1-A)[2-4\sqrt{2}] \\ &= [2-4\sqrt{2}] + A[2+4\sqrt{2}-2+4\sqrt{2}] \\ &= [2-4\sqrt{2}] + 8\sqrt{2}A,\end{aligned}$$

so $A = (9+4\sqrt{2})/(8\sqrt{2}) = (8+9\sqrt{2})/16$, and $B = 1 - A = (8-9\sqrt{2})/16$.

Consequently,

$$a_n = [(8+9\sqrt{2})/16][2+4\sqrt{2}]^n + [(8-9\sqrt{2})/16][2-4\sqrt{2}]^n, \quad n \geq 0.$$

15. Here we find that

$a_0 = 1, \qquad a_1 = 2, \qquad a_2 = 2, \qquad a_3 = 2^2, \qquad a_4 = 2^3, \qquad a_5 = 2^5, \qquad a_6 = 2^8,$

and, in general, $a_n = 2^{F_n}$, where F_n is the nth Fibonacci number for $n \geq 0$.

17. (a) From the previous exercise the number of compositions of $n+3$ with no 1's as summands is F_{n+2}.

(b) (i) The number that start with 2 is the number of compositions of $n+1$ with no 1's as summands. This is F_n.

(ii) F_{n-1}

(iii) The number that start with k, for $2 \leq k \leq n+1$, is the number of compositions of $(n+3)-k$ with no 1s as summands. This is $F_{(n+3)-k-1} = F_{n-k+2}$, $2 \leq k \leq n+1$.

(c) If the composition starts with $n+2$ then there is only one remaining summand – namely, 1. But here we are not allowed to use 1 as a summand, so there are no such compositions that start with $n+2$.

The one-summand composition '$n+3$' is the only composition here that starts (and ends) with $n+3$.

(d) These results provide a combinatorial proof that

$F_{n+2} = \sum_{k+2}^{n+1} F_{n-k+2} + 1 = (F_n + F_{n-1} + \cdots + F_2 + F_1) + 1$, or

$F_{n+2} - 1 = \sum_{i=1}^{n} F_i = \sum_{i=0}^{n} F_i$, since $F_0 = 0$.

19. From $1+\frac{1}{x} = x$ we learn that $x+1 = x^2$, or $x^2 - x - 1 = 0$. So $x = (1 \pm \sqrt{5})/2$ and the points of intersection are $((1+\sqrt{5})/2, (1+\sqrt{5})/2) = (\alpha, \alpha)$ and $((1-\sqrt{5})/2, (1-\sqrt{5})/2) = (\beta, \beta)$.

21. Proof (By the Alternative Form of the Principle of Mathematical Induction):

(a) $F_3 = 2 = (1+\sqrt{9})/2 > (1+\sqrt{5})/2 = \alpha = \alpha^{3-2}$,

$F_4 = 3 = (3+\sqrt{9})/2 > (3+\sqrt{5})/2 = \alpha^2 = \alpha^{4-2}$,

so the result is true for these first two cases (where $n = 3, 4$). This establishes the basis step. Assuming the truth of the statement for $n = 3, 4, 5, \ldots, k (\geq 4)$, where k is a fixed (but arbitrary) integer, we continue now with $n = k+1$:

$$\begin{aligned} F_{k+1} &= F_k + F_{k-1} \\ &> \alpha^{k-2} + \alpha^{(k-1)-2} \\ &= \alpha^{k-2} + \alpha^{k-3} = \alpha^{k-3}(\alpha+1) \\ &= \alpha^{k-3} \cdot \alpha^2 = \alpha^{k-1} = \alpha^{(k+1)-2} \end{aligned}$$

Consequently, $F_n > \alpha^{n-2}$ for all $n \geq 3$ – by the Alternative Form of the Principle of Mathematical Induction.

(b) $F_3 = 2 = (3+\sqrt{1})/2 < (3+\sqrt{5})/2 = \alpha^2 = \alpha^{3-1}$,

$F_4 = 3 = 2+1 < 2+\sqrt{5} = \alpha^3 = \alpha^{4-1}$,

so this result is true for these first two cases (where $n = 3, 4$). This establishes the basis step. Assuming the truth of the statement for $n = 3, 4, 5, \ldots, k (\geq 4)$, where k is a fixed (but arbitrary) integer, we continue now with $n = k+1$:

$$\begin{aligned} F_{k+1} &= F_k + F_{k-1} \\ &< \alpha^{k-1} + \alpha^{(k-1)-1} \\ &= \alpha^{k-1} + \alpha^{k-2} = \alpha^{k-2}(\alpha+1) \\ &= \alpha^{k-2} \cdot \alpha^2 = \alpha^k = \alpha^{(k+1)-1} \end{aligned}$$

Consequently, $F_n < \alpha^{n-1}$ for all $n \geq 3$ – by the Alternative Form of the Principle of Mathematical Induction.

23. Here we shall use auxiliary variables. For $n \geq 1$, let $a_n^{(0)}$ count the number of ternary strings of length n where there are no consecutive 1s and no consecutive 2s and the nth symbol is 0. We define $a_n^{(1)}$ and $a_n^{(2)}$ analogously. Then

$$\begin{aligned} a_n &= a_n^{(0)} + a_n^{(1)} + a_n^{(2)} \\ &= a_{n-1} + [a_{n-1} - a_{n-1}^{(1)}] + [a_{n-1} - a_{n-1}^{(2)}] \\ &= 2a_{n-1} + [a_{n-1} - a_{n-1}^{(1)} - a_{n-1}^{(2)}] \\ &= 2a_{n-1} + a_{n-1}^{(0)} = 2a_{n-1} + a_{n-2} \end{aligned}$$

Letting $a_n = cr^n$, $c \neq 0$, $r \neq 0$, we find that $r^2 - 2r - 1 = 0$, so the characteristic roots are $1 \pm \sqrt{2}$. Consequently, $a_n = c_1(1+\sqrt{2})^n + c_2(1-\sqrt{2})^n$. Here $a_1 = 3$, for the three one-symbol ternary strings 0, 1, and 2. Since we cannot use the two-symbol ternary strings 11 and 22, we have $a_2 = 3^2 - 2 = 7$. Extending the recurrence relation so that we can use $n = 0$, we have $a_2 = 2a_1 + a_0$ so $a_0 = a_2 - 2a_1 = 7 - 2 \cdot 3 = 1$. With

$$\begin{aligned} 1 = a_0 &= c_1 + c_2, \text{ and} \\ 3 = a_1 &= c_1(1+\sqrt{2}) + c_2(1-\sqrt{2}) \\ &= (c_1 + c_2) + \sqrt{2}(c_1 - c_2), \end{aligned}$$

we now have $1 = c_1 + c_2$ and $\sqrt{2} = c_1 - c_2$, so $c_1 = (1+\sqrt{2})/2$ and $c_2 = (1-\sqrt{2})/2$. Consequently,

$$a_n = (1/2)(1+\sqrt{2})^{n+1} + (1/2)(1-\sqrt{2})^{n+1}, \quad n \geq 0.$$

25. Let a_n count the number of ways one can tile a $2 \times n$ chessboard using these colored dominoes and square tiles. Here $a_1 = 4$, $a_2 = 4^2 + 4^2 + 5 = 37$, and, for $n \geq 3$, $a_n = 4a_{n-1} + 16a_{n-2} + 5a_{n-2} = 4a_{n-1} + 21a_{n-2}$. The characteristic equation is $x^2 - 4x - 21 = 0$ and this gives $x = 7$, $x = -3$ as the characteristic roots. Consequently, $a_n = c_1(7)^n + c_2(-3)^n$, $n \geq 1$.

Here $a_0 = (1/21)(a_2 - 4a_1) = 1$ can be introduced to simplify the calculations for c_1, c_2. From $1 = a_0 = c_1 + c_2$ and $4 = 7c_1 - 3c_2$ we learn that $c_1 = 7/10$, $c_2 = 3/10$, so $a_n = (7/10)(7)^n + (3/10)(-3)^n$, $n \geq 0$.

When $n = 10$ we find that the 2×10 chessboard can be tiled in $(7/10)(7)^{10} + (3/10)(-3)^{10} =$ 197,750,389 ways.

27. There is $a_1 = 1$ string of length 1 (namely, 0) in A^*, and $a_2 = 2$ strings of length 2 (namely, 00 and 01) and $a_3 = 5$ strings of length 3 (namely, 000, 001, 010, 011 and 111). For $n \geq 4$ we consider the entry from A at the (right) end of the string.
(1) 0: there are a_{n-1} strings.
(2) 01: there are a_{n-2} strings.
(3) 011, 111: there are a_{n-3} strings in each of these two cases.

Consequently,

$$a_n = a_{n-1} + a_{n-2} + 2a_{n-3}, \quad n \geq 4, \quad a_1 = 1, \quad a_2 = 2, \quad a_3 = 5.$$

From the characteristic equation $r^3 - r^2 - r - 2 = 0$, we find that $(r-2)(r^2+r+1) = 0$ and the characteristic roots are 2 and $(-1 \pm i\sqrt{3})/2$. Since $(-1+i\sqrt{3})/2 = \cos 120° + i \sin 120° = \cos(\frac{2\pi}{3}) + i\sin(\frac{2\pi}{3})$, we have

$$a_n = c_1(2)^n + c_2 \cos\frac{2n\pi}{3} + c_3 \sin\frac{2n\pi}{3}, \quad n \geq 1.$$

From

$$1 = a_1 = 2c_1 - c_2/2 + c_3(\sqrt{3}/2)$$
$$2 = a_2 = 4c_1 - c_2/2 - c_3(\sqrt{3}/2)$$
$$5 = a_3 = 8c_1 + c_2,$$

we learn that $c_1 = 4/7$, $c_2 = 3/7$, and $c_3 = \sqrt{3}/21$, so

$$a_n = (4/7)(2)^n + (3/7)\cos(2n\pi/3) + (\sqrt{3}/21)\sin(2n\pi/3), \quad n \geq 1.$$

[Note that a_n also counts the number of ways one can tile a $1 \times n$ chessboard using 1×1 square tiles of one color, 1×2 rectangular tiles of one color, and 1×3 rectangular tiles that come in *two* colors.]

29. $x_{n+2} - x_{n+1} = 2(x_{n+1} - x_n)$, $n \geq 0$, $x_0 = 1$, and $x_1 = 5$.

$x_{n+2} - 3x_{n+1} + 2x_n = 0$

For $n \geq 0$, let $x_n = cr^n$, where $c, r \neq 0$. Then we get the characteristic equation $r^2 - 3r + 2 = 0 = (r-2)(r-1)$, so $x_n = A(2^n) + B(1^n) = A(2^n) + B$.

$x_0 = 1 = A + B$

$x_1 = 5 = 2A + B$

Hence $4 = A$, $B = -3$, and $x_n = 4(2^n) - 3 = 2^{n+2} - 3$, for $n \geq 0$.

31. Let $b_n = a_n^2$, $b_0 = 16$, $b_1 = 169$.

This yields the linear relation $b_{n+2} - 5b_{n+1} + 4b_n = 0$ with characteristic roots $r = 4, 1$, so $b_n = A(1)^n + B(4)^n$.

$b_0 = 16$, $b_1 = 169 \Longrightarrow A = -35$, $B = 51$ and $b_n = 51(4)^n - 35$. Hence $a_n = \sqrt{51(4)^n - 35}$, $n \geq 0$.

33. Since $\gcd(F_1, F_0) = 1 = \gcd(F_2, F_1)$, consider $n \geq 2$. Then

$$F_3 = F_2 + F_1 (= 1)$$
$$F_4 = F_3 + F_2$$
$$F_5 = F_4 + F_3$$
$$\vdots$$
$$F_{n+1} = F_n + F_{n-1}.$$

Reversing the order of these equations we have the steps in the Euclidean Algorithm for computing the gcd of F_{n+1} and F_n, for $n \geq 2$. Since the last nonzero remainder is $F_1 = 1$, it follows that $\gcd(F_{n+1}, F_n) = 1$ for all $n \geq 2$.

Section 10.3

1. (a) $a_{n+1} - a_n = 2n + 3,\ n \geq 0,\ a_0 = 1$
$a_1 = a_0 + 0 + 3$
$a_2 = a_1 + 2 + 3 = a_0 + 2 + 2(3)$
$a_3 = a_2 + 2(2) + 3 = a_0 + 2 + 2(2) + 3(3)$
$a_4 = a_3 + 2(3) + 3 = a_0 + [2 + 2(2) + 2(3)] + 4(3)$
$\vdots$
$a_n = a_0 + 2[1+2+3+\ldots+(n-1)] + n(3) = 1 + 2[n(n-1)/2] + 3n = 1 + n(n-1) + 3n = n^2 + 2n + 1 = (n+1)^2,\ n \geq 0.$

(b) $a_{n+1} - a_n = 3n^2 - n,\ n \geq 0,\ a_0 = 3$
$a_{n+1} - a_n = 0$
Let $a_n = cr^n$, $c, r \neq 0$. Then the characteristic equation $r - 1 = 0$ gives us the characteristic root $r = 1$ and $a_n^{(h)} = c(1^n) = c$.
$a_n^{(p)} = n(An^2 + Bn + D) = An^3 + Bn^2 + Dn$
$A(n+1)^3 + B(n+1)^2 + D(n+1) - An^3 - Bn^2 - Dn = 3n^2 - n$
$An^3 + 3An^2 + 3An + A + Bn^2 + 2Bn + B + Dn + D - An^3 - Bn^2 - Dn = 3n^2 - n$
$3A = 3;\ A = 1$
$3A + 2B = -1;\ B = -2$
$A + B + D = 0;\ D = 1$
$a_n^{(p)} = n^3 - 2n^2 + n = n(n^2 - 2n + 1) = n(n-1)^2$
$a_n = c + n(n-1)^2$
$3 = a_0 = c$
So $a_n = 3 + n(n-1)^2,\ n \geq 0$.

(c) $a_{n+1} - 2a_n = 5,\ n \geq 0,\ a_0 = 1$
$a_1 = 2a_0 + 5 = 2 + 5$
$a_2 = 2a_1 + 5 = 2^2 + 2 \cdot 5 + 5$
$a_3 = 2a_2 + 5 = 2^3 + (2^2 + 2 + 1)5$
$\vdots$
$a_n = 2^n + 5(1 + 2 + 2^2 + \ldots + 2^{n-1}) = 2^n + 5(2^n - 1) = 6(2^n) - 5,\ n \geq 0.$

(d) $a_{n+1} - 2a_n = 2^n,\ n \geq 0,\ a_0 = 1$
$a_{n+1} - 2a_n = 0$
Let $a_n = cr^n$, $c, r \neq 0$. Then the characteristic equation $r - 2 = 0$ gives us the characteristic root $r = 2$ and $a_n^{(h)} = c(2^n)$.
$a_n^{(p)} = An(2^n)$.
$A(n+1)(2^{n+1}) - 2An(2^n) = 2^n$
$A(n+1)(2) - 2An = 1$
$2An + 2A - 2An = 1$
$2A = 1;\ A = 1/2$
$a_n^{(p)} = (1/2)n(2^n) = n(2^{n-1})$
$a_n = c(2^n) + n(2^{n-1})$
$1 = a_0 = c$

So $a_n = 2^n(2^{n-1}), n \geq 0$.

3. (a) Let $a_n =$ the number of regions determined by the n lines under the conditions specified. When the n-th line is drawn there are $n-1$ points of intersection and n segments are formed on the line. Each of these segments divides a region into two regions and this increases the number of previously existing regions, namely a_{n-1}, by n.
$a_n = a_{n-1} + n,\ n \geq 1,\ a_0 = 1.$
$a_n^{(h)} = A,\ a_n^{(p)} = Bn + Cn^2$
$Bn + Cn^2 = B(n-1) + C(n-1)^2 + n$
$Bn + Cn^2 - Bn + B - Cn^2 + 2Cn - C = n.$
By comparing the coefficients on like powers of n we have $B = C = 1/2$ and $a_n = A + (1/2)n + (1/2)n^2$.
$1 = a_0 = A$ so $a_n = 1 + (1/2)(n)(n+1),\ n \geq 0$.

(b) Let $b_n =$ the number of infinite regions that result for n such lines. When the nth line is drawn it is divided into n segments. The first and nth segments each create a new infinite region. Hence $b_n = b_{n-1} + 2,\ n \geq 2,\ b_1 = 2$. The solution of this recurrence relation is $b_n = 2n,\ n \geq 1,\ b_0 = 1$.

5. (a) $a_{n+2} + 3a_{n+1} + 2a_n = 3^n,\ n \geq 0,\ a_0 = 0,\ a_1 = 1.$
With $a_n = cr^n,\ c, r \neq 0$, the characteristic equation $r^2 + 3r + 2 = 0 = (r+2)(r+1)$ yields the characteristic roots $r = -1, -2$.
Hence $a_n^{(h)} = A(-1)^n + B(-2)^n$, while $a_n^{(p)} = C(3)^n$.
$C(3)^{n+2} + 3C(3)^{n+1} + 2C(3)^n = 3^n \Longrightarrow 9C + 9C + 2C = 1 \Longrightarrow C = 1/20.$
$a_n = A(-1)^n + B(-2)^n + (1/20)(3)^n$
$0 = a_0 = A + B + (1/20)$
$1 = a_1 = -A - 2B + (3/20)$
Hence $1 = a_0 + a_1 = -B + (4/20)$ and $B = -4/5$. Then $A = -B - (1/20) = 3/4$.
$a_n = (3/4)(-1)^n + (-4/5)(-2)^n + (1/20)(3)^n,\ n \geq 0$

(b) $a_n = (2/9)(-2)^n - (5/6)(n)(-2)^n + (7/9),\ n \geq 0$

7. Here the characteristic equation is $r^3 - 3r^2 + 3r - 1 = 0 = (r-1)^3$, so $r = 1, 1, 1$ and $a_n^{(h)} = A + Bn + Cn^2,\ a_n^{(p)} = Dn^3 + En^4$.
$D(n+3)^3 + E(n+3)^4 - 3D(n+2)^3 - 3E(n+2)^4 + 3D(n+1)^3 + 3E(n+1)^4 - Dn^3 - En^4 = 3 + 5n \Longrightarrow D = -3/4,\ E = 5/24.$
$a_n = A + Bn + Cn^2 - (3/4)n^3 + (5/24)n^4,\ n \geq 0.$

9. From Example 10.29, $P = (Si)[1 - (1+i)^{-T}]^{-1}$, where P is the payment, S is the loan (\$2500), T is the number of payments (24) and i is the interest rate per month (1%).

$$P = (2500)(0.01)[1 - (1.01)^{-24}]^{-1} = \$117.68.$$

11. (a) Let $a_n^2 = b_n, n \geq 0$
$b_{n+2} - 5b_{n+1} + 6b_n = 7n$

$b_n^{(h)} = A(3^n) + B(2^n),\ b_n^{(p)} = Cn + D$
$C(n+2) + D - 5[C(n+1) + D] + 6(Cn + D) = 7n \Longrightarrow C = 7/2,\ D = 21/4$
$b_n = A(3^n) + B(2^n) + (7n/2) + (21/4)$
$b_0 = a_0^2 = 1,\ b_1 = a_1^2 = 1$
$1 = b_0 = A + B + 21/4$
$1 = b_1 = 3A + 2B + 7/2 + 21/4$
$3A + 2B = -31/3$
$2A + 2B = -34/4$
$A = 3/4,\ B = -5$
$a_n = [(3/4)(3)^n - 5(2)^n + (7n/2) + (21/4)]^{1/2},\ n \geq 0$
(b) $a_n^2 - 2a_{n-1} = 0,\ n \geq 1,\ a_0 = 2$
$a_n^2 = 2a_{n-1}$
$\log_2 a_n^2 = \log_2(2a_{n-1}) = \log_2 2 + \log_2 a_{n-1}$
$2\log_2 a_n = 1 + \log_2 a_{n-1}$
Let $b_n = \log_2 a_n$.
The solution of the recurrence relation $2b_n = 1 + b_{n-1}$ is $b_n = A(1/2)^n + 1$.
$b_0 = \log_2 a_0 = \log_2 2 = 1$, so $1 = b_0 = A + 1$ and $A = 0$.
Consequently, $b_n = 1,\ n \geq 0$, and $a_n = 2,\ n \geq 0$.

13. (a) Consider the 2^n binary strings of length n. Half of these strings (2^{n-1}) end in 0 and the other half (2^{n-1}) in 1. For the 2^{n-1} binary strings of length $(n-1)$, there are t_{n-1} runs. When we append 0 to each of these strings we get $t_{n-1} + (\frac{1}{2})(2^{n-1})$ runs, where the additional $(\frac{1}{2})(2^{n-1})$ runs arise when we append 0 to the $(\frac{1}{2})(2^{n-1})$ strings of length $(n-1)$ that end in 1. Upon appending 1 to each of the 2^{n-1} binary strings of length $n-1$, we get the remaining $t_{n-1} + (\frac{1}{2})(2^{n-1})$ runs. Consequently we find that

$$t_n = 2t_{n-1} + 2^{n-1},\quad n \geq 2,\quad t_1 = 2.$$

Here $t_n^{(h)} = c(2^n)$, so $t_n^{(p)} = An(2^n)$. Substituting $t_n^{(p)}$ into the recurrence relation we have

$$\begin{aligned} An(2^n) &= 2A(n-1)2^{n-1} + 2^{n-1} \\ &= An(2^n) - A(2^n) + 2^{n-1} \end{aligned}$$

By comparison of coefficients for 2^n and $n2^n$ we learn that $A = \frac{1}{2}$. Consequently, $t_n = t_n^{(h)} + t_n^{(p)} = c(2^n) + n(2^{n-1})$, and $2 = t_1 = c(2) + 1 \Rightarrow c = \frac{1}{2}$, so $t_n = (\frac{1}{2})(2^n) + n(2^{n-1}) = (n+1)(2^{n-1}),\ n \geq 1$.
(b) Here there are 4^n quaternary strings of length n and 4^{n-1} of these end in each of the one symbol suffices 0,1,2, and 3. In this case

$$t_n = 4[t_{n-1} + (\frac{3}{4})4^{n-1}] = 4t_{n-1} + 3(4^{n-1}),\quad n \geq 2,\quad t_1 = 4.$$

Comparable to the solution for part (a), here $t_n^{(h)} = c(4^n)$ and $t_n^{(p)} = An(4^n)$. So $An4^n = 4A(n-1)4^{n-1} + (3)(4^{n-1}) = An4^n - A(4^n) + (\frac{3}{4})4^n$, and $A = \frac{3}{4}$. Consequently, $t_n = c(4^n) + (\frac{3}{4})n4^n$ and $4 = t_1 = 4c + (\frac{3}{4})(4) \Rightarrow c = \frac{1}{4}$, so $t_n = (\frac{1}{4})4^n + (\frac{3}{4})n4^n = 4^{n-1}(1 + 3n)$, $n \geq 1$.

(c) For an alphabet Σ, where $|\Sigma| = r \geq 1$, there are r^n strings of length n and these r^n strings determine a total of $r^{n-1}[1 + (r-1)n]$ runs. [Note: This formula includes the case where $r = 1$.]

15.

```
Program Towers_of_Hanoi (input, output);
Var
    number: integer;    {number = number of disks}

Procedure Move_The_Disks (n: integer; start, inter, finish; char);
{This procedure will move n disks from the start peg to the finish peg using inter as
    the intermediary peg.}
Begin
    If n=1 then
        Writeln ('Move disk from  ', start, '  to  ', finish, '.')
    Else
        Begin
            Move_The_Disks (n-1, start, finish, inter);
            Move_The_Disks (1, start, ' ', finish);
            Move_The_Disks (n-1, inter, start, finish)
        End {else}
End; {procedure}
Begin {main program}
    Write ('How many disks are there?  ');
    Readln (number);
    If number < 1 then
        Writeln ('Your input is not appropriate.')
    Else
        Move_The_Disks (number, '1','2','3')
End.
```

Section 10.4

1. (a) $a_{n+1} - a_n = 3^n \quad n \geq 0,\ a_0 = 1$
Let $f(x) = \sum_{n=0}^{\infty} a_n x^n$.
$\sum_{n=0}^{\infty} a_{n+1}x^{n+1} - \sum_{n=0}^{\infty} a_n x^{n+1} = \sum_{n=0}^{\infty} 3^n x^{n+1}$
$[f(x) - a_0] - xf(x) = x\sum_{n=0}^{\infty}(3x)^n = x/(1-3x)$
$f(x) - 1 - xf(x) = x/(1-3x)$
$f(x) = 1/(1-x) + x/((1-x)(1-3x)) = 1/(1-x) + (-1/2)/(1-x) + (1/2)/(1-3x) =$ $(1/2)/(1-x) + (1/2)(1-3x)$, and $a_n = (1/2)[1+3^n], n \geq 0$.

(b) $a_{n+1} - a_n = n^2, n \geq 0, a_0 = 1$.

Let $f(x) = \sum_{n=0}^{\infty} a_n x^n$.
$\sum_{n=0}^{\infty} a_{n+1}x^{n+1} - \sum_{n=0}^{\infty} a_n x^{n+1} = \sum_{n=0}^{\infty} n^2 x^{n+1}$
$\sum_{n=1}^{\infty} a_n x^n - x\sum_{n=0}^{\infty} a_n x^n = x\sum_{n=0}^{\infty} n^2 x^n$
$[f(x) - a_0] - xf(x) = x[\frac{x(x+1)}{(1-x)^3}]$ (from Example 9.5(d))
$f(x)(1-x) = 1 + \frac{x^2(x+1)}{(1-x)^3}$
$f(x) = \frac{1}{1-x} + \frac{x^2(x+1)}{(1-x)^4}$
$\frac{x^2(x+1)}{(1-x)^4} = \frac{x^3+x^2}{(1-x)^4} = \frac{A}{1-x} + \frac{B}{(1-x)^2} + \frac{C}{(1-x)^3} + \frac{D}{(1-x)^4}$
$x^3 + x^2 = A(1-x)^3 + B(1-x)^2 + C(1-x) + D = A(1-3x+3x^2-x^3) + B(1-2x+x^2) + C(1-x) + D$
$1 = -A;\ A = -1$
$1 = 3A + B;\ B = 4$
$0 = -3A - 2B - C;\ C = -5$
$0 = A + B + C + D;\ D = 2$
$f(x) = \frac{1}{1-x} + [\frac{(-1)}{1-x} + \frac{4}{(1-x)^2} + \frac{(-5)}{(1-x)^3} + \frac{2}{(1-x)^4}]$
From formula (8) in Table 9.2 we have
$a_n = 1 + [(-1) + 4\binom{n+1}{n} + (-5)\binom{n+2}{n} + 2\binom{n+3}{n}]$
$= 1 + [(-1) + 4n + 4 - (\frac{5}{2})(n^2+3n+2) + (\frac{1}{3})(n^3+6n^2+11n+6)]$
$= 1 + [(\frac{1}{3})n^3 - (\frac{1}{2})n^2 + (\frac{1}{6})n] = 1 + (\frac{1}{6})(n)(2n^2-3n+1)$
$= 1 + [n(n-1)(2n-1)/6],\ n \geq 0$

(c) $a_{n+2} - 3a_{n+1} + 2a_n = 0,\ n \geq 0,\ a_0 = 1,\ a_1 = 6$

$$\sum_{n=0}^{\infty} a_{n+2}x^{n+2} - 3\sum_{n=0}^{\infty} a_{n+1}x^{n+2} + 2\sum_{n=0}^{\infty} a_n x^{n+2} = 0$$

$$\sum_{n=0}^{\infty} a_{n+2}x^{n+2} - 3x\sum_{n=0}^{\infty} a_{n+1}x^{n+1} + 2x^2\sum_{n=0}^{\infty} a_n x^n = 0$$

Let $f(x) = \sum_{n=0}^{\infty} a_n x^n$. Then

$(f(x)-1-6x) - 3x(f(x)-1) + 2x^2 f(x) = 0$, and $f(x)(1-3x+2x^2) = 1+6x-3x = 1+3x$. Consequently,

$$f(x) = \frac{1+3x}{(1-2x)(1-x)} = \frac{5}{(1-2x)} + \frac{(-4)}{(1-x)} = 5\sum_{n=0}^{\infty}(2x)^n - 4\sum_{n=0}^{\infty} x^n,$$

and $a_n = 5(2^n) - 4,\ n \geq 0$.

(d) $a_{n+2} - 2a_{n+1} + a_n = 2^n,\ n \geq 0,\ a_0 = 1,\ a_1 = 2$
$\sum_{n=0}^{\infty} a_{n+2}x^{n+2} - 2\sum_{n=0}^{\infty} a_{n+1}x^{n+2} + \sum_{n=0}^{\infty} a_n x^{n+2} = \sum_{n=0}^{\infty} 2^n x^{n+2}$
Let $f(x) = \sum_{n=0}^{\infty} a_n x^n$. Then
$[f(x) - a_0 - a_1 x] - 2x[f(x) - a_0] + x^2 f(x) = x^2\sum_{n=0}^{\infty}(2x)^n$
$f(x) - 1 - 2x - 2xf(x) + 2x + x^2 f(x) = x^2/(1-2x)$

$(x^2 - 2x + 1)f(x) = 1 + x^2/(1-2x) \Longrightarrow f(x) = 1/(1-x)^2 +$
$x^2/((1-2x)(1-x)^2) = (1-2x+x^2)/((1-x)^2(1-2x)) = 1/(1-2x) = 1 + 2x + (2x)^2 + \ldots,$
so $a_n = 2^n,\ n \geq 0.$

3. (a) $a_{n+1} = -2a_n - 4b_n$
$b_{n+1} = 4a_n + 6b_n$
$n \geq 0,\ a_0 = 1,\ b_0 = 0.$
Let $f(x) = \sum_{n=0}^{\infty} a_n x^n,\ g(x) = \sum_{n=0}^{\infty} b_n x^n.$
$\sum_{n=0}^{\infty} a_{n+1}x^{n+1} = -2\sum_{n=0}^{\infty} a_n x^{n+1} - 4\sum_{n=0}^{\infty} b_n x^{n+1}$
$\sum_{n=0}^{\infty} b_{n+1}x^{n+1} = 4\sum_{n=0}^{\infty} a_n x^{n+1} + 6\sum_{n=0}^{\infty} b_n x^{n+1}$
$f(x) - a_0 = -2xf(x) - 4xg(x)$
$g(x) - b_0 = 4xf(x) + 6xg(x)$
$f(x)(1+2x) + 4xg(x) = 1$
$f(x)(-4x) + (1-6x)g(x) = 0$

$$f(x) = \frac{\begin{vmatrix} 1 & 4 \\ 0 & (1-6x) \end{vmatrix}}{\begin{vmatrix} (1+2x) & 4x \\ -4x & (1-6x) \end{vmatrix}} = (1-6x)/(1-2x)^2 =$$

$(1-6x)(1-2x)^{-2} = (1-6x)[\binom{-2}{0} + \binom{-2}{1}(-2x) + \binom{-2}{2}(-2x)^2 + \ldots]$
$a_n = \binom{-2}{n}(-2)^n - 6\binom{-2}{n-1}(-2)^{n-1} = 2^n(1-2n),\ n \geq 0$
$f(x)(-4x) + (1-6x)g(x) = 0 \Longrightarrow g(x) = (4x)f(x)(1-6x)^{-1} \Longrightarrow g(x) =$
$4x(1-2x)^{-2}$ and $b_n = 4\binom{-2}{n-1}(-2)^{n-1} = n(2^{n+1}),\ n \geq 0.$

(b) $a_{n+1} = 2a_n - b_n + 2$
$b_{n+1} = -a_n + 2b_n - 1$
$n \geq 0,\ a_0 = 0,\ b_0 = 1$
Let $f(x) = \sum_{n=0}^{\infty} a_n x^n,\ g(x) = \sum_{n=0}^{\infty} b_n x^n.$
$\sum_{n=0}^{\infty} a_{n+1}x^{n+1} = 2\sum_{n=0}^{\infty} a_n x^{n+1} - \sum_{n=0}^{\infty} b_n x^{n+1} + \sum_{n=0}^{\infty} 2x^{n+1}$
$\sum_{n=0}^{\infty} b_{n+1}x^{n+1} = -\sum_{n=0}^{\infty} a_n x^{n+1} + 2\sum_{n=0}^{\infty} b_n x^{n+1} - \sum_{n=0}^{\infty} x^{n+1}$
$f(x) - a_0 = 2xf(x) - xg(x) + 2x(\frac{1}{1-x})$
$g(x) - b_0 = -xf(x) + 2xg(x) - x(\frac{1}{1-x})$
$f(x)[1-2x] + xg(x) = \frac{2x}{1-x}$
$xf(x) + [1-2x]g(x) = \frac{1-2x}{1-x}$
$g(x) = \frac{2x^2-4x+1}{(1-3x)(1-x)^2} = \frac{(-1/4)}{1-3x} + \frac{(3/4)}{1-x} + \frac{(1/2)}{(1-x)^2}$
$b_n = (-1/4)(3^n) + (3/4) + (1/2)(n+1),\ n \geq 0$
$a_n = -b_{n+1} + 2b_n - 1 = (-1/4) + (1/4)(3^n) + (1/2)n,\ n \geq 0.$

Section 10.5

1. $b_4 = b_0b_3 + b_1b_2 + b_2b_1 + b_3b_0 = 2(5+2) = 14$

$b_n = [(2n)!/((n+1)!(n!))]$, $b_4 = 8!/(5!4!) = 14$

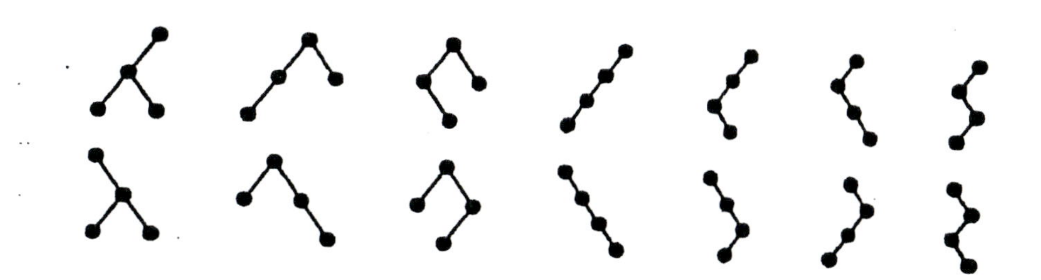

3. $\binom{2n-1}{n} - \binom{2n-1}{n-2} = \left[\frac{(2n-1)!}{n!(n-1)!}\right] - \left[\frac{(2n-1)!}{(n-2)!(n+1)!}\right] =$

$\left[\frac{(2n-1)!(n+1)}{(n+1)!(n-1)!}\right] - \left[\frac{(2n-1)!(n-1)}{(n-1)!(n+1)!}\right] = \left[\frac{(2n-1)!}{(n+1)!(n-1)!}\right][(n+1)-(n-1)] =$

$\frac{(2n-1)!(2)}{(n+1)!(n-1)!} = \frac{(2n-1)!(2n)}{(n+1)!n!} = \frac{(2n)!}{(n+1)(n!)(n!)} = \frac{1}{(n+1)}\binom{2n}{n}$

5.

(a) $(1/9)\binom{16}{8}$ (b) $[(1/4)\binom{6}{3}]^2$

(c) $[(1/6)\binom{10}{5}][(1/3)\binom{4}{2}]$ (d) $(1/6)\binom{10}{5}$

7. (a)

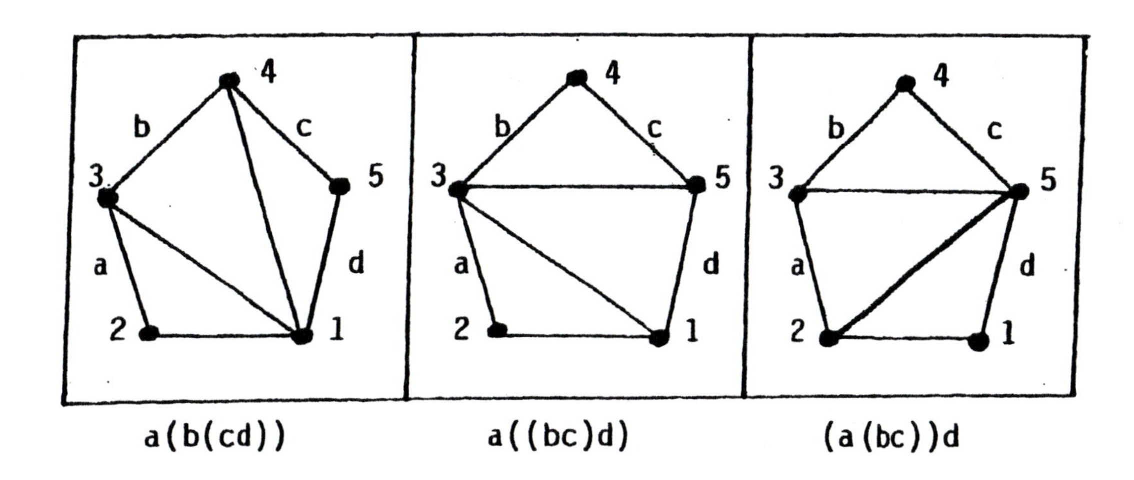

(b) (iii) $(((ab)c)d)e$ (iv) $(ab)(c(de))$

9. In Fig. 10.23 note how vertex 1 is always paired with an even numbered vertex. This must be the case for each $n \geq 0$, otherwise we end up with intersecting chords.
For each $n \geq 1$, let $1 \leq k \leq n$, so that $2 \leq 2k \leq 2n$. Drawing the chord connecting vertex 1 with vertex $2k$, we divide the circumference of the circle into two segments – one containing the vertices $2, 3, \ldots, 2k-1$, and the other containing the vertices $2k+1, 2k+2, \ldots, 2n$. These vertices can be connected by nonintersecting chords in $a_{k-1}a_{n-k}$ ways, so

$$a_n = a_0 a_{n-1} + a_1 a_{n-2} + a_2 a_{n-3} + \ldots + a_{n-2} a_1 + a_{n-1} a_0.$$

Since $a_0 = 1, a_1 = 1, a_2 = 2$, and $a_3 = 5$, we find that $a_n = b_n$, the nth Catalan number.

11.

(a)

x	$f_1(x)$	$f_2(x)$	$f_3(x)$	$f_4(x)$	$f_5(x)$
1	1	3	2	2	1
2	2	3	2	3	3
3	3	3	3	3	3

(b) The functions in part (a) correspond with the following paths from $(0, 0)$ to $(3, 3)$.

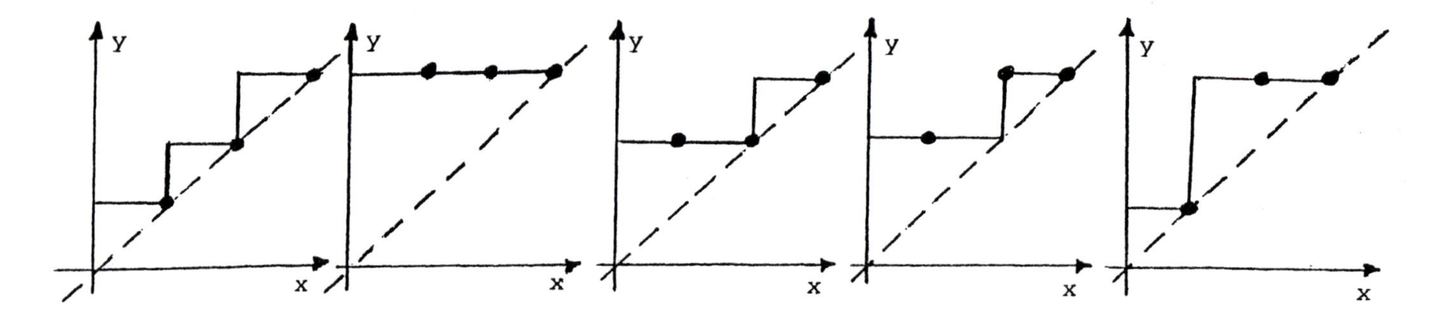

(c) The mountain ranges in Fig. 10.24 of the text.
(d) For $n \in \mathbf{Z}^+$, the number of monotone increasing functions $f : \{1, 2, 3, \ldots, n\} \rightarrow \{1, 2, 3, \ldots, n\}$ where $f(i) \geq i$ for all $1 \leq i \leq n$, is $b_n = (1/(n+1))\binom{2n}{n}$, the n-th Catalan number. This follows from Exercise 3 in Section 1.5. There is a one-to-one correspondence between the paths described in that exercise and the functions being dealt with here.

13. For $n \in \mathbf{N}$, let a_n count the number of these arrangements for a row of n contiguous pennies. Here $a_0 = 1$, $a_1 = 1$, $a_2 = 2$, and $a_3 = 5$. For the general situation, let $n \in \mathbf{N}$ and consider a contiguous row of $n+1$ pennies. These $n+1$ pennies provide n possible locations for placing a penny on the second level. There are two cases to consider:
(1) The first location (as the second level is scanned from left to right) that is empty is at position i, where $1 \leq i \leq n$. So there are $i-1$ pennies (above the first i pennies in the bottom contiguous row) in the positions to the left of position i. These $i-1$ contiguous pennies provide a_{i-1} possible arrangements. The $n - [(i-1)+1] = n - i$ positions (on the second level) to the right of position i are determined by a row of $n-i+1$ contiguous pennies at the bottom level and these $n-i+1$ contiguous pennies provide a_{n-i+1} arrangements. As i goes from 1 to n we get a total of

$\sum_{i=1}^{n} a_{i-1}a_{n-i+1} = a_0a_n + a_1a_{n-1} + a_2a_{n-2} + \cdots + a_{n-1}a_1$ arrangements.
(2) The only situation not covered in case 1 occurs when there is no empty position on the second level. So we have a row of $n+1$ contiguous pennies on the bottom level and n contiguous pennies on the second level — and above these $2n+1$ pennies there are $a_n (= a_n a_0)$ possible arrangements.

From cases (1) and (2) we have $a_{n+1} = a_0a_n + a_1a_{n-1} + a_2a_{n-2} + \cdots + a_{n-1}a_1 + a_na_0$, $a_0 = 1$, so $a_n = b_n = (1/(n+1))\binom{2n}{n}$, the nth Catalan numbers.

15.

(a) 1 3 2, 2 3 1: $E_3 = 2$

(b) 1 3 2 5 4   3 4 1 5 2
1 4 2 5 3   3 4 2 5 1
1 4 3 5 2   3 5 1 4 2
1 5 2 4 3   3 5 2 4 1
1 5 3 4 2

2 3 1 5 4   4 5 1 3 2
2 4 1 5 3   4 5 2 3 1
2 4 3 5 1
2 5 1 4 3
2 5 3 4 1 $E_5 = 16$

(c) For each rise/fall permutation, n cannot be in the first position (unless $n = 1$); n is the second component of a rise in such a permutation. Consequently, n must be at position 2 or 4 ... or $2\lfloor n/2 \rfloor$.

(d) Consider the location of n in a rise/fall permutation $x_1x_2x_3\ldots x_{n-1}x_n$ of $1, 2, 3, \ldots, n$. The number n is in position $2i$ for some $1 \leq i \leq \lfloor n/2 \rfloor$. Here there are $2i-1$ numbers that precede n. These can be selected in $\binom{n-1}{2i-1}$ ways and give rise to E_{2i-1} rise/fall permutations. The $(n-1)-(2i-1) = n-2i$ numbers that follow n give rise to E_{n-2i} rise/fall permutations. Consequently, $E_n = \sum_{i=1}^{\lfloor n/2 \rfloor} \binom{n-1}{2i-1} E_{2i-1}E_{n-2i}$, $n \geq 2$.

(e) Comparable to part (c), here we realize that for $n \geq 2$, 1 is at the end of the permutation or is the first component of a rise in such a permutation. Therefore, 1 must be at position 1 or 3 or ... or $2\lfloor (n-1)/2 \rfloor + 1$.

(f) As in part (d) look now for 1 in a rise/fall permutation of $1, 2, 3, \ldots, n$. We find 1 is position $2i+1$ for some $0 \leq i \leq \lfloor (n-1)/2 \rfloor$. Here there are $2i$ numbers that precede 1. These can be selected in $\binom{n-1}{2i}$ ways and give rise to E_{2i} rise/fall permutations. The remaining $(n-1)-2i = n-2i-1$ numbers that follow 1 give rise to E_{n-2i-1} rise/fall permutations. Therefore, $E_n = \sum_{i=0}^{\lfloor (n-1)/2 \rfloor} \binom{n-1}{2i} E_{2i}E_{n-2i-1}$, $n \geq 1$.

(g) From parts (d) and (f) we have:

(d) $E_n = \binom{n-1}{1}E_1E_{n-2} + \binom{n-1}{3}E_3E_{n-4} + \cdots + \binom{n-1}{2\lfloor n/2\rfloor-1}E_{2\lfloor n/2\rfloor-1}E_{n-2\lfloor n/2\rfloor}$

(f) $E_n = \binom{n-1}{0}E_0E_{n-1} + \binom{n-1}{2}E_2E_{n-3} + \cdots + \binom{n-1}{2\lfloor(n-1)/2\rfloor}E_{2\lfloor(n-1)/2\rfloor}E_{n-2\lfloor(n-1)/2\rfloor-1}$

Adding these equations we find that $2E_n = \sum_{i=0}^{n-1}\binom{n-1}{i}E_iE_{n-i-1}$ or $E_n = (1/2)\sum_{i=0}^{n-1}\binom{n-1}{i}E_iE_{n-i-1}$.

(h)

$$\begin{aligned}
E_6 &= (1/2)\textstyle\sum_{i=0}^{5}\binom{5}{i}E_iE_{5-i}\\
&= (1/2)[\binom{5}{0}E_0E_5 + \binom{5}{1}E_1E_4 + \binom{5}{2}E_2E_3 + \binom{5}{3}E_3E_2 + \binom{5}{4}E_4E_1 + \binom{5}{5}E_5E_0]\\
&= (1/2)[1\cdot1\cdot16 + 5\cdot1\cdot5 + 10\cdot1\cdot2 + 10\cdot2\cdot1 + 5\cdot5\cdot1 + 1\cdot16\cdot1]\\
&= (1/2)[16+25+20+20+25+16] = 61
\end{aligned}$$

$$\begin{aligned}
E_7 &= (1/2)\textstyle\sum_{i=0}^{6}\binom{6}{i}E_iE_{6-i}\\
&= (1/2)[1\cdot1\cdot61 + 6\cdot1\cdot16 + 15\cdot1\cdot5 + 20\cdot2\cdot2 + 15\cdot5\cdot1 + 6\cdot16\cdot1 + 1\cdot61\cdot1]\\
&= 272
\end{aligned}$$

(i) Consider the Maclaurin series expansions
$\sec x = 1 + x^2/2! + 5x^4/4! + 61x^6/6! + \cdots$ and
$\tan x = x + 2x^3/3! + 16x^5/5! + 272x^7/7! + \cdots$
One finds that $\sec x + \tan x$ is the exponential generating function of the sequence 1,1,1,2,5,16,61,272,... – namely, the sequence of Euler numbers.

Section 10.6

1. (a) $f(n) = (5/3)(4n^{\log_3 4} - 1)$ and $f \in O(n^{\log_3 4})$ for $n \in \{3^i | i \in \mathbf{N}\}$

(b) $f(n) = 7(\log_5 n + 1)$ and $f \in O(\log_5 n)$ for $n \in \{5^i | i \in \mathbf{N}\}$

3. (a) $f \in O(\log_b n)$ on $\{b^k | k \in \mathbf{N}\}$

(b) $f \in O(n^{\log_b a})$ on $\{b^k | k \in \mathbf{N}\}$

5. (a) $f(1) = 0$
$f(n) = 2f(n/2) + 1$
From Exercise 2(b), $f(n) = n - 1$.

(b) The equation $f(n) = f(n/2) + (n/2)$ arises as follows: There are $(n/2)$ matches played in the first round. Then there are $(n/2)$ players remaining, so we need $f(n/2)$ additional matches to determine the winner.

7. $O(1)$

9. (a)

$$\begin{array}{rcll} f(n) & \le & af(n/b) & + \; cn \\ af(n/b) & \le & a^2 f(n/b^2) & + \; ac(n/b) \\ a^2 f(n/b^2) & \le & a^3 f(n/b^3) & + \; a^2 c(n/b^2) \\ a^3 f(n/b^3) & \le & a^4 f(n/b^4) & + \; a^3 c(n/b^3) \\ & \vdots & & \vdots \\ a^{k-1} f(n/b^{k-1}) & \le & a^k f(n/b^k) & + \; a^{k-1} c(n/b^{k-1}) \end{array}$$

Hence $f(n) \le a^k f(n/b)^k + cn[1 + (a/b) + (a/b)^2 + \ldots + (a/b)^{k-1}] = a^k f(1) + cn[1 + (a/b) + (a/b)^2 + \ldots + (a/b)^{k-1}]$, since $n = b^k$. Since $f(1) \le c$ and $(n/b^k) = 1$, we have $f(n) \le cn[1 + (a/b) + (a/b)^2 + \ldots + (a/b)^{k-1} + (a/b)^k] = (cn)\sum_{i=0}^{k}(a/b)^i$.

(b) When $a = b$, $f(n) \le (cn)\sum_{i=0}^{k} 1^i = (cn)(k+1)$, where $n = b^k$, or $k = \log_b n$. Hence $f(n) \le (cn)(\log_b n + 1)$ so $f \in O(n \log_b n) = O(n \log n)$, for any base greater than 1.

(c) For $a \ne b$, $cn\sum_{i=0}^{k}(a/b)^i = cn\left[\frac{1-(a/b)^{k+1}}{1-(a/b)}\right]$

$$= (c)(b^k)\left[\frac{1-(a/b)^{k+1}}{1-(a/b)}\right] = c\left[\frac{b^k - (a^{k+1}/b)}{1-(a/b)}\right] = c\left[\frac{b^{k+1} - a^{k+1}}{b-a}\right] =$$

$$= c\left[\frac{a^{k+1} - b^{k+1}}{a-b}\right].$$

(d) From part (c), $f(n) \le (c/(a-b))[a^{k+1} - b^{k+1}]$ $= (ca/(a-b))a^k - (cb/(a-b))b^k$. But $a^k = a^{\log_b n} = n^{\log_b a}$ and $b^k = n$, so $f(n) \le (ca/(a-b))n^{\log_b a} - (cb/(a-b))n$.

(i) When $a < b$, then $\log_b a < 1$, and $f \in O(n)$ on $\mathbf{Z}^+$.
(ii) When $a > b$, then $\log_b a > 1$, and $f \in O(n^{\log_b a})$ on $\mathbf{Z}^+$.

Supplementary Exercises

1. $\binom{n}{k+1} = \frac{n!}{(k+1)!(n-k-1)!} = \frac{(n-k)}{(k+1)}\frac{n!}{k!(n-k)!} = \binom{n-k}{k+1}\binom{n}{k}$

3. There are two cases to consider. Case 1: (1 is a summand) – Here there are $p(n-1, k-1)$ ways to partition $n-1$ into exactly $k-1$ summands. Case 2: (1 is not a summand) – Here each summand $s_1, s_2, \ldots, s_k > 1$. For $1 \le i \le k$, let $t_i = s_i - 1 \ge 1$. Then $t_1, t_2, \ldots, t_k$ provide a partition of $n-k$ into exactly k summands. These cases are exhaustive and disjoint, so by the rule of sum $p(n,k) = p(n-1,k-1) + p(n-k,k)$.

5. (a)

$$A^2 = \begin{bmatrix} 2 & 1 \\ 1 & 1 \end{bmatrix} = \begin{bmatrix} F_3 & F_2 \\ F_2 & F_1 \end{bmatrix}, \quad A^3 = \begin{bmatrix} 3 & 2 \\ 2 & 1 \end{bmatrix} = \begin{bmatrix} F_4 & F_3 \\ F_3 & F_2 \end{bmatrix},$$

$$A^4 = \begin{bmatrix} 5 & 3 \\ 3 & 2 \end{bmatrix} = \begin{bmatrix} F_5 & F_4 \\ F_4 & F_3 \end{bmatrix}.$$

(b) Conjecture: For $n \in \mathbf{Z}^+$, $A^n = \begin{bmatrix} F_{n+1} & F_n \\ F_n & F_{n-1} \end{bmatrix}$,

where F_n denotes the nth Fibonacci number.

Proof: For $n = 1$, $A = A^1 = \begin{bmatrix} 1 & 1 \\ 1 & 0 \end{bmatrix} = \begin{bmatrix} F_2 & F_1 \\ F_1 & F_0 \end{bmatrix}$, so the result is true in

this first case. Assume the result true for $n = k \geq 1$, i.e.,

$A^k = \begin{bmatrix} F_{k+1} & F_k \\ F_k & F_{k-1} \end{bmatrix}$. For $n = k+1$, $A^n = A^{k+1} = A^k \cdot A = \begin{bmatrix} F_{k+1} & F_k \\ F_k & F_{k-1} \end{bmatrix} \begin{bmatrix} 1 & 1 \\ 1 & 0 \end{bmatrix}$

$$= \begin{bmatrix} F_{k+1} + F_k & F_{k+1} \\ F_k + F_{k-1} & F_k \end{bmatrix} = \begin{bmatrix} F_{k+2} & F_{k+1} \\ F_{k+1} & F_k \end{bmatrix}.$$

Consequently, the result is true for all $n \in \mathbf{Z}^+$ by the Principle of Mathematical Induction.

7. From $x^2 - 1 = 1 + \frac{1}{x}$ we find that $x^3 - x = x + 1$, or $x^3 - 2x - 1 = 0$. Since $(-1)^3 - 2(-1) - 1 = -1 + 2 - 1 = 0$, it follows that -1 is a root of $x^3 - 2x - 1$. Consequently, $x - (-1) = x + 1$ is a factor and we have $x^3 - 2x - 1 = (x+1)(x^2 - x - 1)$. So the roots of $x^3 - 2x - 1$ are $-1, (1+\sqrt{5})/2$, and $(1 - \sqrt{5})/2$.
For $x = -1$, $y = (-1)^2 - 1 = 0$.
For $x = (1+\sqrt{5})/2$, $y = [(1+\sqrt{5})/2]^2 - 1 = (1/4)(6 + 2\sqrt{5}) - 1 = [(3+\sqrt{5})/2] - 1 = (1+\sqrt{5})/2$.
For $x = (1-\sqrt{5})/2$, $y = [(1-\sqrt{5})/2]^2 - 1 = (1/4)(6 - 2\sqrt{5}) - 1 = [(3-\sqrt{5})/2] - 1 = (1-\sqrt{5})/2$.

So the points of intersection are $(-1, 0)$, $((1+\sqrt{5})/2, (1+\sqrt{5})/2) = (\alpha, \alpha)$, and $((1-\sqrt{5})/2, (1-\sqrt{5})/2) = (\beta, \beta)$.

9. (a) Since $\alpha^2 = \alpha + 1$, it follows that $\alpha^2 + 1 = 2 + \alpha$ and $(2+\alpha)^2 = 4 + 4\alpha + \alpha^2 = 4(1+\alpha) + \alpha^2 = 5\alpha^2$.
(b) Since $\beta^2 = \beta + 1$ we find that $\beta^2 + 1 = \beta + 2$ and $(2+\beta)^2 = 4 + 4\beta + \beta^2 = 4(1+\beta) + \beta^2 = 5\beta^2$.

(c) $\sum_{k=0}^{2n} \binom{2n}{k} F_{2k+m} = \sum_{k=0}^{2n} \binom{2n}{k} \left[\frac{\alpha^{2k+m} - \beta^{2k+m}}{\alpha - \beta}\right]$

$$= (1/(\alpha-\beta))\left[\sum_{k=0}^{2n} \binom{2n}{k} (\alpha^2)^k \alpha^m - \sum_{k=0}^{2n} \binom{2n}{k} (\beta^2)^k \beta^m\right]$$

$$= (1/(\alpha-\beta))[\alpha^m(1+\alpha^2)^{2n} - \beta^m(1+\beta^2)^{2n}]$$

$$= (1/(\alpha-\beta))[\alpha^m(2+\alpha)^{2n} - \beta^m(2+\beta)^{2n}]$$

$$= (1/(\alpha-\beta))[\alpha^m((2+\alpha)^2)^n - \beta^m((2+\beta)^2)^n]$$

$$= (1/(\alpha-\beta))[\alpha^m(5\alpha^2)^n - \beta^m(5\beta^2)^n]$$

$$= 5^n(1/(\alpha-\beta))[\alpha^{2n+m} - \beta^{2n+m}] = 5^n F_{2n+m}.$$

11. Consider the case where n is even. (The argument for n odd is similar.) For the fence $\mathcal{F}_n = \{a_1, a_2, \ldots, a_n\}$, there are c_{n-1} order-preserving functions $f : \mathcal{F}_n \to \{1, 2\}$ where $f(a_n) = 2$. [Note that $(\{1,2\}, \leq)$ is the same partial order as $\mathcal{F}_2$.] When such a function satisfies $f(a_n) = 1$, then we must have $f(a_{n-1}) = 1$, and there are c_{n-2} of these order-preserving functions. Consequently, since these two cases have nothing in common and cover all possibilities, we find that

$$c_n = c_{n-1} + c_{n-2}, \quad c_1 = 2,\ c_2 = 3.$$

So $c_n = F_{n+2}$, the $(n+2)$nd Fibonacci number.

13. (a) For $n \geq 1$, let a_n count the number of ways one can tile a $1 \times n$ chessboard using the 1×1 white tiles and 1×2 blue tiles. Then $a_1 = 1$ and $a_2 = 2$.

For $n \geq 3$, consider the nth square (at the right end) of the $1 \times n$ chessboard. Two situations are possible here:
(1) This square is covered by a 1×1 white tile, so the preceding $n-1$ squares (of the $1 \times n$ chessboard) can be covered in a_{n-1} ways;
(2) This square and the preceding ($(n-1)$st) square are both covered by a 1×2 blue tile, so the preceding $n-2$ squares (of the $1 \times n$ chessboard) can be covered in a_{n-2} ways. These two situations cover all possibilities and are disjoint, so we have

$$a_n = a_{n-1} + a_{n-2}, \quad n \geq 3, \quad a_1 = 1, \quad a_2 = 2.$$

Consequently, $a_n = F_{n+1}$, the $(n+1)$st Fibonacci number.
(b) (i) There is only $1 = \binom{n}{0} = \binom{n-0}{n-2\cdot 0}$ way to tile the $1 \times n$ chessboard using all white squares.
(ii) Consider the equation $x_1 + x_2 + \cdots + x_{n-1} = n-1$, where $x_i = 1$ for $1 \leq i \leq n-1$. We can select one of the x_i, where $1 \leq i \leq n-1$, in $\binom{n-1}{1} = \binom{n-1}{n-2} = \binom{n-1}{n-2\cdot 1}$ ways. Increase the value of this x_i to 2 and we have

$$x_1 + x_2 + \cdots + x_{n-1} = n.$$

In terms of our tilings we have $i-1$ white tiles, then the one blue tile, and then $n-i-1$ white tiles on the right of the blue tile – for a total of $(i-1)+1+(n-i-1)=n-1$ tiles.
(iii) There are $\binom{n-2}{2}=\binom{n-2}{n-4}=\binom{n-2}{n-2\cdot 2}$ tilings where we have exactly two blue tiles and $n-4$ white ones.
(iv) Likewise we have $\binom{n-3}{3}=\binom{n-3}{n-6}=\binom{n-3}{n-2\cdot 3}$ tilings that use 3 blue tiles and $n-6$ white ones.
(v) For $0 \leq k \leq \lfloor n/2 \rfloor$, there are $\binom{n-k}{k}=\binom{n-k}{n-2k}$ tilings with k blue tiles and $n-2k$ white ones.
(c) $F_{n+1}=\sum_{k=0}^{\lfloor n/2 \rfloor}\binom{n-k}{k}=\sum_{k=0}^{\lfloor n/2 \rfloor}\binom{n-k}{n-2k}$. [Compare this result with the formula presented in the previous exercise.]

15. (a) For each derangement, 1 is placed in position i, $2 \leq i \leq n$. Two things then occur.

Case 1: (i is in position 1) – Here the other $n-2$ integers are deranged in d_{n-2} ways. With $n-1$ choices for i this results in $(n-1)d_{n-2}$ such derangements.

Case 2: (i is not in position 1 (or position i)). Here we consider 1 as the new natural position for i, so there are $n-1$ elements to derange. With $n-1$ choices for i we have $(n-1)d_{n-1}$ derangements. Since the two cases are exhaustive and disjoint, the result follows from the rule of sum.

(b) $d_0=1$ (c) $d_n-nd_{n-1}=d_{n-2}-(n-2)d_{n-3}$

(d) $d_n-nd_{n-1}=(-1)^i[d_{n-i}-(n-i)d_{n-i-1}]$
Let $i=n-2$.
$d_n-nd_{n-1}=(-1)^{n-2}[d_2-2d_1]=(-1)^{n-2}=(-1)^n$

(e) $d_n-nd_{n-1}=(-1)^n$
$(d_n-nd_{n-1})(x^n/n!)=(-1)^n(x^n/n!)$
$\sum_{n=2}^{\infty}(d_n-nd_{n-1})(x^n/n!)=\sum_{n=2}^{\infty}(-x)^n/n!=e^{-x}-1+x$
$\sum_{n=2}^{\infty}d_nx^n/n!-x\sum_{n=2}^{\infty}d_{n-1}x^{n-1}/(n-1)!=e^{-x}-1+x$
$[f(x)-d_1x-d_0]-x[f(x)-d_0]=e^{-x}-1+x$
$f(x)-1-xf(x)+x=e^{-x}-1+x$ and $f(x)=e^{-x}/(1-x)$

17. (a) $a_n=\binom{2n}{n}$

(b) $(r+sx)^t=r^t(1+(s/r)x)^t=r^t[\binom{t}{0}+\binom{t}{1}(s/r)x+\binom{t}{2}(s/r)^2x^2+\ldots]=a_0+a_1x+a_2x^2+\ldots=1+2x+6x^2+\ldots$
$r^t=1 \Longrightarrow r=1$
$\binom{t}{1}s=2=ts$, $\binom{t}{2}s^2=6=s^2[t(t-1)/2]=s(t-1)=2-s$, so $s=-4$, $t=-1/2$, and $(1-4x)^{-1/2}$ generates $\binom{2n}{n}$, $n \geq 0$.

(c) Let a coin be tossed $2n$ times with the sequence of H's and T's counted in a_n. For $1 \leq i \leq n$, there is a smallest i where the number of H's equals the number of T's for the first time after $2i$ tosses. This sequence of $2i$ tosses is counted in b_i; the given sequence of $2n$ tosses is counted in $a_{n-i}b_i$. Since $b_0=0$, as i varies from 0 to n, $a_n=\sum_{i=0}^{n}a_ib_{n-i}$.

(d) Let $g(x) = \sum_{n=0}^{\infty} b_n x^n$, $f(x) = \sum_{n=0}^{\infty} a_n x^n = (1-4x)^{-1/2}$.
$\sum_{n=1}^{\infty} a_n x^n = \sum_{n=1}^{\infty}(a_0 b_n + a_1 b_{n-1} + \ldots + a_n b_0)x^n \Longrightarrow f(x) - a_0 = f(x)g(x)$ or $g(x)$
$= 1 - [1/f(x)] = 1 - (1-4x)^{1/2}$.
$(1-4x)^{1/2} = [\binom{1/2}{0} + \binom{1/2}{1}(-4x) + \binom{1/2}{2}(-4x)^2 + \ldots]$

The coefficient of x^n in $(1-4x)^{1/2}$ is $\binom{1/2}{n}(-4)^n =$

$$\frac{(1/2)((1/2)-1)((1/2)-2)\cdots((1/2)-n+1)}{n!}(-4)^n = \frac{(-1)(1)(3)(5)\cdots(2n-3)}{n!}(2^n) =$$

$$\frac{(-1)(1)(3)\cdots(2n-3)(2)(4)\cdots(2n-2)(2n)}{n!n!} = \frac{(-1)}{(2n-1)}\frac{(2n)!}{n!n!} = [-1/(2n-1)]\binom{2n}{n}.$$

Consequently, the coefficient of x^n in $g(x)$ is $b_n = [1/(2n-1)]\binom{2n}{n}$, $n \geq 1$, $b_0 = 0$.

19. For $x, y, z \in \mathbf{R}$,

$f(f(x,y),z) = f(a+bxy+c(x+y),z) = a+b[(a+bxy+c(x+y))z]+c[(a+bxy+c(x+y))+z)]$
$= a + ac + c^2x + bcxy + b^2xyz + bcxz + c^2y + bcyz + abz + cz$, and

$f(x, f(y,z)) = f(x, a + byz + c(x+y))$
$= a + b[x(a + byz + c(y+z)) + c[x + (a + byz + c(y+z))]$
$= a + ac + abx + cx + c^2y + c^2z + b^2xyz + bcxy + bcxz + bcyz$.

f associative $\Rightarrow f(f(x,y),z) = f(x, f(y,z)) \Rightarrow c^2x + (ab+c)z = abx + cx + c^2z$. With $ab = 1$ it follows that

$$c^2x + z + cz = x + c^2z + cx, \quad \text{or} \quad (c^2 - c - 1)x = (c^2 - c - 1)z.$$

Since x, z are arbitrary, we have $c^2 - c - 1 = 0$. Consequently, $c = \alpha$ or $c = \beta$.

21. Since $A \cap B = \emptyset$, $Pr(S) = Pr(A \cup B) = Pr(A) + Pr(B)$. Consequently, we have $1 = p + p^2$, so $p^2 + p - 1 = 0$ and $p = (-1 \pm \sqrt{5})/2$. Since $(-1-\sqrt{5})/2 < 0$ it follows that $p = (-1+\sqrt{5})/2 = -\beta$.

23. Here $a_1 = 1$ (for the string 0) and $a_2 = 2$ (for the strings 00, 11). For $n \geq 3$, consider the nth bit of a binary string (of length n) where there is no run of 1's of odd length.
(i) If this bit is 0 then the preceding $n-1$ bits can arise in a_{n-1} ways; and
(ii) If this bit is 1, then the $(n-1)$st bit must also be 1 and the preceding $n-2$ bits can arise in a_{n-2} ways.

Since the situations in (i) and (ii) have nothing in common and cover all cases we have

$$a_n = a_{n-1} + a_{n-2}, \quad n \geq 3, a_1 = 1, a_2 = 2.$$

Here $a_n = F_{n+1}, n \geq 1$, and so we have another instance where the Fibonacci numbers arise.

25.

(a) $(n=0)$ $F_1^2 - F_0F_1 - F_0^2 = 1^2 - 0\cdot 1 - 0^2 = 1$
$(n=1)$ $F_2^2 - F_1F_2 - F_1^2 = 1^2 - 1\cdot 1 - 1^2 = -1$
$(n=2)$ $F_3^2 - F_2F_3 - F_2^2 = 2^2 - 1\cdot 2 - 1^2 = 1$
$(n=3)$ $F_4^2 - F_3F_4 - F_3^2 = 3^2 - 2\cdot 3 - 2^2 = -1$

(b) Conjecture: For $n \geq 0$,

$$F_{n+1}^2 - F_nF_{n+1} - F_n^2 = \begin{cases} 1, & n \text{ even} \\ -1, & n \text{ odd} \end{cases}$$

(c) Proof: The result is true for $n = 0, 1, 2, 3$, by the calculations in part (a). Assume the result true for $n = k(\geq 3)$. There are two cases to consider – namely, k even and k odd. We shall establish the result for k even, the proof for k odd being similar. Our induction hypothesis tells us that $F_{k+1}^2 - F_kF_{k+1} - F_k^2 = 1$. When $n = k+1(\geq 4)$ we find that $F_{k+2}^2 - F_{k+1}F_{k+2} - F_{k+1}^2 = (F_{k+1}+F_k)^2 - F_{k+1}(F_{k+1}+F_k) - F_{k+1}^2 = F_{k+1}^2 + 2F_{k+1}F_k + F_k^2 - F_{k+1}^2 - F_{k+1}F_k - F_{k+1}^2 = F_{k+1}F_k + F_k^2 - F_{k+1}^2 = -[F_{k+1}^2 - F_kF_{k+1} - F_k^2] = -1$. The result follows for all $n \in \mathbf{N}$, by the Principle of Mathematical Induction.

27.

(a) $r(C_1, x) = 1 + x$ $r(C_4, x) = 1 + 4x + 3x^2$
$r(C_2, x) = 1 + 2x$ $r(C_5, x) = 1 + 5x + 6x^2 + x^3$
$r(C_3, x) = 1 + 3x + x^2$ $r(C_6, x) = 1 + 6x + 10x^2 + 4x^3$

In general, for $n \geq 3$, $r(C_n, x) = r(C_{n-1}, x) + xr(C_{n-2}, x)$.

(b) $r(C_1, 1) = 2$ $r(C_3, 1) = 5$ $r(C_5, 1) = 13$
$r(C_2, 1) = 3$ $r(C_4, 1) = 8$ $r(C_6, 1) = 21$

[Note: For $1 \leq i \leq n$, if one "straightens out" the chessboard C_i in Fig. 10.28, the result is a $1 \times i$ chessboard – like those studied in the previous exercise.]

29. (a) The partitions counted in $f(n, m)$ fall into two categories:

(1) Partitions where m is a summand. These are counted in $f(n-m, m)$, for m may occur more than once.

(2) Partitions where m is not a summand – so that $m-1$ is the largest possible summand. These partitions are counted in $f(n, m-1)$.

Since these two categories are exhaustive and mutually disjoint it follows that $f(n, m) = f(n-m, m) + f(n, m-1)$.

(b)

```
Program Summands(input,output);
Var
     n: integer;

Function f(n,m: integer): integer;
Begin
     If n=0 then
          f := 1
```

```
     Else if (n < 0) or (m < 1) then
                f := 0
          Else f := f(n,m-1) + f(n-m,m)
End; {of function f}

Begin
     Writeln ('What is the value of n?');
     Readln (n);
     Writeln ('What is the value of m?');
     Readln (m);
     Write ('There are  ', f(n,m):0, '  partitions of  ');
     Write (n:0, '  where  ', m:0, '  is the largest  ');
     Writeln ('summand possible.')
End.
```

(c)

```
Program  Partitions(input,output);
Var
     n: integer;

Function  f(n,m: integer): integer;
Begin
     If n=0 then
          f := 1
     Else if (n < 0) or (m < 1) then
                f := 0
          Else f := f(n,m-1) + f(n-m,m)
End; {of function f}

Begin
     Writeln ('What is the value of n?');
     Readln (n);
     Write ('For n = ', n :0, '  the number of  ');
     Write ('partitions p(', n:0, ') is  ', f(n,n):0, '.')
End.
```

31. The following program will print out the units digit of the first 130 Fibonacci numbers: $F_0 - F_{129}$.

```
Program  Units(input, output);
Var
     FibUnit: array[0..129] of integer;
```

```
        i,j: integer;

Begin
    FibUnit[0] := 0;
    FibUnit[1] := 1;
    For i := 2 to 129 do
        FibUnit[i] := (FibUnit[i-1] + FibUnit[i-2]) Mod 10;
    For i :=0 to 12 do
        For j := 0 to 9 do
            If j < 9 then
                Write (FibUnit[10 * i + j]: 4)
            Else {j = 9}
                Writeln (FibUnit[10 * i + 9]: 4)
End.
```

PART 3

GRAPH THEORY AND APPLICATIONS

CHAPTER 11
AN INTRODUCTION TO GRAPH THEORY

Section 11.1

1. (a) To represent the air routes traveled among a certain set of cities by a particular airline.
(b) To represent an electrical network. Here the vertices can represent switches, transistors, etc., and an edge (x, y) indicates the existence of a wire connecting x to y.
(c) Let the vertices represent a set of job applicants and a set of open positions in a corporation. Draw an edge (A,b) to denote that applicant A is qualified for position b. Then all open positions can be filled if the resulting graph provides a matching between the applicants and open positions.

3. There are six paths from b to f.
(1) $b \to e \to f$
(2) $b \to e \to g \to f$
(3) $b \to c \to d \to e \to f$
(4) $b \to c \to d \to e \to g \to f$
(5) $b \to a \to c \to d \to e \to f$
(6) $b \to a \to c \to d \to e \to g \to f$

5. Each path from a to h must include the edge $\{b, g\}$. There are three paths (in G) from a to b and three paths (in G) from g to h. Consequently, there are nine paths from a to h in G.

There is only one path of length 3, two of length 4, three of length 5, two of length 6, and one of length 7.

7. (a)

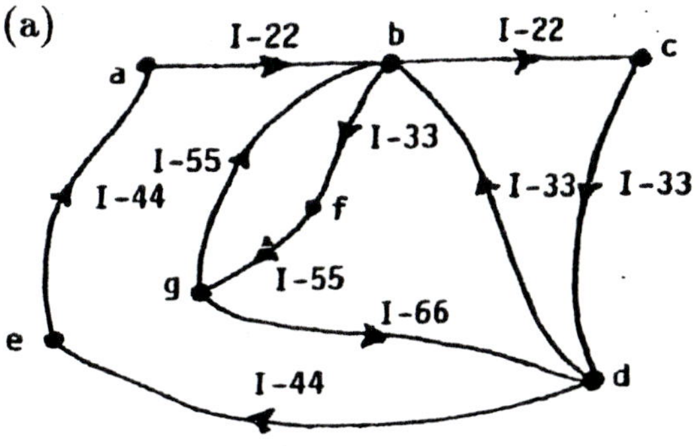

(b) $\{(g, d), (d, e), (e, a)\}$;
$\{(g, b), (b, c), (c, d), (d, e), (e, a)\}$.
(c) Two: One of $\{(b, c), (c, d)\}$ and one of $\{(b, f), (f, g), (g, d)\}$.
(d) No
(e) Yes: Travel the path $\{(c, d), (d, e), (e, a), (a, b), (b, f), (f, g)\}$.

(f) Yes: Travel the path $\{(g, b), (b, f), (f, g), (g, d), (d, b), (b, c), (c, d), (d, e), (e, a), (a, b)\}$.

9. If $\{a, b\}$ is not part of a cycle, then its removal disconnects a and b (and G). If not, there is a path P from a to b and P, together with $\{a, b\}$, provides a cycle containing $\{a, b\}$.
Conversely, if the removal of $\{a, b\}$ from G disconnects G then there exist $x, y \in V$ such that the only path P from x to y contains $e = \{a, b\}$. If e were part of a cycle C, then the edges in $(P - \{e\}) \cup (C - \{e\})$ would provide a second path connecting x to y.

11. (a) Yes (b) No (c) $n - 1$

13. This relation is reflexive, symmetric and transitive, so it is an equivalence relation. The partition of V induced by $\mathcal{R}$ yields the (connected) components of G.

15. For $n \geq 1$, let a_n count the number of closed $v - v$ walks of length n (where, in this case, we allow such a walk to contain or consist of one or more loops). Here $a_1 = 1$ and $a_2 = 2$. For $n \geq 3$ there are a_{n-1} $v - v$ walks where the last edge is the loop $\{v, v\}$ and a_{n-2} $v - v$ walks where the last two edges are both $\{v, w\}$. Since these two cases are exhaustive and have nothing in common we have $a_n = a_{n-1} + a_{n-2}$, $n \geq 3$, $a_1 = 1$, $a_2 = 2$.

We find that $a_n = F_{n+1}$, the $(n + 1)$st Fibonacci number.

Section 11.2

1. (a) Three: (1) $\{b, a\}, \{a, c\}, \{c, d\}, \{d, a\}$
(2) $\{f, c\}, \{c, a\}, \{a, d\}, \{d, c\}$
(3) $\{i, d\}, \{d, c\}, \{c, a\}, \{a, d\}$

(b) G_1 is the subgraph induced by $U = \{a, b, d, f, g, h, i, j\}$
$G_1 = G - \{c\}$

(c) G_2 is the subgraph induced by $W = \{b, c, d, f, g, i, j\}$
$G_2 = G - \{a, h\}$

(d)

b c d f i j

(e)

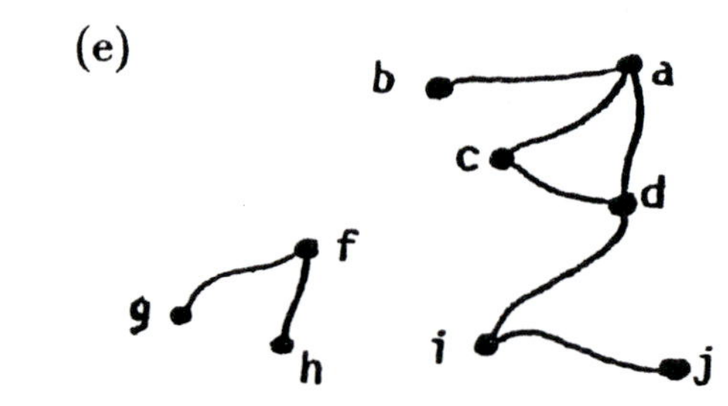

3. (a) For each of the nine edges we have two choices – (1) include it in the subgraph; or, (2) exclude it from the subgraph. So by the rule of product there are $2^9 = 512$ spanning subgraphs.
(b) Four of the spanning subgraphs in part (a) are connected. The graph itself and the

three subgraphs obtained by deleting one of the edges in the cycle $\{a,d\},\{d,c\},\{c,a\}$.
(c) 2^6

5. G is (or is isomorphic to) the complete graph K_n, where $n=|V|$.

7. (a)

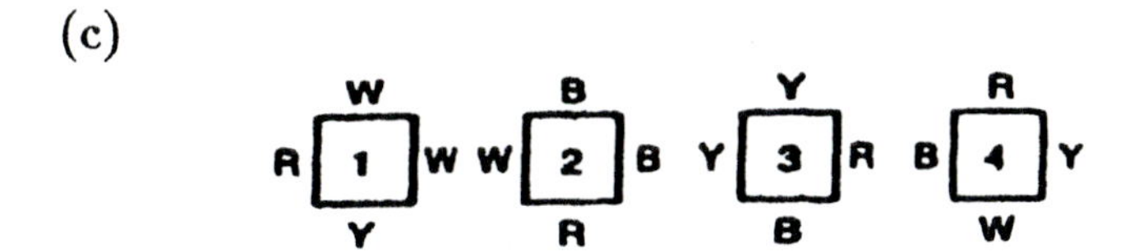

(b) No solution.

(c)

W
R 1 W
Y
B
W 2 B
R
Y
Y 3 R
B
R
B 4 Y
W

9. (a) Each graph has four vertices that are incident with three edges. In the second graph these vertices (w,x,y,z) form a cycle. This is not so for the corresponding vertices (a,b,g,h) in the first graph. Hence the graphs are *not* isomorphic.

(b) In the first graph the vertex d is incident with four edges. No vertex in the second graph has this property, so the graphs are *not* isomorphic.

11. (a) If $G_1=(V_1,E_1)$ and $G_2=(V_2,E_2)$ are isomorphic, then there is a function $f:V_1\longrightarrow V_2$ that is one-to-one and onto and preserves adjacencies. If $x,y\in V_1$ and $\{x,y\}\notin E_1$, then $\{f(x),f(y)\}\notin E_2$. Hence the same function f preserves adjacencies for $\overline{G}_1,\overline{G}_2$ and can be used to define an isomorphism for $\overline{G}_1,\overline{G}_2$. The converse follows in a similar way.

(b) They are not isomorphic. The complement of the graph containing vertex a is a cycle of length 8. The complement of the other graph is the disjoint union of two cycles of length 4.

13. If G is the cycle with edges $\{a,b\},\{b,c\},\{c,d\},\{d,e\}$ and $\{e,a\}$, then $\overline{G}$ is the cycle with edges $\{a,c\}$, $\{c,e\}$, $\{e,b\}$, $\{b,d\}$, $\{d,a\}$. Hence, G and $\overline{G}$ are isomorphic. Conversely, if G is a cycle on n vertices and $G,\overline{G}$ are isomorphic, then $n=(1/2)\binom{n}{2}$, or $n=(1/4)(n)(n-1)$, and $n=5$.

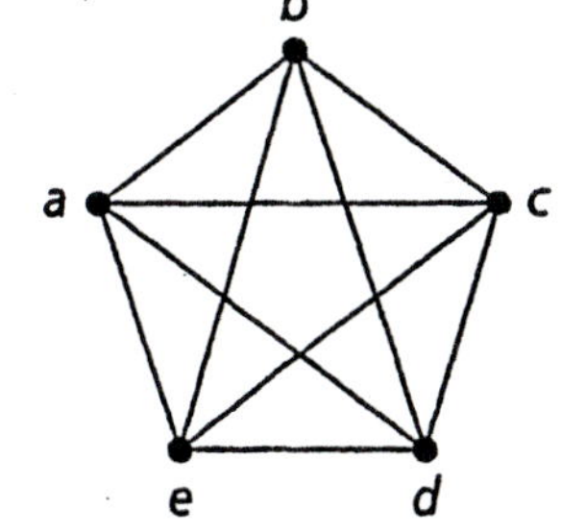

15. (a) Here f must also maintain directions. So if $(a,b)\in E_1$, then $(f(a),f(b))\in E_2$.

(b) They are not isomorphic. Consider vertex a in the first graph. It is incident to one vertex and incident from two other vertices. No vertex in the other graph has this property.

17. There are two cases to consider:

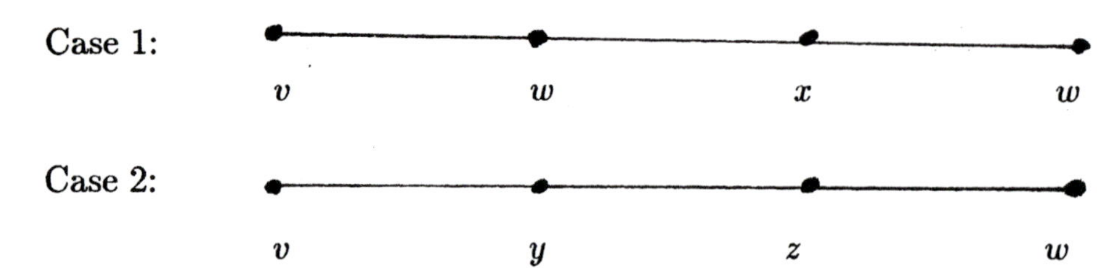

Here there are $n-2$ choices for y – namely, any vertex other than v, w – and there are $n-2$ choices for z – namely, any vertex other than w or the vertex selected for z.

Consequently, there are $(n-1)+(n-2)^2 = n^2 - 3n + 3$ walks of length 3 from v to w.

Section 11.3

1. (a) Since $18 = 2|E| = \sum_{v \in V} \deg(v) = 3|V|$, it follows that $|V| = 6$.
(b) Here $30 = 2|E| = \sum_{v \in V} \deg(v) = k|V|$, where k is an integer-constant. Consequently, $|V|$ is a divisor of 30, so $|V| = 1$ or 2 or 3 or 5 or 6 or 10 or 15 or 30. [In the first four cases G must be a multigraph; when $|V| = 30$, G is disconnected.]
(c) From $20 = 2|E| = \sum_{v \in V} \deg(v) = 2(4)+(|V|-2)3$, it follows that $3|V| = 20-8+6 = 18$, so $|V| = 6$.

3. Since $38 = 2|E| = \sum_{v \in V} \deg(v) \geq 4|V|$, the largest possible value for $|V|$ is 9. We can have (i) seven vertices of degree 4 and two of degree 5; or (ii) eight vertices of degree 4 and one of degree 6. The graph in part (a) of the figure is an example for case (i); an example for case (ii) is provided in part (b) of the figure.

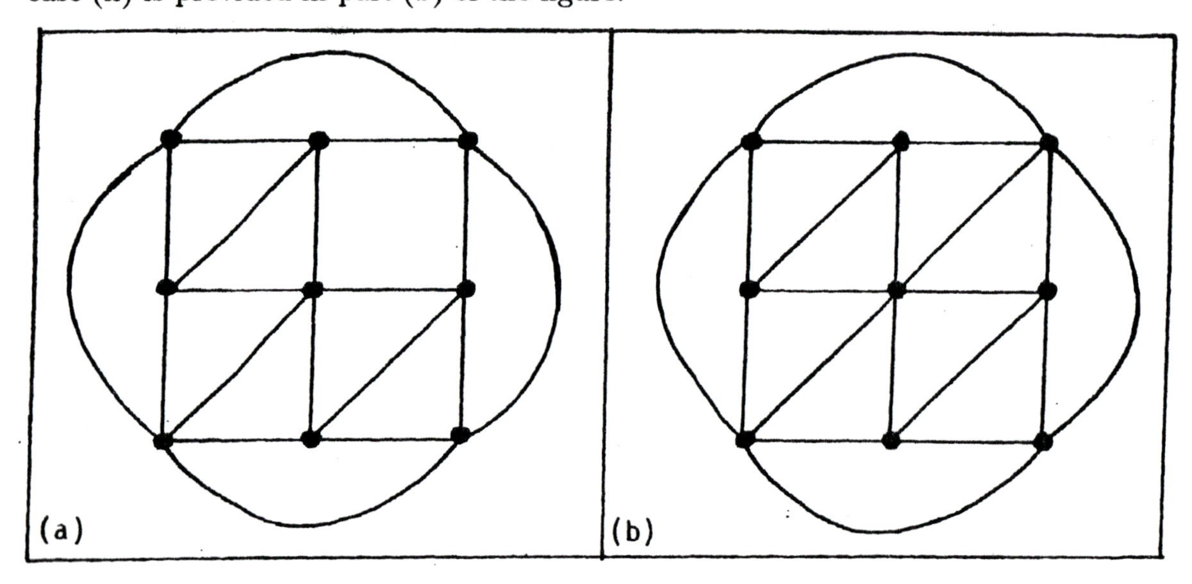

5. (a) $|V_1| = 8 = |V_2|$; $|E_1| = 14 = |E_2|$.
(b) For V_1 we find that $\deg(a) = 3$, $\deg(b) = 4$, $\deg(c) = 4$, $\deg(d) = 3$, $\deg(e) = 3$,

$\deg(f) = 4$, $\deg(g) = 4$, and $\deg(h) = 3$. For V_2 we have $\deg(s) = 3$, $\deg(t) = 4$, $\deg(u) = 4$, $\deg(v) = 3$, $\deg(w) = 4$, $\deg(x) = 3$, $\deg(y) = 3$, and $\deg(z) = 4$. Hence each of the two graphs has four vertices of degree 3 and four of degree 4.
(c) Despite the results in parts (a) and (b) the graphs G_1 and G_2 are *not* isomorphic.

In the graph G_2 the four vertices of degree 4 — namely, t, u, w, and z — are on a cycle of length 4. For the graph G_1 the vertices b, c, f, and g — each of degree 4 — do not lie on a cycle of length 4.

A second way to observe that G_1 and G_2 are not isomorphic is to consider once again the vertices of degree 4 in each graph. In G_1 these vertices induce a disconnected subgraph consisting of the two edges $\{b, c\}$ and $\{f, g\}$. The four vertices of degree 4 in graph G_2 induce a connected subgraph that has five edges — every possible edge except $\{u, z\}$.

7. a) Let v be any vertex in the graph. If $\deg(v) \geq 2$, then for each pair of vertices adjacent to v there is a path of length two, where v is the unique vertex of degree 2 for that path. Hence the number of paths of length two is $\binom{3}{2}$ (for a) $+\binom{3}{2}$ (for b) $+\binom{4}{2}$ (for c) $+\binom{3}{2}$ (for d) $+\binom{2}{2}$ (for e) $+\binom{3}{2}$ (for f) $= 3+3+6+3+1+3 = 19$.
b) $\sum_{i=1}^{n} \binom{d_i}{2}$ [Note: No assumption about connectedness is made here.]

9. From Example 11.12 we know that the hypercube Q_n has $n2^{n-1}$ edges. Consequently,
a) $n \cdot 2^{n-1} = 524,288 \Rightarrow n = 16$; and
b) $n \cdot 2^{n-1} = 4,980,736 \Rightarrow n = 19$, so there are $2^{19} = 524,288$ vertices in this hypercube.

11. The number of edges in K_n is $\binom{n}{2} = n(n-1)/2$. If the edges of K_n can be partitioned into such cycles of length 4, then 4 divides $\binom{n}{2}$ and $\binom{n}{2} = 4t$ for some $t \in \mathbf{Z}^+$. For each vertex v that appears in a cycle, there are two edges (of K_n) incident to v. Consequently, each vertex v of K_n has even degree, so n is odd. Therefore, $n-1$ is even and as $4t = \binom{n}{2} = n(n-1)/2$, it follows that $8t = n(n-1)$. So 8 divides $n(n-1)$, and since n is odd, it follows (from the Fundamental Theorem of Arithmetic) that 8 divides $n-1$. Hence $n-1 = 8k$, or $n = 8k+1$, for some $k \in \mathbf{Z}^+$.

13. $\delta|V| \leq \sum_{v \in V} \deg(v) \leq \Delta|V|$. Since $2|E| = \sum_{v \in V} \deg(v)$, it follows that
$\delta|V| \leq 2|E| \leq \Delta|V|$ so $\delta \leq 2(e/n) \leq \Delta$.

15. Proof: Start with a cycle $v_1 \to v_2 \to v_3 \to \ldots \to v_{2k-1} \to v_{2k} \to v_1$. Then draw the k edges $\{v_1, v_{k+1}\}, \{v_2, v_{k+2}\}, \ldots, \{v_i, v_{i+k}\}, \ldots, \{v_k, v_{2k}\}$. The resulting graph has $2k$ vertices each of degree 3.

17. (Corollary 11.1) Let $V = V_1 \cup V_2$ where V_1 (V_2) contains all vertices of odd (even) degree. Then $2|E| - \sum_{v \in V_2} \deg(v) = \sum_{v \in V_1} \deg(v)$ is an even integer. For $|V_1|$ odd, $\sum_{v \in V_1} \deg(v)$ is odd.
(Corollary 11.2) For the converse let $G = (V, E)$ have an Euler trail with a, b as the starting and terminating vertices, respectively. Add the edge $\{a, b\}$ to G to form the

graph $G' = (V, E')$, where G' has an Euler circuit. Hence G' is connected and each vertex has even degree. Removing edge $\{a, b\}$ the vertices in G will have the same even degree except for a, b. $\deg_G(a) = \deg_{G'}(a) - 1, \deg_G(b) = \deg_{G'}(b) - 1$, so the vertices a, b have odd degree in G. Also, since the edges in G form an Euler trail, G is connected.

19. (a) Let $a, b, c, x, y \in V$ with $\deg(a) = \deg(b) = \deg(c) = 1$, $\deg(x) = 5$, and $\deg(y) = 7$. Since $\deg(y) = 7$, y is adjacent to all of the other (seven) vertices in V. Therefore vertex x is not adjacent to any of the vertices a, b, and c. Since x cannot be adjacent to itself, unless we have loops, it follows that $\deg(x) \leq 4$, and we cannot draw a graph for the given conditions.

(b)

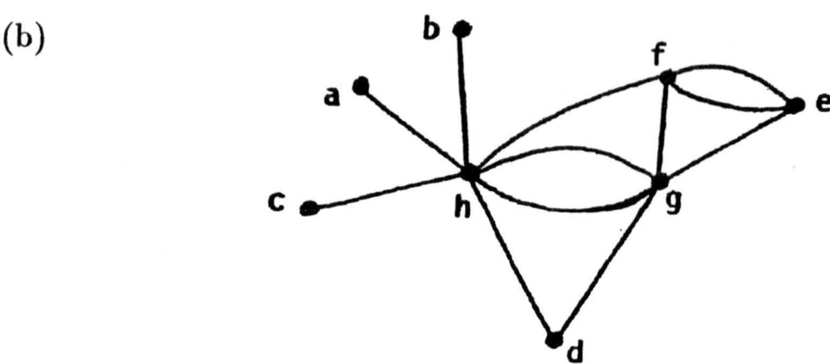

21. n odd; $n = 2$

23. Yes. Model the situation with a graph where there is a vertex for each room and the surrounding corridor. Draw an edge between two vertices if there is a door common to both rooms, or a room and the surrounding corridor. The resulting multigraph is connected with every vertex of even degree.

25. (a) (i) Let the vertices of K_6 be $v_1, v_2, v_3, v_4, v_5, v_6$, where $\deg(v_i) = 5$ for all $1 \leq i \leq 6$. Consider the subgraph S of K_6 obtained (from K_6) by deleting the edges $\{v_2, v_5\}$ and $\{v_3, v_6\}$. Then S is connected with $\deg(v_1) = \deg(v_4) = 5$, and $\deg(v_i) = 4$ for $i \in \{2, 3, 5, 6\}$. Hence S has an Euler trail that starts at v_1 (or v_4) and terminates at v_4 (or v_1). This Euler trail in S is then a trail of maximum length in K_6, and its length is $\binom{6}{2} - (1/2)[6 - 2] = 15 - 2 = 13$.

(ii) $\binom{8}{2} - (1/2)[8 - 2] = 28 - 3 = 25$

(iii) $\binom{10}{2} - (1/2)[10 - 2] = 45 - 4 = 41$

(iv) $\binom{2n}{2} - (1/2)[2n - 2] = n(2n - 1) - (n - 1) = 2n^2 - 2n + 1$.

(b) (i) Label the vertices of K_6 as in section (i) of part (a) above. Now consider the subgraph T of K_6 obtained (from K_6) by deleting the edges $\{v_1, v_4\}$, $\{v_2, v_5\}$, and $\{v_3, v_6\}$. Then T is connected with $\deg(v_i) = 4$ for all $1 \leq i \leq n$. Hence T has an Euler circuit and this Euler circuit for T is then a circuit of maximum length in K_6. The length of the circuit is $\binom{6}{2} - (1/2)(6) = 15 - 3 = 12$.

(ii) $\binom{8}{2} - (1/2)(8) = 28 - 4 = 24$

(iii) $\binom{10}{2} - (1/2)(10) = 45 - 5 = 40$

(iv) $\binom{2n}{2} - (1/2)(2n) = n(2n-1) - n = 2n^2 - 2n = 2n(n-1)$.

27. From Exercise 24 we know that $\sum_{v \in V} \text{id}(v) = e = \sum_{v \in V} \text{od}(v)$, where $e = \binom{n}{2}$, the number of edges in G. Consequently, we now find that $\sum_{v \in V}[\text{od}(v) - \text{id}(v)] = 0$. For each $v \in V, \text{od}(v) + \text{id}(v) = n - 1$, so $0 = (n-1) \cdot 0 = \sum_{v \in V}(n-1)[\text{od}(v) - \text{id}(v)] = \sum_{v \in V}[\text{od}(v) + \text{id}(v)][\text{od}(v) - \text{id}(v)] = \sum_{v \in V}[(\text{od}(v))^2 - (\text{id}(v))^2]$, and the result follows.

29. (a) and (b) (c)

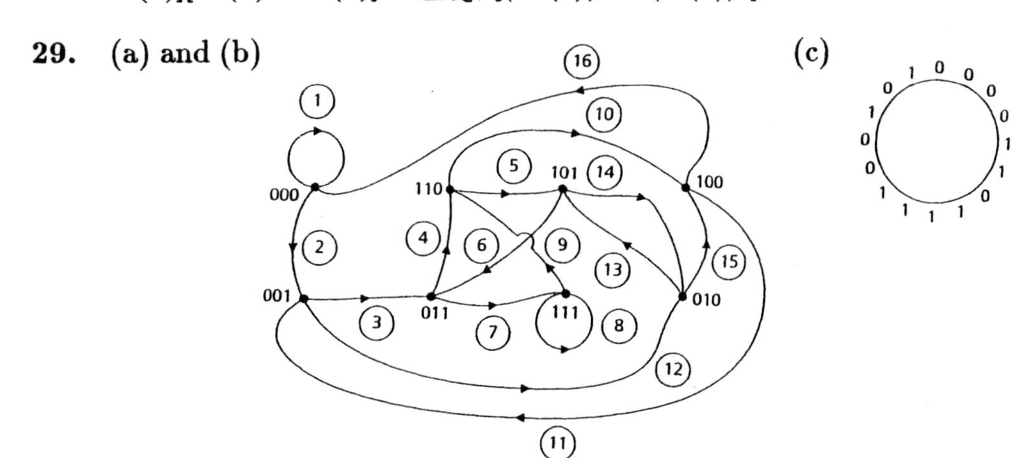

31. Let $|V| = n \geq 2$. Since G is loop-free and connected, for all $x \in V$ we have $1 \leq \deg(x) \leq n - 1$. Apply the pigeonhole principle with the n vertices as the pigeons and the $n - 1$ possible degrees as the pigeonholes.

33. (a) Label the rows and columns of the first matrix with a, b, c. Then the graph for this adjacency matrix is a path of two edges where $\deg(a) = \deg(b) = 1$ and $\deg(c) = 2$.

Now label the rows and columns of the second matrix with x, y, z. The graph for this adjacency matrix is a path of two edges where $\deg(y) = \deg(z) = 1$ and $\deg(x) = 2$.

Define $f : \{a, b, c\} \to \{x, y, z\}$ by $f(a) = y$, $f(b) = z$, $f(c) = x$. This function provides an isomorphism for these two graphs.

Alternatively, if we start with the first matrix and interchange rows 1 and 3 and then interchange columns 1 and 3 (on the resulting matrix), we obtain the second matrix. This also shows us that the graphs (corresponding to these adjacency matrices) are isomorphic.

(b) Yes. Here each graph is (isomorphic to) a cycle of length 4 with an additional edge connecting two nonadjacent vertices on the cycle.
(c) No. In the first graph there is a vertex of degree 3 and one of degree 1. For the second graph all the vertices have degree 2.

35. No. Let each person represent a vertex for a graph. If v, w represent two of these people, draw the edge $\{v, w\}$ if the two shake hands. If the situation were possible, then we would have a graph with 15 vertices, each of degree 3. So the sum of the degrees of the vertices would be 45, an odd integer. This contradicts Theorem 11.2.

37. Assign the Gray code $\{00, 01, 11, 10\}$ to the four horizontal levels: top – 00; second (from the top) – 01; second from the bottom – 11; bottom – 10. Likewise, assign the same code to the four vertical levels: left (or, first) – 00; second – 01; third – 11; right (or, fourth) – 10. This provides the labels for $p_1, p_2, \ldots, p_{16}$, where, for instance, p_1 has the label $(00, 00)$, p_2 has the label $(01, 00), \ldots, p_7$ has the label $(11, 01), \ldots, p_{11}$ has the label $(11, 10)$, and p_{16} has the label $(10, 10)$.

Define the function f from the set of 16 vertices of this grid to the vertices of Q_4 by $f((ab, cd)) = abcd$. Here $f((ab, cd)) = f((a_1b_1, c_1d_1)) \Rightarrow abcd = a_1b_1c_1d_1 \Rightarrow a = a_1, b = b_1, c = c_1, d = d_1 \Rightarrow (ab, cd) = (a_1b_1, c_1d_1) \Rightarrow f$ is one-to-one. Since the domain and codomain of f both contain 16 vertices, it follows from Theorem 5.11 that f is also onto. Finally, let $\{(ab, cd), (wx, yz)\}$ be an edge in the grid. Then either $ab = wx$ and cd, yz differ in one component or $cd = yz$ and ab, wx differ in one component. Suppose that $ab = wx$ and $c = y$, but $d \neq z$. Then $\{abcd, wxyz\}$ is an edge in Q_4. The other cases follow in a similar way. Conversely, suppose that $\{f((a_1b_1, c_1d_1)), f((w_1x_1, y_1z_1))\}$ is an edge in Q_4. Then $a_1b_1c_1d_1, w_1x_1y_1z_1$ differ in exactly one component – say the first. Then in the grid, there is an edge for the vertices $(0b_1, c_1d_1)$, $(1b_1, c_1d_1)$. The arguments are similar for the other three components. Consequently, f establishes an isomorphism between the three-by-three grid and a subgraph of Q_4.

[Note: The three-by-three grid has 24 edges while Q_4 has 32 edges.]

Section 11.4

1.

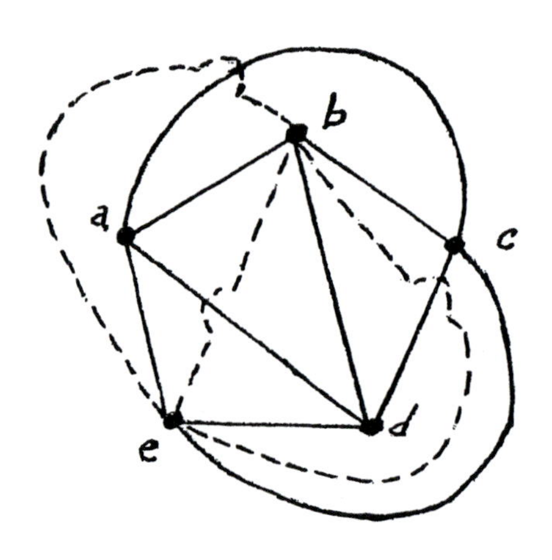

In this situation vertex b is in the region formed by the edges {a,d}, {d,c}, {c,a} and vertex e is outside of this region. Consequently the edge {b,e} will cross one of the edges {a,d}, {d,c}, {a,c} (as shown).

3. (a)

Graph	Number of vertices	Number of edges
$K_{4,7}$	11	28
$K_{7,11}$	18	77
$K_{m,n}$	$m+n$	mn

(b) $m = 6$

5. (a) Let $V_1 = \{a, d, e, h\}$ and $V_2 = \{b, c, f, g\}$. Then every vertex of G is in $V_1 \cup V_2$ and $V_1 \cap V_2 = \emptyset$. Also every edge in G may be written as $\{x, y\}$ where $x \in V_1$ and $y \in V_2$. Consequently, the graph G in part (a) of the figure is bipartite.
(b) Let $V_1' = \{a, b, g, h\}$ and $V_2' = \{c, d, e, f\}$. Then every vertex of G' is in $V_1' \cup V_2'$ and $V_1' \cap V_2' = \emptyset$. Since every edge of G' may be written as $\{x, y\}$, with $x \in V_1'$ and $y \in V_2'$, it follows that this graph is bipartite. In fact G' is (isomorphic to) the complete bipartite graph $K_{4,4}$.
(c) This graph is *not* bipartite. If $G'' = (V'', E'')$ were bipartite, let the vertices of G'' be partitioned as $V_1'' \cup V_2''$, where each edge in G'' is of the form $\{x, y\}$ with $x \in V_1''$ and $y \in V_2''$. We assume vertex a is in V_1''. Now consider the vertices b, c, d, and e. Since $\{a, b\}$ and $\{a, c\}$ are edges of G'' we must have b, c in V_2''. Also, $\{b, d\}$ is an edge in the graph, so d is in V_1''. But then $\{d, e\} \in E'' \Rightarrow e \in V_2''$, while $\{c, e\} \in E'' \Rightarrow e \in V_1''$.

7. The vertices in $K_{m,n}$ may be partitioned as $V_1 \cup V_2$ where $|V_1| = m$, $|V_2| = n$, and each edge of the graph has the form $\{x, y\}$ where $x \in V_1$ and $y \in V_2$.

(a) In order to obtain a cycle of length four we need to select two vertices from each of V_1 and V_2. This can be done in $\binom{m}{2}\binom{n}{2}$ ways — each resulting in a distinct cycle of length four.
[Note: Say we select vertices a, b from V_1 and vertices c, d from V_2. We do *not* distinguish the cycles $a \to c \to b \to d \to a$ and $a \to d \to b \to c \to a$.]
(b) For a path of length two there is one vertex of (path) degree 2 and two vertices of (path) degree 1. If the vertex of (path) degree 2 is in V_1 then there are $m\binom{n}{2}$ such paths. There are $n\binom{m}{2}$ such paths when the vertex of (path) degree 2 is in V_2. Hence there are $m\binom{n}{2} + n\binom{m}{2} = (1/2)(mn)[m + n - 2]$ paths of length 2 in $K_{m,n}$.
(c) Here a path of length 3 has the form $a \to b \to c \to d$ where $a, c \in V_1$ and $b, d \in V_2$. By the rule of product there are $(m)(n)(m-1)(n-1) = 4\binom{m}{2}\binom{n}{2}$ such paths in $K_{m,n}$.

9. (a) $\binom{4}{2} = 6$
(b) $(1/2)(7)(3)(6)(2)(5)(1)(4) = 2520$
(c) $(1/2)(12)(7)(11)(6)(10)(5)(9)(4)(8)(3)(7)(2)(6)(1)(5) = 50,295,168,000$
(d) $(1/2)(n)(m)(n-1)(m-1)(n-2)\cdots(2)(n-(m+1))(1)(n-m)$

11. Partition V as $V_1 \cup V_2$ with $|V_1| = m$, $|V_2| = v - m$. If G is bipartite, then the maximum number of edges that G can have is $m(v - m) = -[m - (v/2)]^2 + (v/2)^2$, a function of m. For a given value of v, when v is even, $m = v/2$ maximizes $m(v - m) = (v/2)[v - (v/2)] = (v/2)^2$. For v odd, $m = (v-1)/2$ or $m = (v+1)/2$ maximizes $m(v-m) = [(v-1)/2][v - ((v-1)/2)] = [(v-1)/2][(v+1)/2] = [(v+1)/2][v - ((v+1)/2)] = (v^2 - 1)/4 = \lfloor (v/2)^2 \rfloor < (v/2)^2$. Hence if $|E| > (v/2)^2$, G cannot be bipartite.

13. (a)

a: {1,2}	f: {4,5}
b: {3,4}	g: {2,5}
c: {1,5}	h: {2,3}
d: {2,4}	i: {1,3}
e: {3,5}	j: {1,4}

(b) G is (isomorphic to) the Petersen graph. (See Fig. 11.52(a)).

15. The result follows if and only if mn is even (that is, at least one of m, n is even).

Suppose, without loss of generality, that m is even — say, $m = 2t$. Let V denote the vertex set of $K_{m,n}$ where $V = V_1 \cup V_2$ and $V_1 = \{v_1, v_2, \ldots, v_t, v_{t+1}, \ldots, v_m\}$, $V_2 = \{w_1, w_1, \ldots, w_n\}$. The mn edges in $K_{m,n}$ are of the form $\{v_i, w_j\}$ where $1 \leq i \leq m$, $1 \leq j \leq n$. Now consider the subgraphs G_1, G_2 of $K_{m,n}$ where G_1 is induced by $\{v_1, v_2, \ldots, v_t\} \cup V_2$ and G_2 is induced by $\{v_{t+1}, v_{t+2}, \ldots, v_m\} \cup V_2$. Each of G_1, G_2 is isomorphic to $K_{t,n}$, and every edge in $K_{m,n}$ is in exactly one of G_1, G_2.

If both m, n are odd, then $K_{m,n}$ has an odd number of edges and cannot be decomposed into two isomorphic subgraphs — since each such subgraph has the same number of edges as the other.

17. (a) There are 17 vertices, 34 edges and 19 regions and $v - e + r = 17 - 34 + 19 = 2$.
(b) Here we find 10 vertices, 24 edges and 16 regions and $v - e + r = 10 - 24 + 16 = 2$.

19. From $32 = 2|E| = \sum_{v \in V} \deg(v) = 4|V|$ we learn that $|V| = 8$. Then with $|V| - |E| + r = 2$, it follows that $r = 2 + 16 - 8 = 10$.

21. If not, $\deg(v) \geq 6$ for all $v \in V$. Then $2e = \sum_{v \in V} \deg(v) \geq 6|V|$, so $e \geq 3|V|$, contradicting $e \leq 3|V| - 6$ (Corollary 11.3.)

23. (a) $2e \geq kr = k(2 + e - v) \Longrightarrow (2 - k)e \geq k(2 - v) \Longrightarrow e \leq [k/(k-2)](v-2)$.

(b) 4

(c) In $K_{3,3}$, $e = 9$, $v = 6$. $[k/(k-2)](v-2) = (4/2)(4) = 8 < 9 = e$. Since $K_{3,3}$ is connected, it must be nonplanar.

(d) Here $k = 5$, $v = 10$, $e = 15$ and $[k/(k-2)](v-2) = (5/3)(8) = (40/3) < 15 = e.$ Since the Petersen graph is connected, it must be nonplanar.

25. (a) The dual for the tetrahedron (Fig. 11.59(b)) is the graph itself. For the graph (cube) in Fig. 11.59(d) the dual is the octahedron, and vice versa. Likewise, the dual of the dodecahedron is the icosahedron, and vice versa.
(b) For $n \in \mathbf{Z}^+$, $n \geq 3$, the dual of the wheel graph W_n is W_n itself.

27.

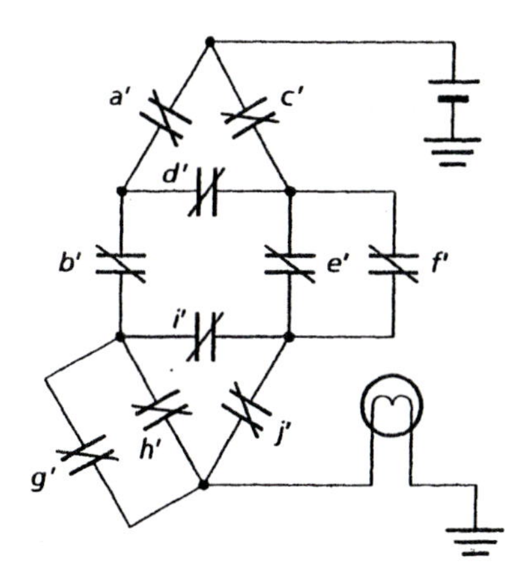

29. Proof:

a) As we mentioned in the remark following Example 11.18, when G_1, G_2 are homeomorphic graphs then they may be regarded as isomorphic except, possibly, for vertices of degree 2. Consequently, two such graphs will have the same number of vertices of odd degree.

b) Now if G_1 has an Euler trail, then G_1 (is connected and) has all vertices of even degree – except two, those being the vertices at the beginning and end of the Euler trail. From part (a) G_2 is likewise connected with all vertices of even degree, except for two of odd degree. Consequently, G_2 has an Euler trail. [The converse follows in a similar way.]

c) If G_1 has an Euler circuit, then G_1 (is connected and) has all vertices of even degree. From part (a) G_2 is likewise connected with all vertices of even degree, so G_2 has an Euler circuit. [The converse follows in a similar manner.]

Section 11.5

1.

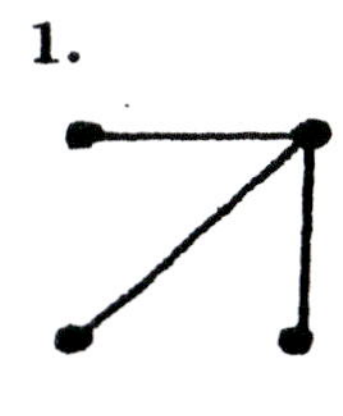

(a)

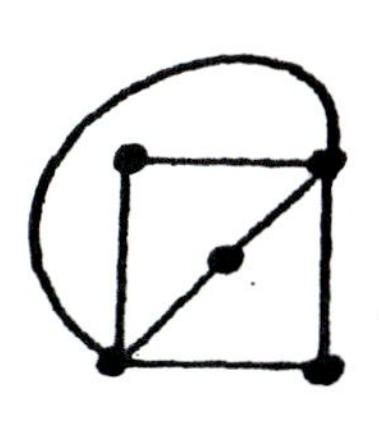

(b)

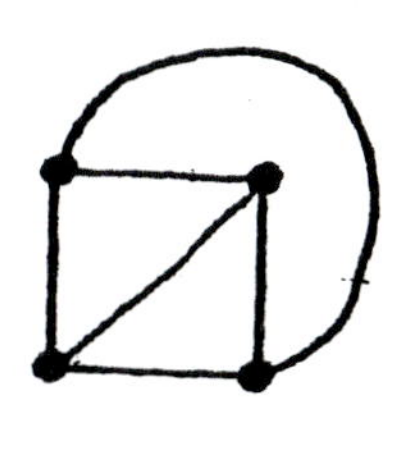

(c)

(d)

3. (a) Hamilton cycle: $a \to g \to k \to i \to h \to b \to c \to d \to j \to f \to e \to a$
(b) Hamilton cycle: $a \to d \to b \to e \to g \to j \to i \to f \to h \to c \to a$

(c) Hamilton cycle: $a \to h \to e \to f \to g \to i \to d \to c \to b \to a$

(d) The edges $\{a,c\}$, $\{c,d\}$, $\{d,b\}$, $\{b,e\}$, $\{e,f\}$, $\{f,g\}$ provide a Hamilton path for the given graph. However, there is no Hamilton cycle, for such a cycle would have to include the edges $\{b,d\}$, $\{b,e\}$, $\{a,c\}$, $\{a,e\}$, $\{g,f\}$, and $\{g,e\}$ – and, consequently, the vertex e will have degree greater than 2.

(e) The path $a \to b \to c \to d \to e \to j \to i \to h \to g \to f \to k \to l \to m \to n \to o$ is one possible Hamilton path for this graph. Another possibility is the path $a \to b \to c \to d \to i \to h \to g \to f \to k \to l \to m \to n \to o \to j \to e$. However, there is no Hamilton cycle. For if we try to construct a Hamilton cycle we must include the edges $\{a,b\}$, $\{a,f\}$, $\{f,k\}$, $\{k,l\}$, $\{d,e\}$, $\{e,j\}$, $\{j,o\}$ and $\{n,o\}$. This then forces us to eliminate the edges $\{f,g\}$ and $\{i,j\}$ from further consideration. Now consider the vertex i. If we use edges $\{d,i\}$ and $\{i,n\}$, then we have a cycle on the vertices d,e,j,o,n and i – and we cannot get a Hamilton cycle for the given graph. Hence we must use only one of the edges $\{d,i\}$ and $\{i,n\}$. Because of the symmetry in this graph let us select edge $\{d,i\}$ – and then edge $\{h,i\}$ so that vertex i will have degree 2 in the Hamilton cycle we are trying to construct. Since edges $\{d,i\}$ and $\{d,e\}$ are now being used, we eliminate edge $\{c,d\}$ and this then forces us to include edges $\{b,c\}$ and $\{c,h\}$ in our construction. Also we must include the edge $\{m,n\}$ since we eliminated edge $\{i,n\}$ from consideration. Next we eliminate edges $\{h,m\}$, $\{h,g\}$ and $\{b,g\}$. Finally we must include edge $\{m,l\}$ and then eliminate edge $\{l,g\}$. But now we have eliminated the four edges $\{b,g\}$, $\{f,g\}$, $\{h,g\}$ and $\{l,g\}$ and g is consequently isolated.

(f) For this graph we find the Hamilton cycle $a \to b \to c \to d \to e \to j \to i \to h \to g \to l \to m \to n \to o \to t \to s \to r \to q \to p \to k \to f \to a$.

5. (a) If we remove any one of the vertices a, b or g, the resulting subgraph has a Hamilton cycle. For example, upon removing vertex a, we find the Hamilton cycle $b \to d \to c \to f \to g \to e \to b$.

(b) The following Hamilton cycle exists if we remove vertex g : $a \to b \to c \to d \to e \to j \to o \to n \to i \to h \to m \to l \to k \to f \to a$. A symmetric situation results upon removing vertex i.

7. (a) We can arrange the n vertices in a cycle in $(n!)/n = (n-1)!$ ways. [As we did in Example 1.16.] Then, as we do not distinguish between a clockwise and a counterclockwise orientation, the final answer is $(1/2)(n-1)!$.

(b) From the argument in Example 11.29, the number of edge-disjoint Hamilton cycles in K_{21} is $(1/2)(21-1) = 10$.

(c) $(1/2)(19-1) = 9$.

9. Let $G = (V,E)$ be a loop-free undirected graph with no odd cycles. We assume that G is connected – otherwise, we work with the components of G. Select any vertex x in V and let $V_1 = \{v \in V | d(x,v)$, the length of a shortest path between x and v, is odd$\}$ and $V_2 = \{w \in V | d(x,w)$, the length of a shortest path between x and w, is even$\}$. Note that (i) $x \in V_2$; (ii) $V = V_1 \cup V_2$; and (iii) $V_1 \cap V_2 = \emptyset$. We claim that each edge $\{a,b\}$ in E has one vertex in V_1 and the other vertex in V_2.

For suppose that $e = \{a, b\} \in E$ with $a, b \in V_1$. (The proof for $a, b \in V_2$ is similar.) Let $E_a = \{\{a, v_1\}, \{v_1, v_2\}, \ldots, \{v_{m-1}, x\}\}$ be the m edges in a shortest path from a to x, and let $E_b = \{\{b, v_1'\}, \{v_1', v_2'\}, \ldots, \{v_{n-1}', x\}\}$ be the n edges in a shortest path from b to x. Note that m, n are both odd. If $\{v_1, v_2, \ldots, v_{m-1}\} \cap \{v_1', v_2', \ldots, v_{n-1}'\} = \emptyset$, then the set of edges $E' = \{\{a, b\}\} \cup E_a \cup E_b$ provides an odd cycle in G. Otherwise, let $w(\neq x)$ be the first vertex where the paths come together, and let $E'' = \{\{a, b\}\} \cup \{\{a, v_1\}, \{v_1, v_2\}, \ldots, \{v_i, w\}\} \cup \{\{b, v_1'\}, \{v_1', v_2'\}, \ldots, \{v_j', w\}\}$, for some $1 \leq i \leq m-1$ and $1 \leq j \leq n-1$. Then either E'' provides an odd cycle for G or $E' - E''$ contains an odd cycle for G.

11. (a)

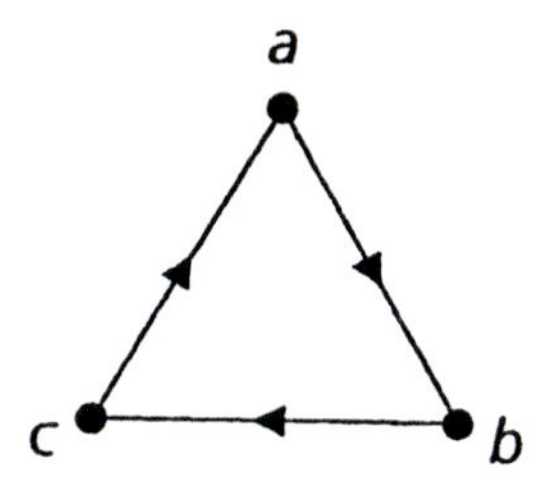

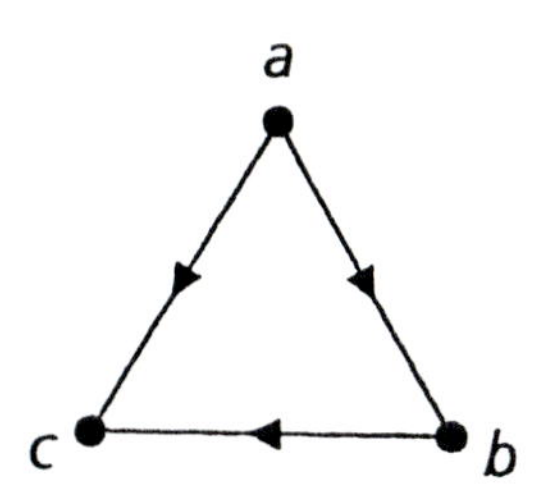

(b)

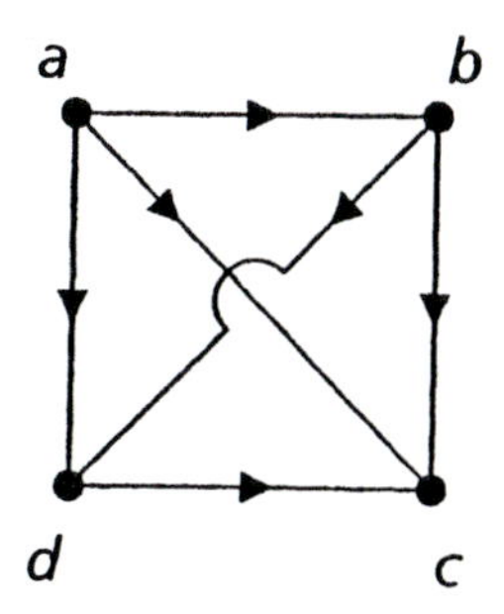

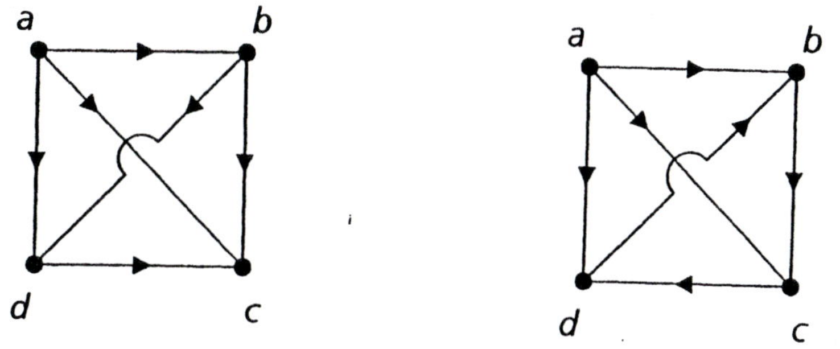

$\text{od}(a) = 3 \quad \text{id}(a) = 0$
$\text{od}(b) = 2 \quad \text{id}(b) = 1$
$\text{od}(c) = 0 \quad \text{id}(c) = 3$
$\text{od}(d) = 1 \quad \text{id}(d) = 2$

$\text{od}(a) = 3 \quad \text{id}(a) = 0$
$\text{od}(b) = 1 \quad \text{id}(b) = 2$
$\text{od}(c) = 1 \quad \text{id}(c) = 2$
$\text{od}(d) = 1 \quad \text{id}(d) = 2$

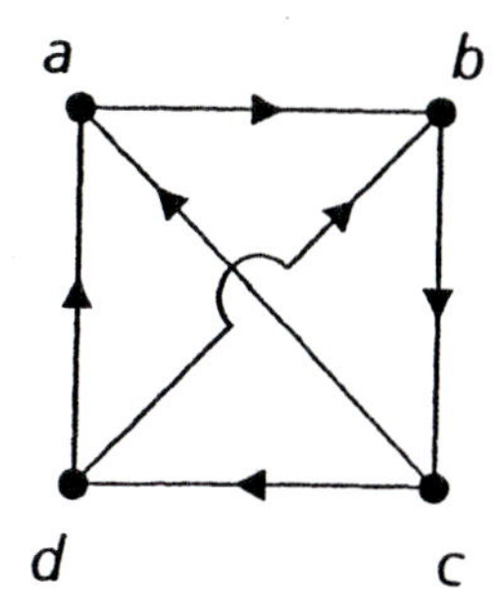

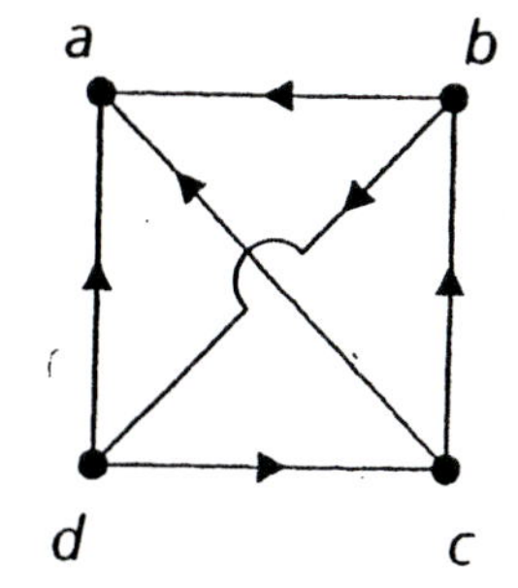

$od(a) = 1$	$id(a) = 2$	$od(a) = 0$	$id(a) = 3$
$od(b) = 1$	$id(b) = 2$	$od(b) = 2$	$id(b) = 1$
$od(c) = 2$	$id(c) = 1$	$od(c) = 2$	$id(c) = 1$
$od(d) = 2$	$id(d) = 1$	$od(d) = 2$	$id(d) = 1$

13. Proof: If not, there exists a vertex x such that $(v, x) \notin E$ and, for all $y \in V$, $y \neq v, x$, if $(v, y) \in E$ then $(y, x) \notin E$. Since $(v, x) \notin E$, we have $(x, v) \in E$, as T is a tournament. Also, for each y mentioned earlier, we also have $(x, y) \in E$. Consequently, $od(x) \geq od(v)+1$ – contradicting $od(v)$ being a maximum!

15.

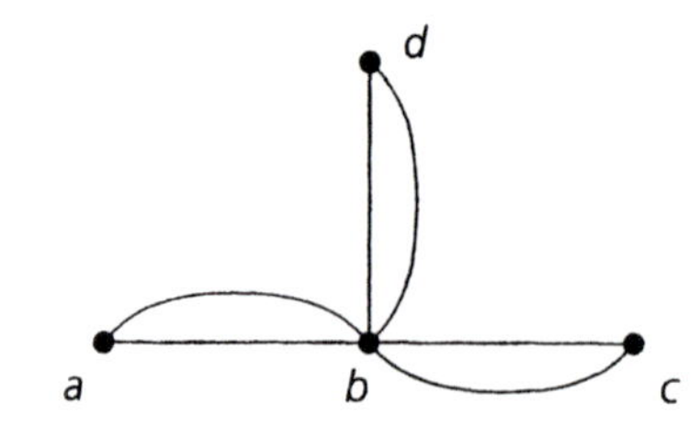

For the multigraph in the given figure, $|V| = 4$ and $\deg(a) = \deg(c) = \deg(d) = 2$ and $\deg(b) = 6$. Hence $\deg(x) + \deg(y) \geq 4 > 3 = 4 - 1$ for all nonadjacent $x, y \in V$, but the multigraph has no Hamilton path.

17. For $n \geq 5$ let $C_n = (V, E)$ denote the cycle on n vertices. Then C_n has (actually is) a Hamilton cycle, but for all $v \in V, \deg(v) = 2 < n/2$.

19. This follows from Theorem 11.9, since for all (nonadjacent) $x, y \in V$, $\deg(x) + \deg(y) = 12 > 11 = |V|$.

21. When $n = 5$ the graphs C_5 and $\overline{C}_5$ are isomorphic, and both are Hamilton cycles on five vertices.

For $n \geq 6$, let u, v denote nonadjacent vertices in $\overline{C}_n$. Since $\deg(u) = \deg(v) = n - 3$ we find that $\deg(u) + \deg(v) = 2n - 6$. Also, $2n - 6 \geq n \iff n \geq 6$, so it follows from Theorem 11.9 that the cocycle $\overline{C}_n$ contains a Hamilton cycle when $n \geq 6$.

23. (a) The path $v \to v_1 \to v_2 \to v_3 \to \ldots \to v_{n-1}$ provides a Hamilton path for H_n. Since $\deg(v) = 1$ the graph cannot have a Hamilton cycle.
(b) Here $|E| = \binom{n-1}{2} + 1$. (So the number of edges required in Corollary 11.6 cannot be decreased.)

25. (a) (i) $\{a, c, f, h\}, \{a, g\}$; (ii) $\{z\}, \{u, w, y\}$
(b) (i) $\beta(G) = 4$; (ii) $\beta(G) = 3$
(c) (i) 3 (ii) 3 (iii) 3 (iv) 4 (v) 6
(vi) The maximum of m and n.
(d) The complete graph on $|I|$ vertices.

Section 11.6

1. Draw a vertex for each species of fish. If two species x, y must be kept in separate aquaria, draw the edge $\{x, y\}$. The smallest number of aquaria needed is then the chromatic number of the resulting graph.

3. We can model this problem with graphs. For either part of the problem draw the undirected graph $G = (V, E)$ where $V = \{1, 2, 3, 4, 5, 6, 7\}$ and $\{i, j\} \in E$ when chemicals i and j require separate storage compartments. For part (a), the graph (in part (a) of the figure) has chromatic number 3, so here Jeannette will need three separate storage compartments to safely store these seven chemicals.

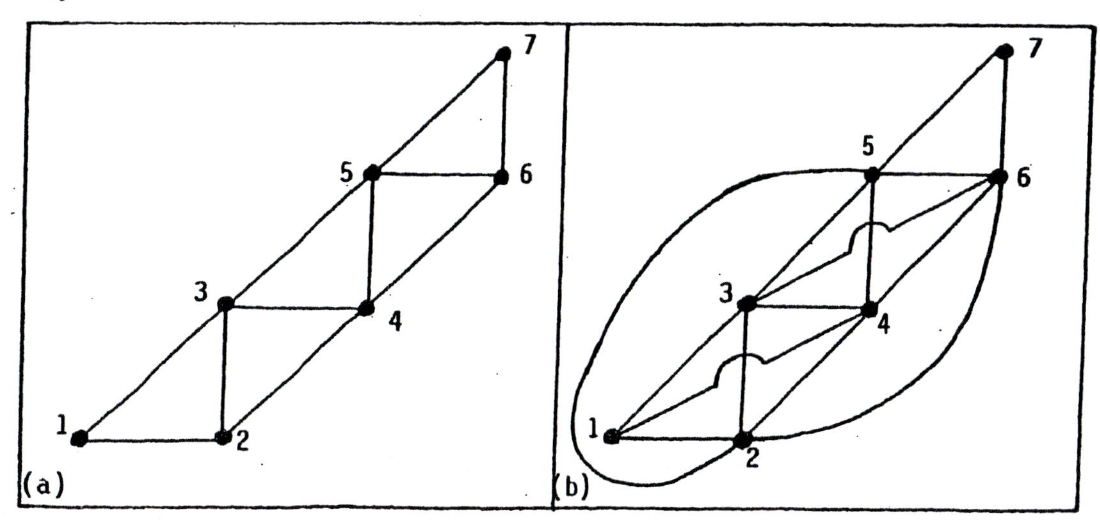

Now consider the graph in part (b) of the figure. Note here that the subgraph induced by the vertices 2,3,4,5,6 is (isomorphic to) K_5. Consequently, with these additional conditions Jeannette will need five separate storage compartments to store these seven chemicals safely.

5. (a) $P(G, \lambda) = \lambda(\lambda - 1)^3$
(b) For $G = K_{1,n}$ we find that $P(G, \lambda) = \lambda(\lambda - 1)^n$.
$\chi(K_{1,n}) = 2$.

7. (a) Since the graph is bipartite its chromatic number is 2.
(b) 2 (n even); 3 (n odd)
(c) Figure 11.59(d): 2; Fig. 11.62(a): 3; Fig. 11.85(i); 2; Fig. 11.85(ii): 3 (d) 2

9. (a) (1) Start at vertex t. There are λ choices for coloring this vertex. And then there are $\lambda - 1$ choices for coloring vertex z and $\lambda - 2$ choices for vertex y. Since vertices t and y have different colors, there are $\lambda - 2$ choices for coloring vertex x. Finally, we can color vertex w with any color except the one we used for vertex x – so there are $\lambda - 1$ choices. By the rule of product it follows that the chromatic polynomial is $\lambda(\lambda - 1)^2(\lambda - 2)^2$.

(2) We use Theorem 11.10. Let G denote the graph and let $e = \{t, w\}$. Then $P(G, \lambda) = P(G_e, \lambda) - P(G'_e, \lambda)$, where G_e is a path with five vertices and G'_e is a cycle

with four vertices. Then $P(G_e, \lambda) = \lambda(\lambda-1)^4$ and $P(G'_e, \lambda) = \lambda(\lambda-1)(\lambda^2-3\lambda+3) = \lambda^4 - 4\lambda^3 + 6\lambda^2 - 3\lambda$ (from Example 11.36). Consequently,

$$\begin{aligned} P(G,\lambda) &= \lambda(\lambda-1)^4 - \lambda(\lambda-1)(\lambda^2-3\lambda-3) \\ &= \lambda(\lambda-1)[\lambda^3 - 3\lambda^2 + 3\lambda - 1 - \lambda^2 + 3\lambda - 3] \\ &= \lambda(\lambda-1)[\lambda^3 - 4\lambda^2 + 6\lambda - 4] = \lambda(\lambda-1)(\lambda-2)(\lambda^2-2\lambda+2). \end{aligned}$$

(3) $\lambda(\lambda-1)(\lambda-2)(\lambda^2-5\lambda+7)$

(b) (1) 3: (2) 3; (3) 3

(c) (1) 720; (2) 1020; (3) 420

11. Let $e = \{v, w\}$ be the deleted edge. There are $\lambda(1)(\lambda-1)(\lambda-2)\cdots(\lambda-(n-2))$ proper colorings of G_n where v, w share the same color and $\lambda(\lambda-1)(\lambda-2)\cdots(\lambda-(n-1))$ proper colorings where v, w are colored with different colors. In total there are $P(G_n, \lambda) = \lambda(\lambda-1)\cdots(\lambda-n+2) + \lambda(\lambda-1)\cdots(\lambda-n+1) = \lambda(\lambda-1)\cdots(\lambda-n+3)(\lambda-n+2)^2$ proper colorings for G_n.

Here $\chi(G_n) = n-1$.

13. (a) $|V| = 2n$; $|E| = (1/2)\sum_{v\in V}\deg(v) = (1/2)[4(2)+(2n-4)(3)] = (1/2)[8+6n-12] = 3n-2$, $n \geq 1$.

(b) For $n = 1$, we find that $G = K_2$ and $P(G,\lambda) = \lambda(\lambda-1) = \lambda(\lambda-1)(\lambda^2-3\lambda+3)^{1-1}$ so the result is true in this first case. For $n = 2$, we have $G = C_4$, the cycle of length 4, and here $P(G,\lambda) = \lambda(\lambda-1)^3 - \lambda(\lambda-1)(\lambda-2) = \lambda(\lambda-1)(\lambda^2-3\lambda+3)^{2-1}$. So the result follows for $n = 2$. Assuming the result true for an arbitrary (but fixed) $n \geq 1$, consider the situation for $n+1$. Write $G = G_1 \cup G_2$, where G_1 is C_4 and G_2 is the ladder graph for n rungs. Then $G_1 \cap G_2 = K_2$, so from Theorem 11.14 we have $P(G,\lambda) = P(G_1,\lambda)\cdot P(G_2,\lambda)/P(K_2,\lambda) = [(\lambda)(\lambda-1)(\lambda^2-3\lambda+3)][(\lambda)(\lambda-1)(\lambda^2-3\lambda+3)^{n-1}]/(\lambda)(\lambda-1) = (\lambda)(\lambda-1)(\lambda^2-3\lambda+3)^n$. Consequently, the result is true for all $n \geq 1$, by the Principle of Mathematical Induction.

15. (a) $\lambda(\lambda-1)(\lambda-2)$ (b) Follows from Theorem 11.10

(c) Follows by the rule of product.

(d)

$$\begin{aligned} P(C_n,\lambda) &= P(P_{n-1},\lambda) - P(C_{n-1},\lambda) = \lambda(\lambda-1)^{n-1} - P(C_{n-1},\lambda) \\ &= [(\lambda-1)+1](\lambda-1)^{n-1} - P(C_{n-1},\lambda) \\ &= (\lambda-1)^n + (\lambda-1)^{n-1} - P(C_{n-1},\lambda) \Longrightarrow \\ & P(C_n,\lambda) - (\lambda-1)^n = (\lambda-1)^{n-1} - P(C_{n-1},\lambda). \end{aligned}$$

Replacing n by $n-1$ yields

$$P(C_{n-1},\lambda) - (\lambda-1)^{n-1} = (\lambda-1)^{n-2} - P(C_{n-2},\lambda) = (-1)[P(C_{n-2},\lambda) - (\lambda-1)^{n-2}].$$

Hence

$$P(C_n,\lambda) - (\lambda-1)^n = P(C_{n-2},\lambda) - (\lambda-1)^{n-2} = (-1)^2[P(C_{n-2},\lambda) - (\lambda-1)^{n-2}].$$

(e) Continuing from part (d),

$$\begin{aligned} P(C_n, \lambda) &= (\lambda-1)^n + (-1)^{n-3}[P(C_3, \lambda) - (\lambda-1)^3] \\ &= (\lambda-1)^n + (-1)^{n-1}[\lambda(\lambda-1)(\lambda-2) - (\lambda-1)^3] \\ &= (\lambda-1)^n + (-1)^n(\lambda-1). \end{aligned}$$

17. From Theorem 11.13, the expansion for $P(G, \lambda)$ will contain exactly one occurrence of the chromatic polynomial of K_n. Since no larger graph occurs this term determines the degree as n and the leading coefficient as 1.

19. (a) For $n \in \mathbf{Z}^+$, $n \geq 3$, let C_n denote the cycle on n vertices.

If n is odd then $\chi(C_n) = 3$. But for each v in C_n, the subgraph $C_n - v$ is a path with $n-1$ vertices and $\chi(C_n - v) = 2$. So for n odd C_n is color-critical.

However, when n is even we have $\chi(C_n) = 2$, and for each v in C_n, the subgraph $C_n - v$ is still a path with $n-1$ vertices and $\chi(C_n - v) = 2$. Consequently, cycles with an even number of vertices are not color-critical.

(b) For every complete graph K_n, where $n \geq 2$, we have $\chi(K_n) = n$, and for each vertex v in K_n, $K_n - v$ is (isomorphic to) K_{n-1}, so $\chi(K_n - v) = n-1$. Consequently, every complete graph with at least one edge is color-critical.

(c) Suppose that G is not connected. Let G_1 be a component of G where $\chi(G_1) = \chi(G)$, and let G_2 be any other component of G. Then $\chi(G_1) \geq \chi(G_2)$ and for all v in G_2 we find that $\chi(G - v) = \chi(G_1) = \chi(G)$, so G is not color-critical.

(d) If not, let $v \in V$ with $\deg(v) \leq k-2$. Since G is color-critical we have $\chi(G-v) \leq k-1$, and so we can properly color the vertices in the subgraph $G - v$ with at most $k-1$ colors. Since $\deg(v) \leq k-2$, we have used at most $k-2$ colors to color all vertices in G adjacent to v. Therefore we do not need a new color (beyond those needed to color the subgraph $G - v$) in order to color v and can color all vertices in G with at most $k-1$ colors. But this contradicts $\chi(G) = k$.

Supplementary Exercises

1. $\binom{n}{2} = 56 + 80 = 136 \Longrightarrow n(n-1) = 272 \Longrightarrow n = 17.$

3. (a) Label the vertices of K_6 with $a, b, \ldots, f$. Of the five edges on a at least three have the same color, say red, and let these edges be $\{a,b\}, \{a,c\}, \{a,d\}$. If the edges $\{b,c\}, \{c,d\}, \{b,c\}$ are all blue, the result follows. If not, one of these edges, say $\{c,d\}$, is red and then $\{a,c\}, \{a,d\}, \{c,d\}$ yield a red triangle.

(b) Consider the six people as vertices. If two people are friends (strangers) draw a red (blue) edge connecting their respective vertices. The result then follows from part (a).

5. (a) We can redraw G_2 as

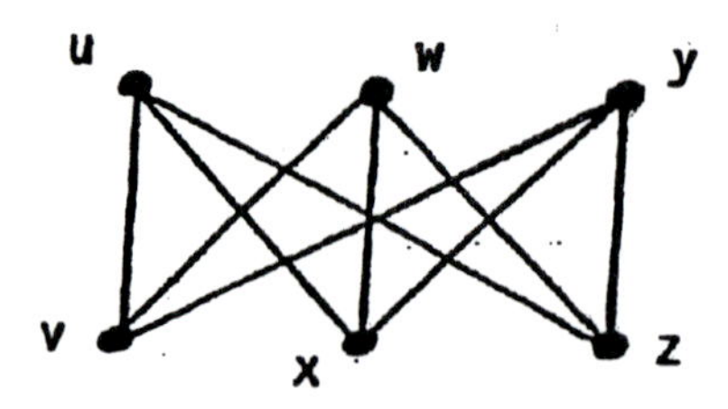

(b) Let $V_1 = \{1,2,3,4,5,6\}$ and $V_2 = \{u,v,w,x,y,z\}$. There are two cases to consider.
(1) We want $f : V_1 \to V_2$ such that $f(\{1,2,3\}) = \{u,w,y\}$ and $f(\{4,5,6\}) = \{v,x,z\}$. There are $(3!)(3!) = 36$ such functions.
(2) Alternately, we want $f : V_1 \to V_2$ such that $f(\{1,2,3\}) = \{v,x,z\}$ and $f(\{4,5,6\}) = \{u,w,y\}$. Again there are 36 such functions. So in total we have 72 different isomorphisms.

7. (a) Let the vertices of $K_{3,7}$ be partitioned as $V_1 \cup V_2$ where $|V_1| = 3$ and $|V_2| = 7$. Then there are $(3)(7)(2)(6)(1)(5) = 1260$ paths of length 5 where each such path contains all three vertices in V_1.
(b) With V_1, V_2 as in part (a) we find that there are $(1/2)(3)(7)(2)(6)(1)$ paths of length 4 that start and end with a vertex in V_1, and there are also $(1/2)(7)(3)(6)(2)(5)$ paths of length 4 that start and end with a vertex in V_2. Consequently, there are $126 + 630 = 756$ paths of length 4 in $K_{3,7}$.
(c) (Case 1: p is odd, $p = 2k + 1$ for $k \in \mathbf{N}$). Here there are mn paths of length $p = 1$ (when $k = 0$) and $(m)(n)(m-1)(n-1)\cdots(m-k)(n-k)$ paths of length $p = 2k+1 \geq 3$. (Case 2: p is even, $p = 2k$ for $k \in \mathbf{Z}^+$). When $p < 2m$ (i.e., $k < m$) the number of paths of length p is $(1/2)(m)(n)(m-1)(n-1)\cdots(n-(k-1))(m-k) + (1/2)(n)(m)(n-1)(m-1)\cdots(m-(k-1))(n-k)$. For $p = 2m$ we find $(1/2)(n)(m)(n-1)(m-1)\cdots(m-(m-1))(n-m)$ paths of (longest) length $2m$.

9. (a) Let I be independent and $\{a,b\} \in E$. If neither a nor b is in $V - I$, then $a, b \in I$, and since they are adjacent, I is not independent. Conversely, if $I \subseteq V$ with $V - I$ a covering of G, then if I is not independent there are vertices $x, y \in I$ with $\{x,y\} \in E$. But $\{x,y\} \in E \Longrightarrow$ either x or y is in $V - I$.

(b) Let I be a largest maximal independent set in G and K a minimal covering. From part (a), $|K| \leq |V - I| = |V| - |I|$ and $|I| \geq |V - K| = |V| - |K|$, or $|K| + |I| \geq |V| \geq |K| + |I|$.

11. Since we are selecting n edges and no two have a common vertex, the selection of n edges will include exactly one occurrence of every vertex. We consider two mutually disjoint and exhaustive cases:

(1) The edge $\{x_n, y_n\}$ is in the selection: Then $\{x_{n-1}, x_n\}$ and $\{y_{n-1}, y_n\}$ are not in the selection and we must select the remaining $n - 1$ edges from the resulting subgraph (a ladder graph with $n - 1$ rungs) in a_{n-1} ways.

(2) The edge $\{x_n, y_n\}$ is not in the selection: Then in order to have x_n and y_n appear in the selection we must include edges $\{x_{n-1}, x_n\}$ and $\{y_{n-1}, y_n\}$. Consequently, we must now select the other $n-2$ edges from the resulting subgraph (a ladder graph with $n-2$ rungs) in a_{n-2} ways.

Hence $a_n = a_{n-1} + a_{n-2}$, $a_0 = 1$, $a_1 = 1$, and $a_n = F_{n+1}$, the $(n+1)$st Fibonacci number.

13. If the vertex y_n is included in the independent subset then we cannot use any of the vertices y_{n-1}, x_{n-1}, or x_n. There are a_{n-2} such subsets — and another a_{n-2} independent subsets where x_n is included. In addition, there are a_{n-1} independent subsets when both x_n and y_n are excluded. This leads us to the recurrence relation

$$a_n = a_{n-1} + 2a_{n-2},$$

with initial conditions $a_1 = 3$, $a_2 = 5$.

To solve this recurrence relation let $a_n = Ar^n$, where $A \neq 0$, $r \neq 0$. This leads to the characteristic equation

$$r^2 - r - 2 = 0,$$

and the characteristic roots -1 and 2. Therefore, $a_n = A_1(-1)^n + A_2(2^n)$, where A_1, A_2 are constants.

$a_1 = 3, a_2 = 5 \Rightarrow 2a_0 = 5 - 3 \Rightarrow a_0 = 1.$

$1 = a_0 = A_1 + A_2.$

$3 = a_1 = -A_1 + 2A_2 = -(1 - A_2) + 2A_2 = -1 + 3A_2$, so $A_2 = 4/3$, and $A_1 = 1 - A_2 = -1/3$.

Consequently, $a_n = (-1/3)(-1)^n + (4/3)(2^n)$, $n \geq 0$ (or $n \geq 1$).

15. (a) $\gamma(G) = 2$; $\beta(G) = 3$; $\chi(G) = 4$.

(b) G has neither an Euler trail nor an Euler circuit; G does have a Hamilton cycle.

(c) G is not bipartite but it is planar.

17. (a) $\chi(G) \geq \omega(G)$ (b) They are equal.

19. (a) The constant term is 3, not 0. This contradicts Theorem 11.11.

(b) The leading coefficient is 3, not 1. This contradicts the result in Exercise 17 of Section 11.6.

(c) The sum of the coefficients in -1, not 0. This contradicts Theorem 11.12.

21. (a) $a_1 = 2$, $a_2 = 3$. For $n \geq 3$ label the vertices of P_n as $v_1, v_2, v_3, \ldots, v_n$ where the edges are $\{v_1, v_2\}, \{v_2, v_3\}, \ldots, \{v_{n-1}, v_n\}$. In constructing an independent subset S from P_n we consider two cases:

(1) $v_n \notin S$: Then S is an independent subset of P_{n-1} and there are a_{n-1} such subsets.

(2) $v_n \in S$: Then $v_{n-1} \notin S$ and $S - \{v_n\}$ is one of the a_{n-2} independent subsets of P_{n-2}.

Hence $a_n = a_{n-1} + a_{n-2}$, $n \geq 3$, $a_1 = 2$, $a_2 = 3$, *or* $a_n = a_{n-1} + a_{n-2}$, $n \geq 2$, $a_0 = 1$, $a_1 = 2$. So $a_n = F_{n+2}$, the $(n+2)$nd Fibonacci number.

(b) Consider the subgraph of G_1 induced by the vertices 1,2,3,4. From part (a) we know that this subgraph determines 8 ($= F_6$, the sixth (nonzero) Fibonacci number) independent subsets of $\{1,2,3,4\}$. Therefore, the graph G_1 has $1 + F_6$ independent subsets of vertices.

Likewise the graph G_2 has $1 + F_7$ independent subsets (of vertices), and the graph G_n determines $1 + F_{n+2}$ such subsets.

(c) $H_1 : 3 + F_6 = (2^2 - 1) + F_6$
$H_2 : 3 + F_7 = (2^2 - 1) + F_7$
$H_3 : 3 + F_{n+2} = (2^2 - 1) + F_{n+2}$

(d) There are $2^s - 1 + m$ independent subsets of vertices for graph $G' = (V', E')$.

CHAPTER 12
TREES

Section 12.1

1. (a)

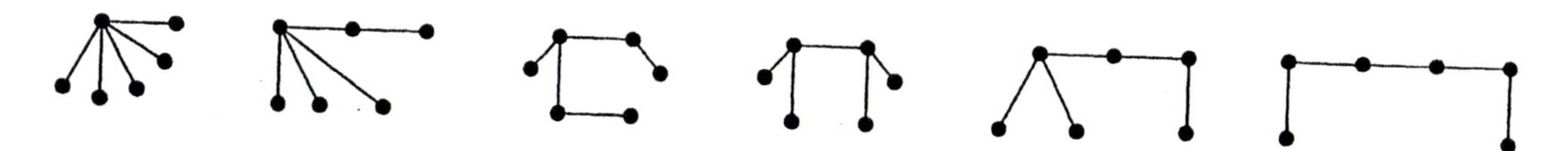

(b) Consider just the carbon atoms.

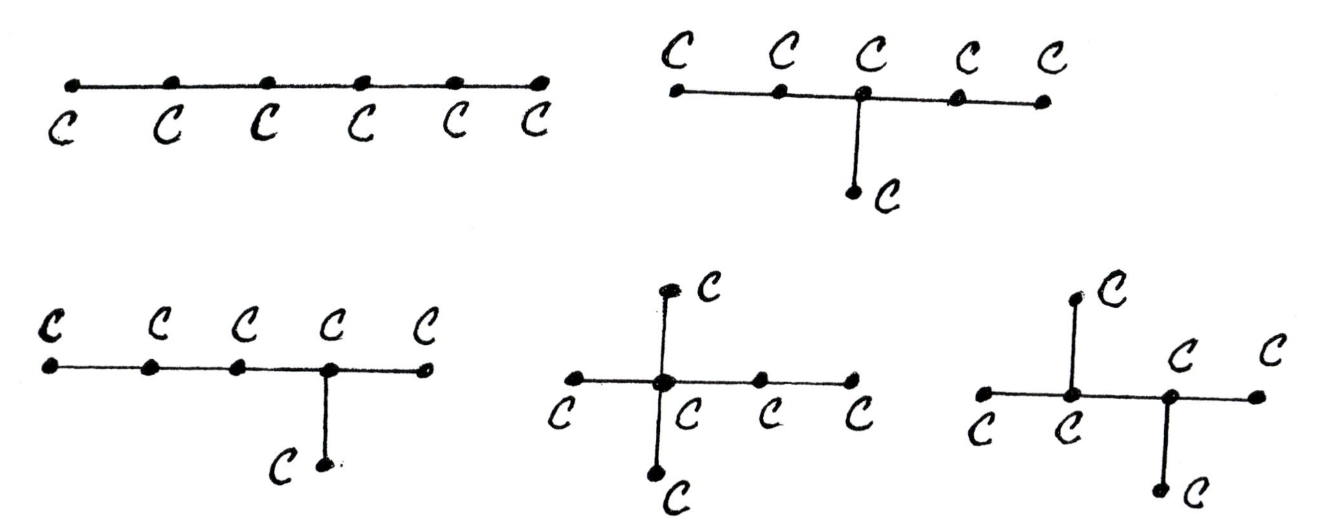

Now add the 14 hydrogen atoms to each "tree" so that each carbon atom (considered as a tree vertex) has degree 4. From the resulting trees we see that hexane has five isomers.

3. (a) Let $e_1, e_2, \ldots, e_7$ denote the numbers of edges for the seven trees, and let $v_1, v_2, \ldots v_7$, respectively, denote the numbers of vertices. Then $v_i = e_i + 1$, for all $1 \leq i \leq 7$, and $|V_1| = v_1 + v_2 + \ldots + v_7 = (e_1 + e_2 + \ldots + e_7) + 7 = 40 + 7 = 47$.

(b) Let n denote the number of trees in F_2. Then if e_i, v_i, $1 \leq i \leq n$, denote the numbers of edges and vertices, respectively, in these trees, it follows that $v_i = e_i + 1$, for all $1 \leq i \leq n$, and $62 = v_1+v_2+\ldots+v_n = (e_1+1)+(e_2+1)+\ldots+(e_n+1) = (e_1+e_2+\ldots+e_n)+n = 51+n$, so $n = 62 - 51 = 11$ trees in F_2.

5. A path is a tree with only two pendant vertices.

7. 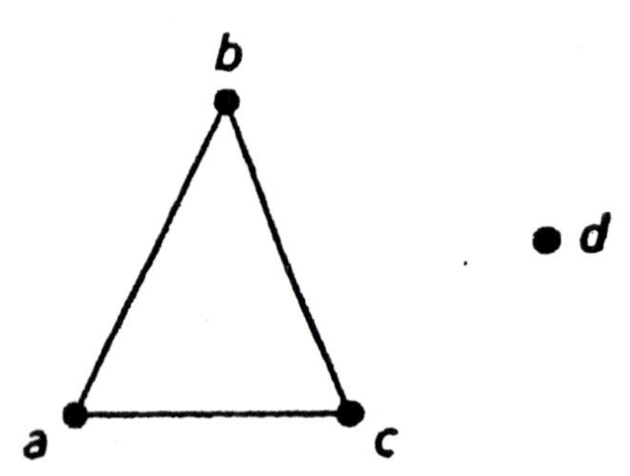

9. If there is a (unique) path between each pair of vertices in G then G is connected. If G contains a cycle then there is a pair of vertices x, y with two distinct paths connecting x and y. Hence, G is a loop-free connected undirected graph with no cycles, so G is a tree.

11. Since T is a tree, there is a unique path connecting any two distinct vertices of T. Hence there are $\binom{n}{2}$ distinct paths in T.

13. (a) In part (i) of the given figure we find the complete bipartite graph $K_{2,3}$. Parts (ii) and (iii) of the figure provide two nonisomorphic spanning trees for $K_{2,3}$.
(b) Up to isomorphism these are the only spanning trees for $K_{2,3}$.

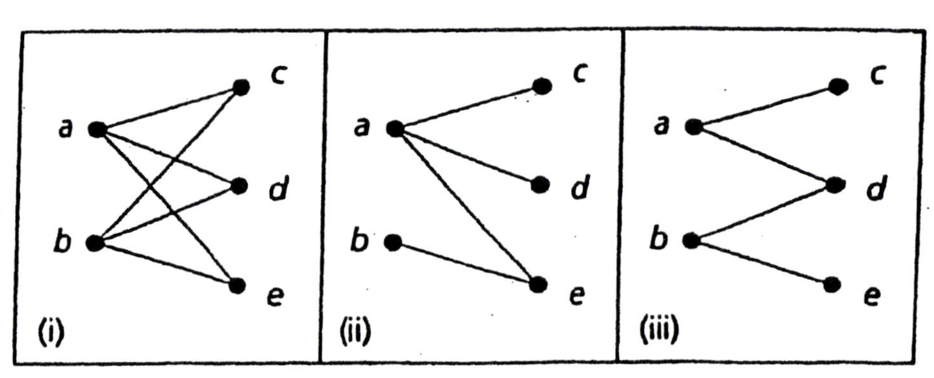

15. (a) 6: Any one of the six spanning trees for C_6 (the cycle on six vertices) together with the path connecting f to k.
(b) $6 \cdot 6 = 36$

17. (a) $n \geq m + 1$
(b) Let k be the number of pendant vertices in T. From Theorem 11.2 and Theorem 12.3 we have

$$2(n-1) = 2|E| = \sum_{v \in V} \deg(v) \geq k + m(n-k).$$

Consequently, $[2(n-1) \geq k + m(n-k)] \Rightarrow [2n - 2 \geq k + mn - mk] \Rightarrow [k(m-1) \geq 2 - 2n + mn = 2 + (m-2)n \geq 2 + (m-2)(m+1) = 2 + m^2 - m - 2 = m^2 - m = m(m-1)]$, so $k \geq m$.

19. (a) If the complement of T contains a cut set, then the removal of these edges disconnects G and there are vertices x, y with no path connecting them. Hence T is not a spanning tree for G.

(b) If the complement of C contains a spanning tree, then every pair of vertices in G has a path connecting them and this path includes no edges of C. Hence the removal of the edges in C from G does not disconnect G, so C is not a cut set for G.

21. (a) (i) 3,4,6,3,8,4 (ii) 3,4,6,6,8,4

(b) No pendant vertex of the given tree appears in the sequence so the result is true for these vertices. When an edge $\{x,y\}$ is removed and y is a pendant vertex (of the tree or one of the resulting subtrees), then the $\deg(x)$ is decreased by 1 and x is placed in the sequence. As the process continues either (i) this vertex x becomes a pendant vertex in a subtree and is removed but not recorded in the sequence, or (ii) the vertex x is left as one of the last two vertices of an edge. In either case x has been listed in the sequence $(\deg(x)-1)$ times.

(c)

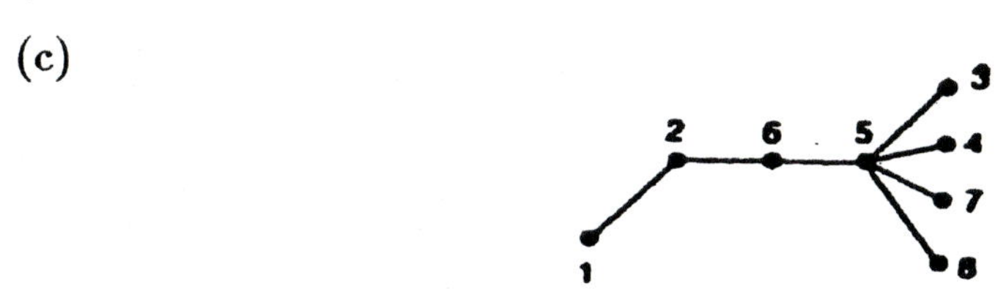

(d) Input: The given Prüfer code $x_1, x_2, \ldots, x_{n-2}$.
Output: The unique tree T with n vertices labeled with $1, 2, \ldots, n$. (This tree T has the Prüfer code $x_1, x_2, \ldots, x_{n-2}$.)

```
C := [x_1, x_2, ..., x_{n-2}]     {Initializes C as a list (ordered set).}
L := [1, 2, ..., n]               {Initializes L as a list (ordered set).}
T := ∅
for i := 1 to n − 2 do
    v := smallest element in L not in C
    w := first entry in C
    T := T ∪ {{v, w}}             {Add the new edge {v, w} to the present forest.}
    delete v from L
    delete the first occurrence of w from C
T := T ∪ {{y, z}}                 {The vertices y, z are the last two remaining entries in L.}
```

23. (a) If the tree contains $n+1$ vertices then it is (isomorphic to) the complete bipartite graph $K_{1,n}$ – often called the *star* graph.
(b) If the tree contains n vertices then it is (isomorphic to) a path on n vertices.

25. Let $E_1 = \{\{a,b\},\{b,c\},\{c,d\},\{d,e\},\{b,h\},\{d,i\},\{f,i\},\{g,i\}\}$ and
$E_2 = \{\{a,h\},\{b,i\},\{h,i\},\{g,h\},\{f,g\},\{c,i\},\{d,f\},\{e,f\}\}$.

Section 12.2

1.

(a) f,h,k,p,q,s,t	(b) a	(c) d
(d) e,f,j,q,s,t	(e) q,t	(f) 2
(g) k,p,q,s,t		

3. (a) $1 + w - x\ y * \pi \uparrow z\ 3$
(b) $/ \uparrow a - b\ c + d * e\ f$ results in $a^{(b-c)}/(d+(e*f))$. With $a = c = d = e = 2$ and $b = f = 4$, $a^{(b-c)}/(d+(e*f)) = 2^2/(2+8) = 4/10 = 0.4$.

5. Preorder: r,j,h,g,e,d,b,a,c,f,i,k,m,p,s,n,q,t,v,w,u
Inorder: h,e,a,b,d,c,g,f,j,i,r,m,s,p,k,n,v,t,w,q,u
Postorder: a,b,c,d,e,f,g,h,i,j,s,p,m,v,w,t,u,q,n,k,r

7. (a)
(i) & (iii) (ii)

(b)
(i) (ii) (iii)

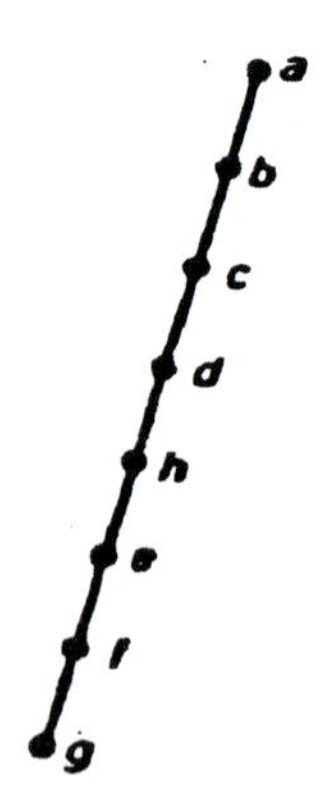

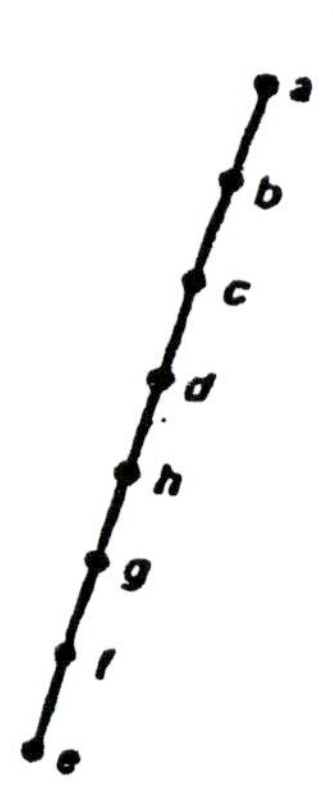

9.

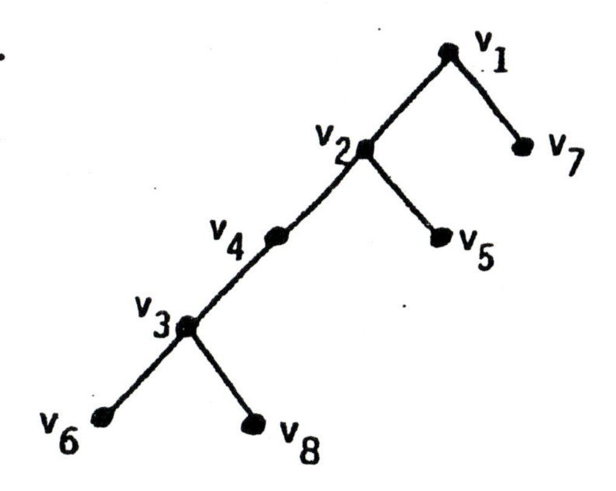

G is connected.

11. Theorem 12.6

(a) Each internal vertex has m children so there are mi vertices that are the children of some other vertex. This accounts for all vertices in the tree except the root. Hence $n = mi + 1$

(b) $\ell + i = n = mi + 1 \Longrightarrow \ell = (m-1)i + 1$

(c) $\ell = (m-1)i + 1 \Longrightarrow i = (\ell - 1)/(m-1)$

$n = mi + 1 \Longrightarrow i = (n-1)/m$.

(Corollary 12.1)

Since the tree is balanced $m^{h-1} < \ell \leq m^h$ by Theorem 12.7.

$m^{h-1} < \ell \leq m^h \Longrightarrow \log_m(m^{h-1}) < \log_m(\ell) \leq \log_m(m^h) \Longrightarrow$

$(h-1) < \log_m \ell \leq h \Longrightarrow h = \lceil \log_m \ell \rceil$.

13. (a) From part (a) of Theorem 12.6 we have $|V|$ = number of vertices in $T = 3i + 1 = 3(34) + 1 = 103$. So T has $103 - 1 = 102$ edges. From part (b) of the same theorem we find that the number of leaves in T is $(3-1)(34) + 1 = 69$. [We can also obtain the number of leaves as $|V| - i = 103 - 34 = 69$.]

(b) It follows from part (c) of Theorem 12.6 that the given tree has $(817-1)/(5-1) = 816/4 = 204$ internal vertices.

15. (a)

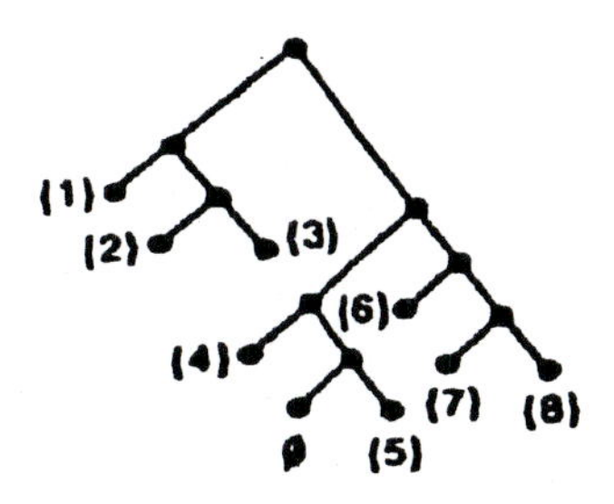

(b) 9; 5

(c) $h(m-1)$; $(h-1) + (m-1)$

17. 21845; $1 + m + m^2 + \ldots + m^{h-1} = (m^h - 1)/(m - 1)$.

19.

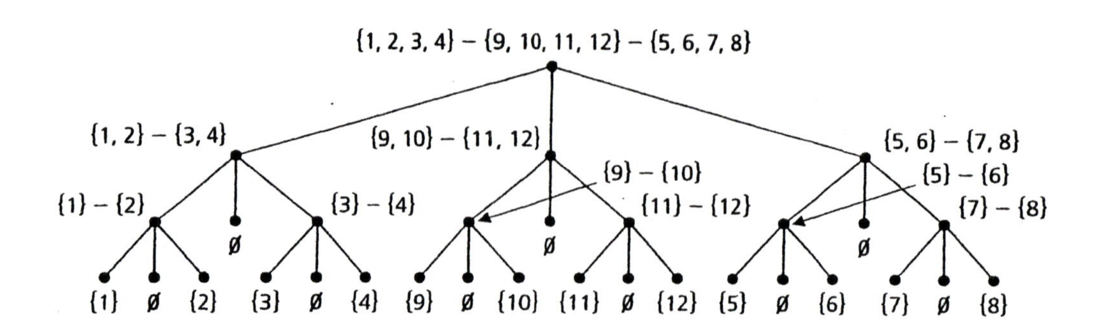

21. Let T be a complete binary tree with 31 vertices. The left and right subtrees of T are then *complete binary trees* on $2k+1$ and $30-(2k+1)$ vertices, respectively, with $0 \leq k \leq 14$.

The number of ways the left subtree can have $11(= 2\cdot5+1)$ vertices is $(\frac{1}{6})\binom{10}{5}$. This leaves $19(= 2\cdot9+1)$ vertices for the right subtree where there are $(\frac{1}{10})\binom{18}{9}$ possibilities. So by the rule of product there are $(\frac{1}{6})\binom{10}{5}(\frac{1}{10})\binom{18}{9} = 204,204$ complete binary trees on 31 vertices with 11 vertices in the left subtree of the root. A similar argument tells us that there are $(\frac{1}{11})\binom{20}{10}(\frac{1}{5})\binom{8}{4} = 235,144$ complete binary trees on 31 vertices with 21 vertices in the right subtree of the root.

23. (a) 1,2,5,11,12,13,14,3,6,7,4,8,9,10,15,16,17
(b) The preorder traversal of the rooted tree.

Section 12.3

1. (a) $L_1: 1,3,5,7,9$ $\quad L_2: 2,4,6,8,10$
(b) $L_1: \; 1,3,5,7,\ldots,2m-3, m+n$
$L_2: \; 2,4,6,8,\ldots,2m-2,2m-1,2m,2m+1,\ldots,m+n-1$

3. (a)

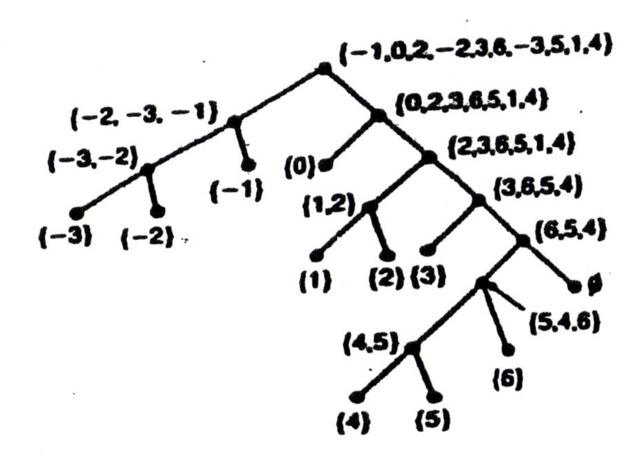

(b)

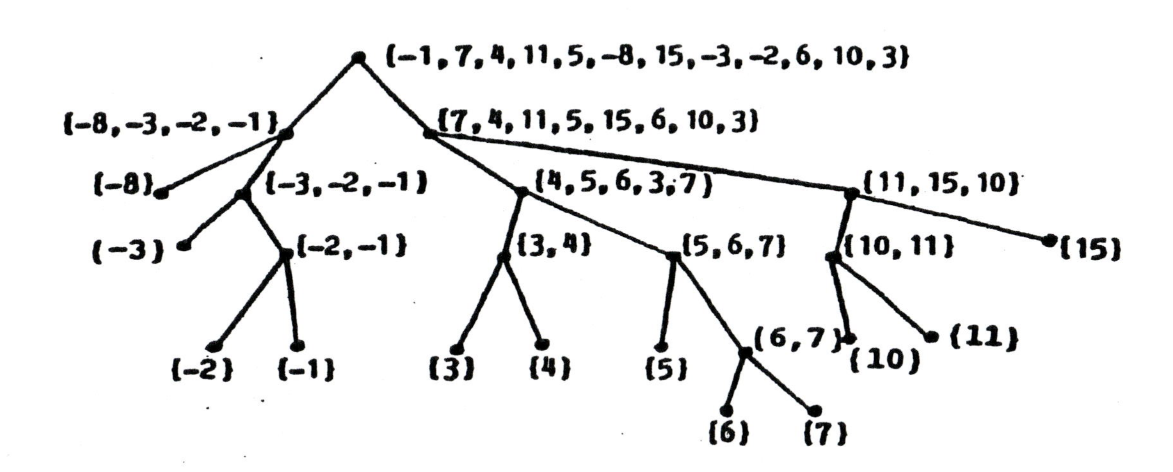

Section 12.4

1. (a) tear (b) tatener (c) rant

3.

a:	111	e:	10	h:	010
b:	110101	f:	0111	i:	00
c:	0110	g:	11011	j:	110100
d:	0001				

5. Since the tree has $m^7 = 279,936$ leaves, it follows that $m = 6$. From part(c) of Theorem 12.6 we find that there are $(m^7 - 1)/(m - 1) = (279,935)/5 = 55,987$ internal vertices.

7.

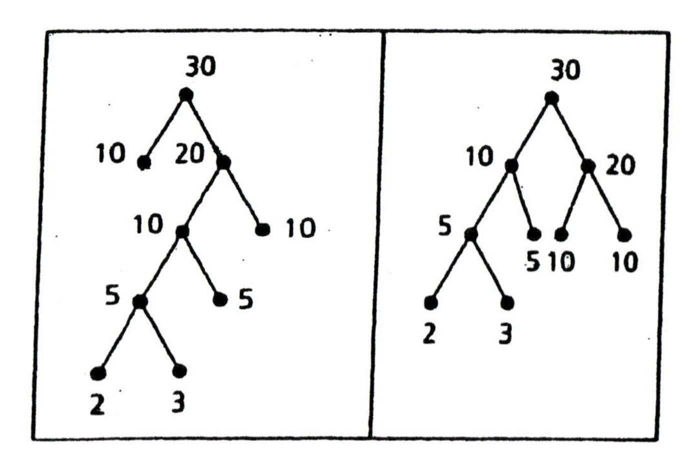

Amend part (a) of Step 2 for the Huffman tree algorithm as follows. If there are $n(> 2)$ such trees with smallest root weights w and w', then

(i) if $w < w'$ and $n-1$ of these trees have root weight w', select a tree (of root weight w') with smallest height; and

(ii) if $w = w'$ (and all n trees have the same smallest root weight), select two trees (of root weight w) of smallest height.

Section 12.5

1. The articulation points are b, e, f, h, j, k. The biconnected components are
$B_1 : \{\{a, b\}\}$; $B_2 : \{\{d, e\}\}$;
$B_3 : \{\{b, c\}, \{c, f\}, \{f, e\}, \{e, b\}\}$; $B_4 : \{\{f, g\}, \{g, h\}, \{h, f\}\}$;
$B_5 : \{\{h, i\}, \{i, j\}, \{j, h\}\}$; $B_6 : \{\{j, k\}\}$;
$B_7 : \{\{k, p\}, \{p, n\}, \{n, m\}, \{m, k\}, \{p, m\}\}$.

3. (a) T can have as few as one or as many as $n-2$ articulation points. If T contains a vertex of degree $(n-1)$, then this vertex is the only articulation point. If T is a path with n vertices and $n-1$ edges, then the $n-2$ vertices of degree 2 are all articulation points.
(b) In all cases, a tree on n vertices has $n-1$ biconnected components. Each edge is a biconnected component.

5. $\chi(G) = \max\{\chi(B_i) | 1 \leq i \leq k\}$.

7. Proof: Suppose that G has a pendant vertex, say x, and that $\{w, x\}$ is the (unique) edge in E incident with x. Since $|V| \geq 3$ we know that $\deg(w) \geq 2$ and that $\kappa(G - w) \geq 2 > 1 = \kappa(G)$. Consequently, w is an articulation point of G.

9.

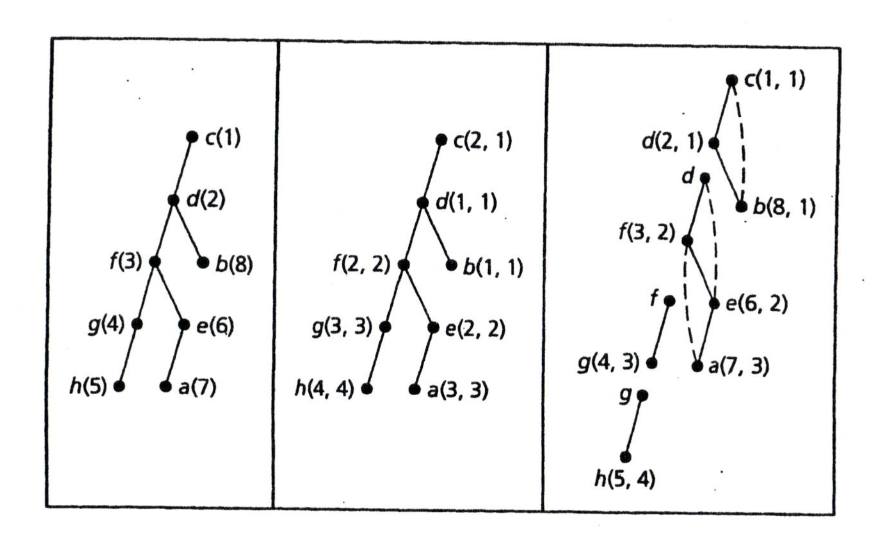

(a) The first tree provides the depth-first spanning tree T for G where the order prescribed for the vertices is reverse alphabetical and the root is c.
(b) The second tree provides $(\text{low}'(v), \text{low}(v))$ for each vertex v of G (and T). These results follow from step (2) of the algorithm.

For the third tree we find $(\text{dfi}(v), \text{low}(v))$ for each vertex v. Applying step (3) of the algorithm we find the articulation points d, f, and g, and the four biconnected components.

11. No! For any loop-free connected undirected graph $G = (V, E)$ where $|V| \geq 2$, we have $\text{low}(x_1) = \text{low}(x_2) = 1$. (Note: Vertices x_1 and x_1 are always on the same biconnected component.)

13. Proof: If not, let $v \in V$ where v is an articulation point of G. Then $\kappa(G - v) > \kappa(G) = 1$. (From Exercise 19 of Section 11.6 we know that G is connected.) Now $G - v$ is disconnected with components $H_1, H_2, \ldots, H_t$, for $t \geq 2$. For $1 \leq i \leq t$, let $v_i \in H_i$. Then $H_i + v$ is a subgraph of $G - v_{i+1}$, and $\chi(H_i + v) \leq \chi(G - v_{i+1}) < \chi(G)$. (Here $v_{t+1} = v_1$.) Now let $\chi(G) = n$ and let $\{c_1, c_2, \ldots, c_n\}$ be a set of n colors. For each subgraph $H_i + v$, $1 \leq i \leq t$, we can properly color the vertices of $H_i + v$ with at most $n - 1$ colors — and can use c_1 to color vertex v for all of these t subgraphs. Then we can join these t subgraphs together at vertex v and obtain a proper coloring for the vertices of G where we use less than $n(= \chi(G))$ colors.

Supplementary Exercises

1. If G is a tree, consider G as a rooted tree. Then there are λ choices for coloring the root of G and $(\lambda - 1)$ choices for coloring each of its descendants. The result then follows by the rule of product.

Conversely, if $P(G, \lambda) = \lambda(\lambda - 1)^{n-1}$, then since the factor λ only occurs once, the graph G is connected. $P(G, \lambda) = \lambda(\lambda - 1)^{n-1} = \lambda^n - (n-1)\lambda^{n-1} + \ldots + (-1)^{n-1}\lambda \Longrightarrow G$ has n vertices and $(n-1)$ edges. Therefore by part (d) of Theorem 12.5, G is a tree.

3. (a) 1011001010100

(b) (i) (ii)

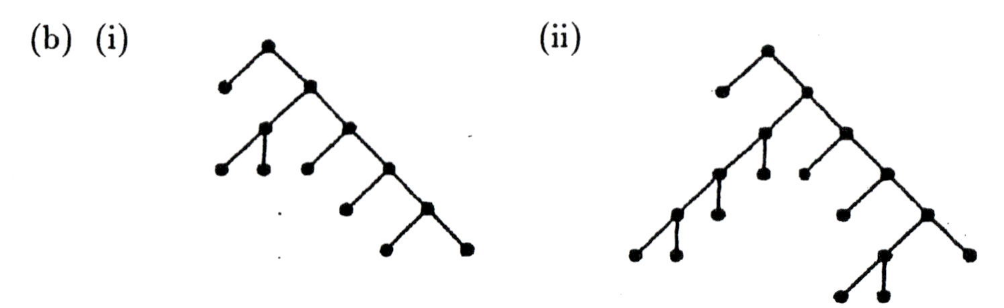

(c) Since the last two vertices visited in a preorder traversal are leaves, the last two symbols in the characteristic sequence of every complete binary tree are 00.

5. We assume that $G = (V, E)$ is connected – otherwise we work with a component of G. Since G is connected, and $\deg(v) \geq 2$ for all $v \in V$, it follows from Theorem 12.4 that G is not a tree. But every connected graph that is not a tree must contain a cycle.

7. For $1 \leq i(< n)$, let $x_i =$ the number of vertices v where $\deg(v) = i$. Then $x_1 + x_2 + \ldots + x_{n-1} = |V| = |E| + 1$, so $2|E| = 2(-1 + x_1 + x_2 + \ldots + x_{n-1})$. But $2|E| = \sum_{v \in V} \deg(v) = (x_1 + 2x_2 + 3x_3 + \ldots + (n-1)x_{n-1})$. Solving $2(-1 + x_1 + x_2 + \ldots + x_{n-1}) = x_1 + 2x_2 + \ldots + (n-1)x_{n-1}$ for x_1, we find that $x_1 = 2 + x_3 + 2x_4 + 3x_5 + \ldots + (n-3)x_{n-1} = 2 + \sum_{\deg(v_i) \geq 3}[\deg(v_i) - 2]$.

9. (a) G^2 is isomorphic to K_5.
(b) G^2 is isomorphic to K_4.
(c) G^2 is isomorphic to K_{n+1}, so the number of new edges is $\binom{n+1}{2} - n = \binom{n}{2}$.

(d) If G^2 has an articulation point x, then there exists $u, v \in V$ such that every path (in G^2) from u to v passes through x. (This follows from Exercise 2 of Section 12.5.) Since G is connected, there exists a path P (in G) from u to v. If x is not on this path (which is also a path in G^2), then we contradict x being an articulation point in G^2. Hence the path P (in G) passes through x, and we can write $P: u \rightarrow u_1 \rightarrow \ldots \rightarrow u_{n-1} \rightarrow u_n \rightarrow x \rightarrow v_m \rightarrow v_{m-1} \rightarrow \ldots \rightarrow v_1 \rightarrow v$. But then in G^2 we add the edge $\{u_n, v_m\}$, and the path P' (in G^2) given by $P': u \rightarrow u_1 \rightarrow \ldots \rightarrow u_{n-1} \rightarrow u_n \rightarrow v_m \rightarrow v_{m-1} \rightarrow \ldots \rightarrow v_1 \rightarrow v$ does not pass through x. So x is not an articulation point of G^2, and G^2 has no articulation points.

11. (a) $\ell_n = \ell_{n-1} + \ell_{n-2}$, for $n \geq 3$ and $\ell_1 = \ell_2 = 1$. Since this is precisely the Fibonacci recurrence relation, we have $l_n = F_n$, the nth Fibonacci number, for $n \geq 1$.
(b) $i_n = i_{n-1} + i_{n-2} + 1$, $n \geq 3$, $i_1 = i_2 = 0$. The summand "+1" arises when we count the root, an internal vertex.
(Homogeneous part of solution):

$$i_n^{(h)} = i_{n-1}^{(h)} + i_{n-2}^{(h)},\ n \geq 3$$

$$i_n^{(h)} = A\alpha^n + B\beta^n, \text{ where } \alpha = (1+\sqrt{5})/2 \text{ and } \beta = (1-\sqrt{5})/2.$$

(Particular part of solution):

$$i_n^{(p)} = C, \text{ a constant}$$

Upon substitution into the recurrence relation $i_n = i_{n-1} + i_{n-2} + 1$, $n \geq 3$, we find that

$$C = C + C + 1,$$

so $C = -1$,
and $i_n = A\alpha^n + B\beta^n - 1$.
With $i_1 = i_2 = 0$ we have

$$0 = i_1 = A\alpha + B\beta - 1$$

$$0 = i_2 = A\alpha^2 + B\beta^2 - 1,$$

and consequently,
$B = (\alpha - 1)/[\beta(\alpha - \beta)] = [((1+\sqrt{5})/2) - 1]/[((1-\sqrt{5})/2)(\sqrt{5})] =$
$[1 + \sqrt{5} - 2]/[(1-\sqrt{5})(\sqrt{5})] = -1/\sqrt{5}$, and $A = [1 - B\beta]/\alpha = 1/\sqrt{5}$. Therefore,

$$i_n = (1/\sqrt{5})\alpha^n - (1/\sqrt{5})\beta^n - 1 = F_n - 1,$$

where F_n denotes the nth Fibonacci number, for $n \geq 1$.
(c) $v_n = \ell_n + i_n$, for all $n \in \mathbf{Z}^+$. Consequently, $v_n = F_n + F_n - 1 = 2F_n - 1$, where, as in parts (a) and (b), F_n denotes the nth Fibonacci number.

13. (a) For the spanning trees of G there are two mutually exclusive and exhaustive cases:
(i) The edge $\{x_1, y_1\}$ is in the spanning tree: These spanning trees are counted in b_n.
(ii) The edge $\{x_1, y_1\}$ is not in the spanning tree: In this case the edges $\{x_1, x_2\}$, $\{y_1, y_2\}$ are both in the spanning tree. Upon removing the edges $\{x_1, x_2\}, \{y_1, y_2\}$, and $\{x_1, y_1\}$, from the original ladder graph, we now need a spanning tree for the resulting smaller ladder graph with $n - 1$ rungs. There are a_{n-1} spanning trees in this case.

(b) Here there are three mutually exclusive and exhaustive cases:
(i) The edges $\{x_1, x_2\}$ and $\{y_1, y_2\}$ are both in the spanning tree: Delete $\{x_1, x_2\}$, $\{y_1, y_2\}$, and $\{x_1, y_1\}$ from the graph. Then b_{n-1} counts those spanning trees for ladders with $n - 1$ rungs where $\{x_2, y_2\}$ is included. For each of these delete $\{x_2, y_2\}$ and add $\{x_1, x_2\}$, $\{y_1, y_2\}$ and $\{x_1, y_1\}$.
(ii) The edge $\{x_1, x_2\}$ is in the spanning tree but the edge $\{y_1, y_2\}$ is not: Now the removal of the edges $\{x_1, y_1\}, \{x_1, x_2\}$, and $\{y_1, y_2\}$ from G results in a subgraph that is a ladder graph on $n - 1$ rungs. This subgraph has a_{n-1} spanning trees.
(iii) Here the edge $\{y_1, y_2\}$ is in the spanning tree but the edge $\{x_1, x_2\}$ is not: As in case (ii) there are a_{n-1} spanning trees.

On the basis of the preceding argument we have $b_n = b_{n-1} + 2a_{n-1}$, $n \geq 2$.

(c) $a_n = a_{n-1} + b_n$
$b_n = b_{n-1} + 2a_{n-1}$
$a_n = a_{n-1} + b_{n-1} + 2a_{n-1} = 3a_{n-1} + b_{n-1}$
$b_n = a_n - a_{n-1}$, so $b_{n-1} = a_{n-1} - a_{n-2}$
$a_n = 3a_{n-1} + a_{n-1} - a_{n-2} = 4a_{n-1} - a_{n-2}$, $n \geq 3$, $a_1 = 1$, $a_2 = 4$
$a_n - 4a_{n-1} + a_{n-2} = 0$
$r^2 - 4r + 1 = 0$
$r = (1/2)(4 \pm \sqrt{16-4}) = 2 \pm \sqrt{3}$
So $a_n = A(2+\sqrt{3})^n + B(2-\sqrt{3})^n$
$a_0 = 0 \Longrightarrow A + B = 0 \Longrightarrow B = -A$.
$a_1 = 1 = A(2+\sqrt{3}) - A(2-\sqrt{3}) = 2A\sqrt{3} \Longrightarrow A = 1/2\sqrt{3}$ and $B = -1/2\sqrt{3}$.
Therefore $a_n = (1/(2\sqrt{3}))[(2+\sqrt{3})^n - (2-\sqrt{3})^n], n \geq 0$.

15. (a) (i) 3 (ii) 5
(b) $a_n = a_{n-1} + a_{n-2}$, $n \geq 5$, $a_3 = 2$, $a_4 = 3$.
$a_n = F_{n+1}$, the $(n+1)$st Fibonacci number.

17. Here the input consists of
(a) the k (≥ 3) vertices of the spine – ordered from left to right as $v_1, v_2, \ldots, v_k$;
(b) $\deg(v_i)$, in the caterpillar, for all $1 \leq i \leq k$; and
(c) n, the number of vertices in the caterpillar, with $n \geq 3$.

If $k = 3$, the caterpillar is the complete bipartite graph (or, star) $K_{1,n-1}$, for some $n \geq 3$. We label v_1 with 1 and the remaining vertices with $2, 3, \ldots, n$. This provides the edge labels (the absolute value of the difference of the vertex labels) $1, 2, 3, \ldots, n-1$ – a graceful labeling.

For $k > 3$ we consider the following.
$\ell := 2$ {ℓ is the largest low label}
$h := n - 1$ {h is the smallest high label}
label v_1 with 1
label v_2 with n
for $i := 2$ **to** $k-1$ **do**
  if $2\lfloor i/2 \rfloor = i$ **then** {i is even}
    begin
      if v_i has unlabeled leaves that are not on the spine **then**
        assign the $\deg(v_i) - 2$ labels from ℓ
        to $\ell + \deg(v_i) - 3$ to these leaves of v_i
      assign the label $\ell + \deg(v_i) - 2$ to v_{i+1}
      $\ell := \ell + \deg(v_i) - 1$
    end
  else
    begin

if v_i has unlabeled leaves that are not on the spine **then**
assign the $\deg(v_i) - 2$ labels from $h - [\deg(v_i) - 3]$
to h to these leaves of v_i
assign the label $h - \deg(v_i) + 2$ to v_{i+1}
$h := h - \deg(v_i) + 1$

end

19. (a) $1, -1, 1, 1, -1, -1$ $\quad$ $1, 1, -1, 1, -1, -1$ $\quad$ $1, -1, 1, -1, 1, -1$

(b)

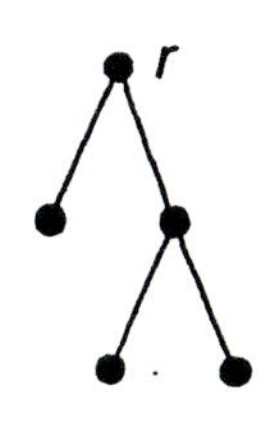

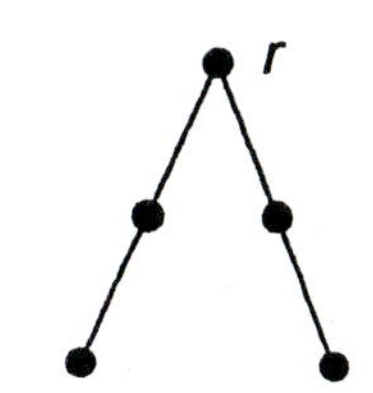

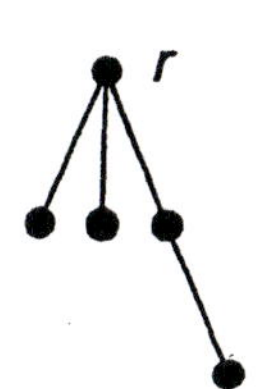

In total there are 14 ordered rooted trees on five vertices.

(c) This is another example where the Catalan numbers arise. There are $(\frac{1}{n+1})\binom{2n}{n}$ ordered rooted trees on $n+1$ vertices.

21. (a) There are $\binom{5}{3} - 2 = 8$ nonidentical (though some are isomorphic) spanning trees for the kite induced by a, b, c, d. Since there are four vertices, a spanning tree has three edges and the only selections of three edges that do not provide a spanning tree are $\{a, c\}, \{b, c\}, \{a, b\}$ and $\{a, b\}, \{a, d\}, \{b, d\}$.

(b) There are $8 \cdot 1 \cdot 8 \cdot 1 \cdot 8 \cdot 1 \cdot 8 = 8^4$ nonidentical (though some are isomorphic) spanning trees of G that do not contain edge $\{c, h\}$. These spanning trees must include the edges $\{g, k\}, \{l, p\}$, and $\{d, o\}$, and there are eight nonidentical (though some are isomorphic) spanning trees for each of the four subgraphs that are kites.

(c) Consider the kite induced by a, b, c, d. There are eight two-tree forests for this kite that have no path between c and d. These forests can be obtained from the five edges of the kite by removing three edges at a time, as follows:

(i) $\{a, b\}, \{a, c\}, \{b, c\}$
(ii) $\{a, c\}, \{b, c\}, \{b, d\}$
(iii) $\{a, c\}, \{a, d\}, \{b, c\}$
(iv) $\{a, b\}, \{a, d\}, \{b, d\}$
(v) $\{a, d\}, \{b, c\}, \{b, d\}$
(vi) $\{a, c\}, \{a, d\}, \{b, d\}$
(vii) $\{a, b\}, \{a, d\}, \{b, c\}$
(viii) $\{a, b\}, \{a, c\}, \{b, d\}$

Vertex c is isolated for (i), (ii), (iii). For (iv), (v), (vi), vertex d is isolated. The forests for (vii), (viii) each contain two disconnected edges: $\{a, c\}, \{b, d\}$ for (vii) and $\{a, d\}, \{b, c\}$ for (viii).

Consequently, there are $4 \cdot 8 \cdot 1 \cdot 8 \cdot 1 \cdot 8 \cdot 1 \cdot 8 \cdot 1 = 4 \cdot 8^4$ nonidentical (though some are isomorphic) spanning trees for G that contain each of the four edges $\{c, h\}, \{g, k\}, \{l, p\}$, and $\{d, o\}$.

(d) In total there are $4 \cdot 8^4 + 4 \cdot 8^4 = 2(4 \cdot 8^4)$ nonidentical (though some are isomorphic) spanning trees for G.

(e) $2n8^n$

CHAPTER 13
OPTIMIZATION AND MATCHING

Section 13.1

1. (a) If not, let $v_i \in \overline{S}$, where $1 \leq i \leq m$ and i is the smallest such subscript. Then $d(v_0, v_i) < d(v_0, v_{m+1})$, and we contradict the choice of v_{m+1} as a vertex v in $\overline{S}$ for which $d(v_0, v)$ is a minimum.
(b) Suppose there is a shorter directed path (in G) from v_0 to v_k. If this path passes through a vertex in $\overline{S}$, then from part (a) we have a contradiction. Otherwise, we have a shorter directed path P'' from v_0 to v_k and P'' only passes through vertices in S. But then $P'' \cup \{(v_k, v_{k+1}), (v_{k+1}, v_{k+2}), \ldots, (v_{m-1}, v_m), (v_m, v_{m+1})\}$ is a directed path (in G) from v_0 to v_{m+1}, and it is shorter than path P.

3. (a) $d(a,b) = 5$; $d(a,c) = 6$; $d(a,f) = 12$; $d(a,g) = 16$; $d(a,h) = 12$
(b) $f: \{(a,c),(c,f)\}$ $g: \{(a,b),(b,h),(h,g)\}$
$h: \{(a,b),(b,h)\}$

5. False – consider the weighted graph

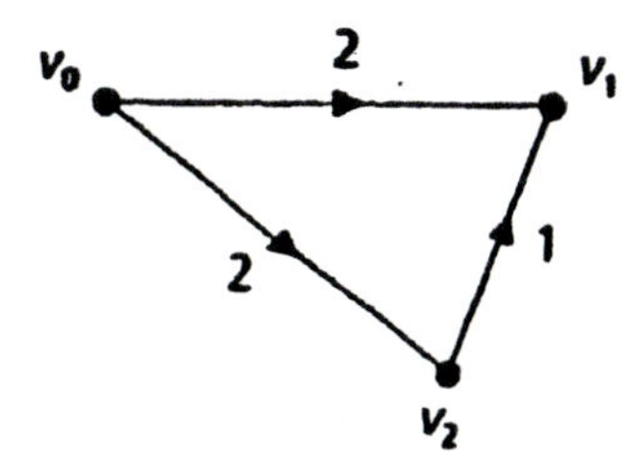

Section 13.2

1. Kruskal's Algorithm generates the following sequence (of forests) which terminates in a minimal spanning tree T of weight 18:

(1) $F_1 = \{\{e,h\}\}$, (2) $F_2 = F_1 \cup \{\{a,b\}\}$,
(3) $F_3 = F_2 \cup \{\{b,c\}\}$, (4) $F_4 = F_3 \cup \{\{d,e\}\}$,
(5) $F_5 = F_4 \cup \{\{e,f\}\}$, (6) $F_6 = F_5 \cup \{\{a,e\}\}$,

(7) $F_7 = F_6 \cup \{\{d, g\}\}$, (8) $F_8 = T = F_7 \cup \{\{f, i\}\}$.

Note: The answer given here is not unique.

3. No! Consider the following counterexample:

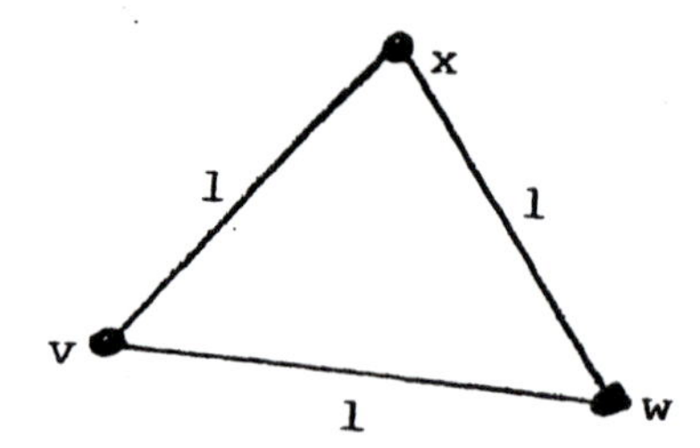

Here $V = \{v, x, w\}$, $E = \{\{v, x\}, \{x, w\}, \{v, w\}\}$ and $E' = \{\{v, x\}, \{x, w\}\}$.

5. (a) Evansville – Indianapolis (168); Bloomington – Indianapolis (51); South Bend – Gary (58); Terre Haute – Bloomington (58); South Bend – Fort Wayne (79); Indianapolis – Forth Wayne (121).

(b) Fort Wayne – Gary (132); Evansville – Indianapolis (168); Bloomington – Indianapolis (51); Gary – South Bend (58); Terre Haute – Bloomington (58); Indianapolis – Fort Wayne (121).

7. (a) To determine an optimal tree of maximal weight replace the two occurrences of "small" in Kruskal's Algorithm by "large".

(b) Use the edges: South Bend – Evansville (303); Fort Wayne – Evansville (290); Gary – Evansville (277); Fort Wayne – Terre Haute (201); Gary – Bloomington (198); Indianapolis – Evansville (168).

9. When the weights of the edges are all distinct, in each step of Kruskal's Algorithm a unique edge is selected.

Section 13.3

1. (a) $s = 2$; $t = 4$; $w = 5$; $x = 9$; $y = 4$ (b) 18

(c) (i) $P = \{a, b, h, d, g, i\}$; $\overline{P} = \{z\}$
(ii) $P = \{a, b, h, d, g\}$; $\overline{P} = \{i, z\}$
(iii) $P = \{a, h\}$; $\overline{P} = \{b, d, g, i, z\}$

3. **(1)**

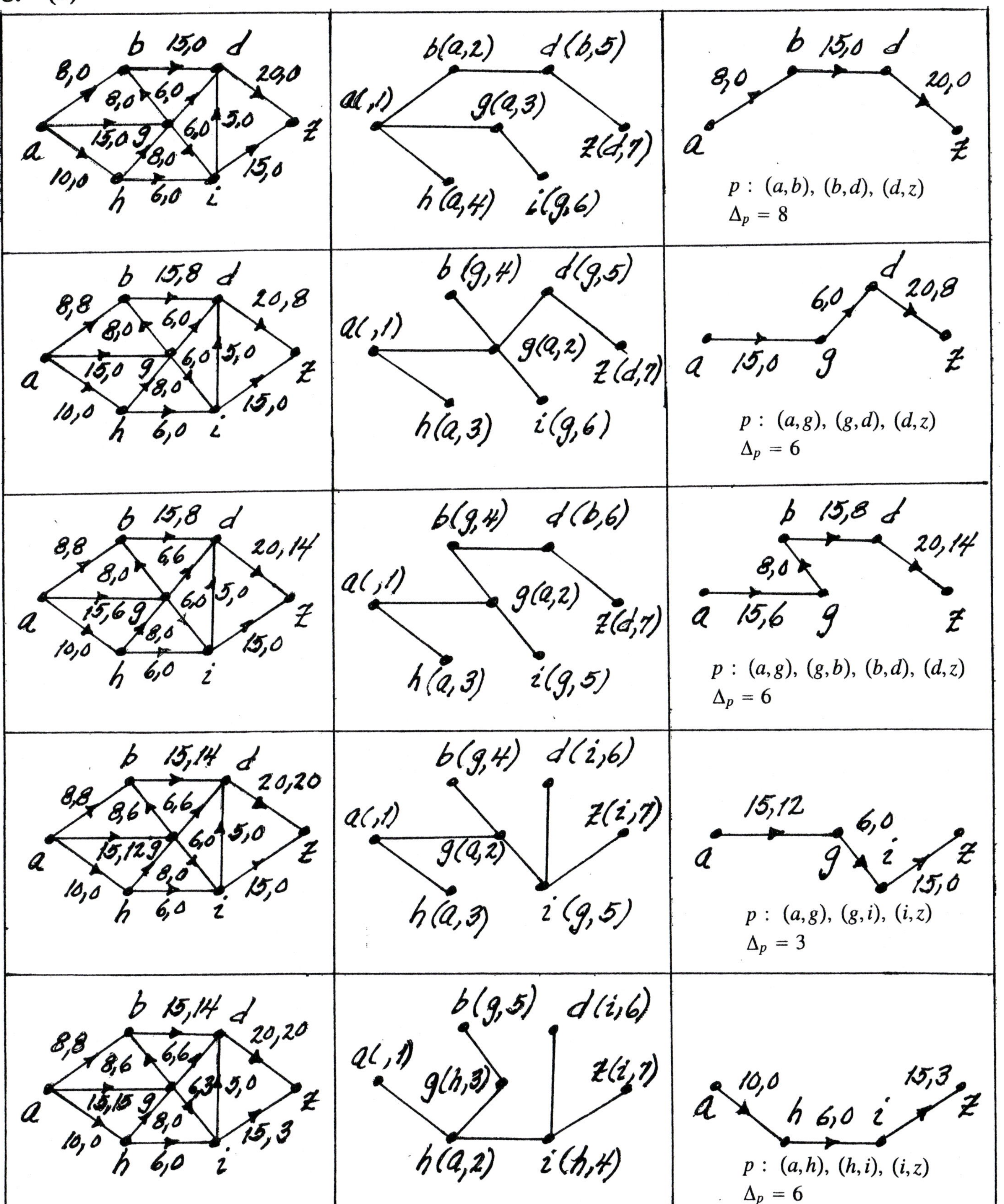

a(–,1)
b(a,2)
d(b,5)
g(a,3)
z(d,7)
h(a,4)
i(g,6)
p : (a,b), (b,d), (d,z)
Δp = 8
b(g,4)
d(g,5)
a(–,1)
g(a,2)
z(d,7)
h(a,3)
i(g,6)
p : (a,g), (g,d), (d,z)
Δp = 6
b(g,4)
d(b,6)
a(–,1)
g(a,2)
z(d,7)
h(a,3)
i(g,5)
p : (a,g), (g,b), (b,d), (d,z)
Δp = 6
b(g,4)
d(i,6)
a(–,1)
z(i,7)
g(a,2)
h(a,3)
i(g,5)
p : (a,g), (g,i), (i,z)
Δp = 3
b(g,5)
d(i,6)
a(–,1)
g(h,3)
z(i,7)
h(a,2)
i(h,4)
p : (a,h), (h,i), (i,z)
Δp = 6

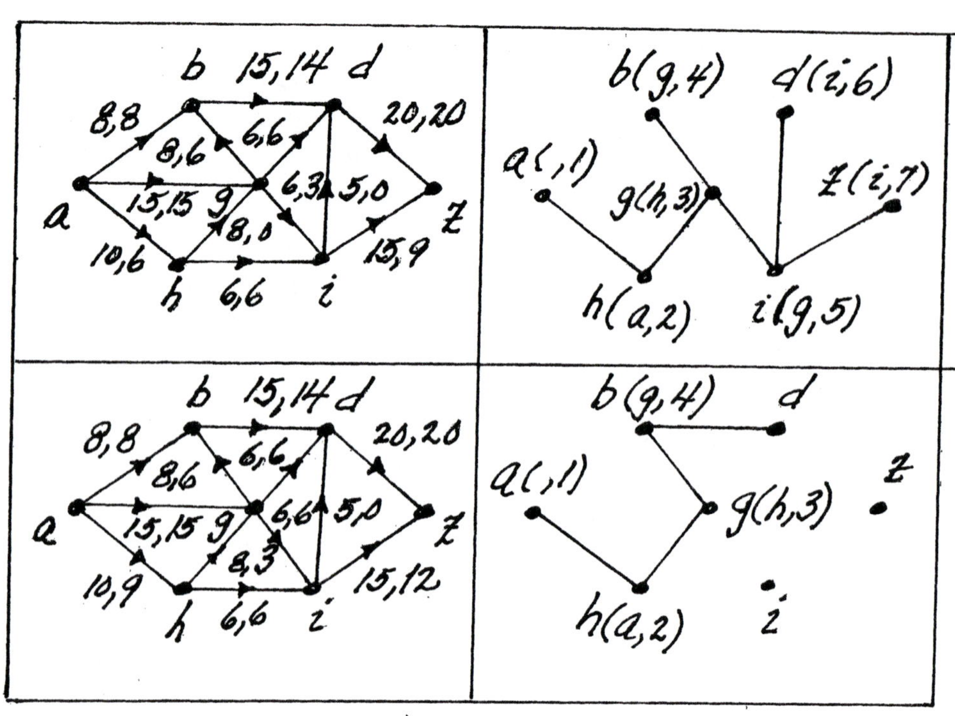

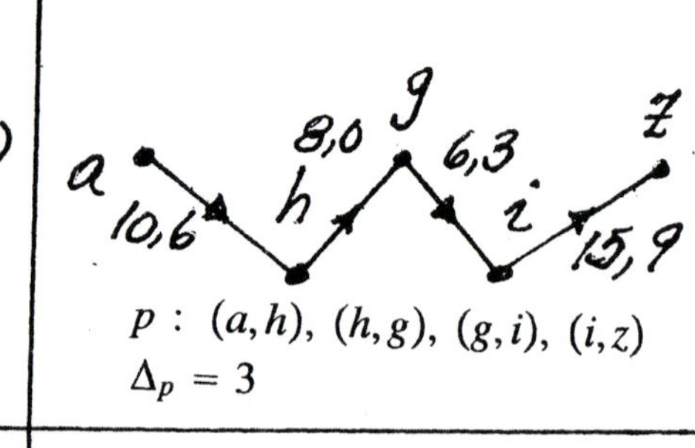

p : (a,h), (h,g), (g,i), (i,z)
$\Delta_p = 3$

The maximal flow is 32, which is $c(P, \overline{P})$ for $P = \{a, b, d, g, h\}$ and $\overline{P} = \{i, z\}$.

(2)

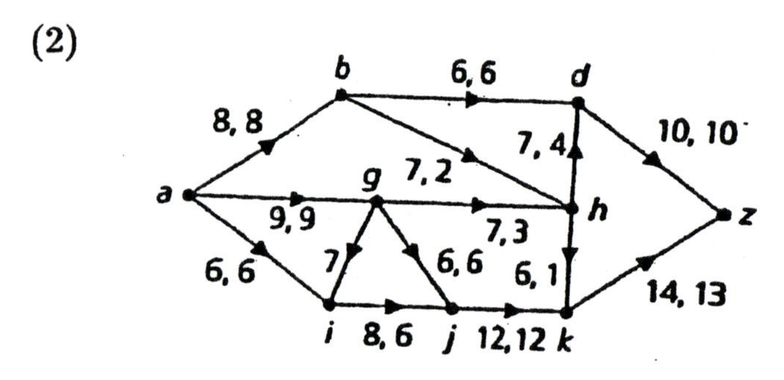

The maximal flow is 23, which is $c(P, \overline{P})$ for $P = \{a\}$ and $\overline{P} = \{b, g, i, j, d, h, k, z\}$.

5. Here $c(e)$ is a positive integer for each $e \in E$ and the initial flow is defined as $f(e) = 0$ for all $e \in E$. The result follows because Δ_p is a positive integer for each application of the Edwards-Karp algorithm and, in the Ford-Fulkerson algorithm, $f(e) - \Delta_p$ will not be negative for a backward edge.

7.

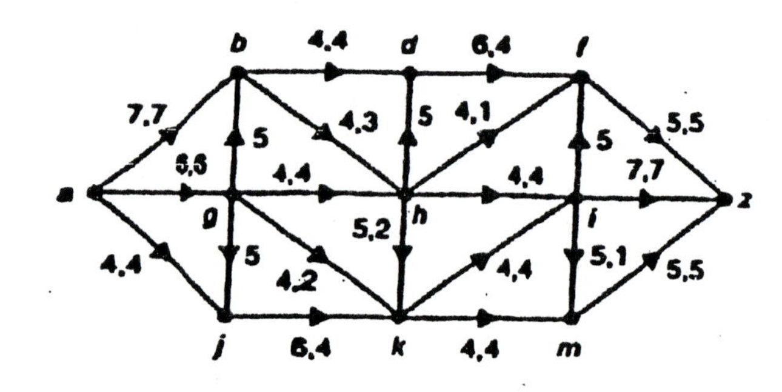

Section 13.4

1. $5/\binom{8}{4} = 1/14$

3. Let the committees be represented as $c_1, c_2, \ldots, c_6$, according to the way they are listed in the exercise.

 (a) Select the members as follows: $c_1 - A$; $c_2 - G$; $c_3 - M$; $c_4 - N$; $c_5 - K$; $c_6 - R$.

 (b) Select the nonmembers as follows: $c_1 - K$; $c_2 - A$; $c_3 - G$; $c_4 - J$; $c_5 - M$; $c_6 - P$.

5. (a) A one-factor for a graph $G = (V, E)$ consists of edges which have no common vertex. So the one-factor contains an even number of vertices, and since it spans G we must have $|V|$ even.
 (b) Consider the Petersen graph as shown in Figure 11.52 (a) of the text. The edges

$$\{e, a\} \qquad \{b, c\} \qquad \{d, i\} \qquad \{g, j\} \qquad \{f, h\}$$

 provide a one-factor for this graph.
 (c) There are $(5)(3) = 15$ one-factors for K_6.
 (d) Label the vertices of K_{2n} with $1, 2, 3, \ldots, 2n-1, 2n$. We can pair vertex 1 with any of the other $2n-1$ vertices, and we are then confronted, in the case where $n \geq 2$, with finding a one-factor for the graph K_{2n-2}. Consequently,

$$a_n = (2n-1)a_{n-1}, \qquad a_1 = 1.$$

 We find that

$$a_n = (2n-1)a_{n-1} = (2n-1)(2n-3)a_{n-2} = (2n-1)(2n-3)(2n-5)a_{n-3} = \ldots$$

$$= (2n-1)(2n-3)(2n-5)\cdots(5)(3)(1) = \frac{(2n)(2n-1)(2n-2)(2n-3)\cdots(4)(3)(2)(1)}{(2n)(2n-2)\cdots(4)(2)}$$

$$= \frac{(2n)!}{2^n(n!)}$$

7. Yes, such an assignment can be made by Fritz. Let X be the set of student applicants and Y the set of part-time jobs. Then for all $x \in X$, $y \in Y$, draw the edge (x, y) if applicant x is qualified for part-time job y. Then $\deg(x) \geq 4 \geq \deg(y)$ for all $x \in X$, $y \in Y$, and the result follows from Corollary 13.6.

9. (a) (1) Select i from A_i, for $1 \leq i \leq 4$.
 (2) Select $i+1$ from A_i, for $1 \leq i \leq 3$, and 1 from A_4.

 (b) 2

11. Proof: For each subset A of X, let G_A be the subgraph of G induced by the vertices in $A \cup R(A)$. If e is the number of edges in G_A, then $e \geq 4|A|$ because $\deg(a) \geq 4$ for all

$a \in A$. Likewise $e \leq 5|R(A)|$ because $\deg(b) \leq 5$ for all $b \in R(A)$. So $5|R(A)| \geq 4|A|$ and $\delta(A) = |A| - |R(A)| \leq |A| - (4/5)|A| = (1/5)|A| \leq (1/5)|X| = 2$. Then since $\delta(G) = \max\{\delta(A)|A \subseteq X\}$ we have $\delta(G) \leq 2$.

13. (a) $\delta(G) = 1$. A maximal matching of X into Y is given by $\{\{x_1, y_4\}, \{x_2, y_2\}, \{x_3, y_1\}, \{x_5, y_3\}\}$.

(b) If $\delta(G) = 0$, there is a complete matching of X into Y, and $\beta(G) = |Y|$, or $|Y| = \beta(G) - \delta(G)$. If $\delta(G) = k > 0$, let $A \subseteq X$ where $|A| - |R(A)| = k$. Then $A \cup (Y - R(A))$ is a largest maximal independent set in G and $\beta(G) = |A| + |Y - R(A)| = |Y| + (|A| - |R(A)|) = |Y| + \delta(G)$, so $|Y| = \beta(G) - \delta(G)$.

(c) Figure 13.30 (a): $\{x_1, x_2, x_3, y_2, y_4, y_5\}$;
Figure 13.32: $\{x_3, x_4, y_2, y_3, y_4\}$.

Supplementary Exercises

1. $d(a,b) = 5$; $d(a,c) = 11$; $d(a,d) = 7$; $d(a,e) = 8$;
$d(a,f) = 19$; $d(a,g) = 9$; $d(a,h) = 14$
(Note that the loop at vertex g and the edges (c,a) of weight 9 and (f,e) of weight 5 are of no significance.)

3. (a) The edge e_1 will always be selected in the first step of Kruskal's Algorithm.

(b) Again using Kruskal's Algorithm, edge e_2 will be selected in the first application of Step (2) unless each of the edges e_1, e_2 is incident with the same two vertices, i.e., the edges e_1, e_2 form a circuit and G is a multigraph.

5.

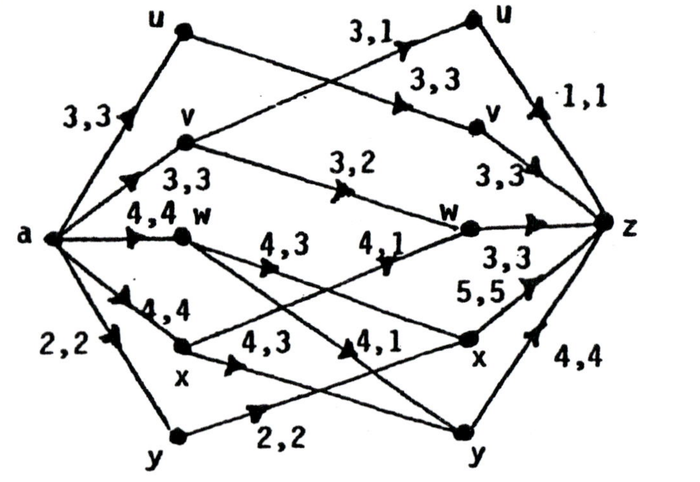

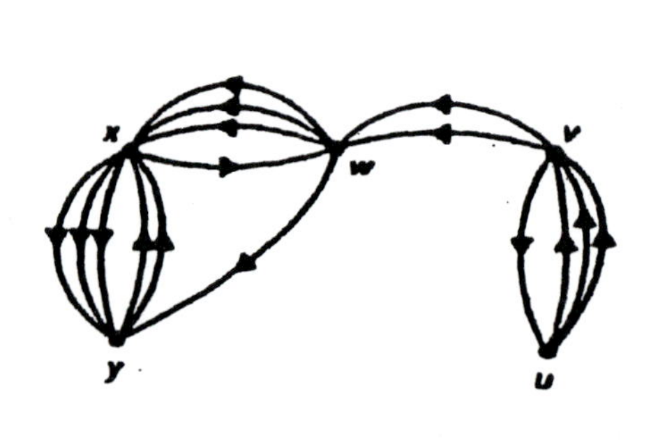

The transport network in the first diagram is determined by using the in degrees of the vertices for the capacities of the edges terminating at the sink z; the out degrees of the vertices are used for the capacities of the edges that originate at the source a.

7. The number of different systems of distinct representatives is d_n, the number of derangements of $\{1, 2, 3, \ldots, n\}$.

9. The vertices (in the line graph $L(G)$) determined by E' form a maximal independent set.

PART 4

MODERN APPLIED ALGEBRA

CHAPTER 14
RINGS AND MODULAR ARITHMETIC

Section 14.1

1. (Example 14.5): $-a = a,\ -b = e,\ -c = d,\ -d = c,\ -e = b$
(Example 14.6): $-s = s,\ -t = y,\ -v = x,\ -w = w,\ -x = v,\ -y = t$

3.

(a)	$(a+b)+c$	$=$	$(b+a)+c$	Commutative Law of $+$
		$=$	$b+(a+c)$	Associative Law of $+$
		$=$	$b+(c+a)$	Commutative Law of $+$
(b)	$d+a(b+c)$	$=$	$d+(ab+ac)$	Distributive Law of $\cdot$ over $+$
		$=$	$(d+ab)+ac$	Associative Law of $+$
		$=$	$(ab+d)+ac$	Commutative Law of $+$
		$=$	$ab+(d+ac)$	Associative Law of $+$
(c)	$c(d+b)+ab$	$=$	$ab+c(d+b)$	Commutative Law of $+$
		$=$	$ab+(cd+cb)$	Distributive Law of $\cdot$ over $+$
		$=$	$ab+(cb+cd)$	Commutative Law of $+$
		$=$	$(ab+cb)+cd$	Associative Law of $+$
		$=$	$(a+c)b+cd$	Distributive Law of $\cdot$ over $+$
(d)	$a(bc)+(ab)d$	$=$	$(ab)c+(ab)d$	Associate Law of $\cdot$
		$=$	$(ab)(c+d)$	Distributive Law of $\cdot$ over $+$
		$=$	$(ab)(d+c)$	Commutative Law of $+$

5. (a) (i) The closed binary operation $\oplus$ is associative. For all $a, b, c \in \mathbf{Z}$ we find that

$$(a \oplus b) \oplus c = (a+b-1) \oplus c = (a+b-1)+c-1 = a+b+c-2,$$

and

$$a \oplus (b \oplus c) = a \oplus (b+c-1) = a+(b+c-1)-1 = a+b+c-2.$$

(ii) For the closed binary operation $\odot$ and all $a, b, c \in \mathbf{Z}$, we have
$(a \odot b) \odot c = (a+b-ab) \odot c = (a+b-ab)+c-(a+b-ab)c = a+b-ab+c-ac-bc+abc = a+b+c-ab-ac-bc+abc$; and
$a \odot (b \odot c) = a \odot (b+c-bc) = a+(b+c-bc)-a(b+c-bc) = a+b+c-bc-ab-ac+abc = a+b+c-ab-ac-bc+abc$.
Consequently, this closed binary operation is also associative.
(iii) Given any integers a, b, c, we find that

$(b \oplus c) \odot a = (b+c-1) \odot a = (b+c-1)+a-(b+c-1)a = b+c-1+a-ba-ca+a = 2a+b+c-1-ba-ca$, and

$(b \odot a) \oplus (c \odot a) = (b+a-ba) \oplus (c+a-ca) = (b+a-ba)+(c+a-ca)-1 = 2a+b+c-1-ba-ca$.

Therefore the second distributive law holds. (The proof for the first distributive law is similar.)

(b) For all $a, b \in \mathbf{Z}$,

$$a \odot b = a+b-ab = b+a-ba = b \odot a,$$

because both ordinary addition and ordinary multiplication are commutative operations for $\mathbf{Z}$. Hence $(\mathbf{Z}, \oplus, \odot)$ is a commutative ring.

(c) Aside from 0 the only other unit is 2, since $2 \odot 2 = 2+2-(2 \cdot 2) = 0$, the unity for $(\mathbf{Z}, \oplus, \odot)$.

(d) This ring is an integral domain, but not a field. For all $a, b \in \mathbf{Z}$ we see that $a \odot b = 1$ (the zero element) $\Rightarrow a+b-ab = 1 \Rightarrow a(1-b) = (1-b) \Rightarrow (a-1)(1-b) = 0 \Rightarrow a = 1$ or $b = 1$, so there are no proper divisors of zero in $(\mathbf{Z}, \oplus, \odot)$.

7. From Exercise 6 for this section we know that we need to determine the condition(s) on k, m for which the Distributive Laws will hold. Since $\odot$ is commutative we can focus on just one of these laws.

If $x, y, z \in \mathbf{Z}$, then

$x \odot (y \oplus z) = (x \odot y) \oplus (x \odot z) \Rightarrow$

$x \odot (y+z-k) = (x+y-mxy) \oplus (x+z-mxz)$

$\Rightarrow x+(y+z-k)-mx(y+z-k) = (x+y-mxy)+(x+z-mxz)-k$

$\Rightarrow x+y+z-k-mxy-mxz+mkx = x+y-mxy+x+z-mxz-k$

$\Rightarrow mkx = x \Rightarrow mk = 1 \Rightarrow m = k = 1$ or $m = k = -1$, since $m, k \in \mathbf{Z}$.

9. (a) We first consider the associative law of addition. For $a, b, c \in \mathbf{Q}$, here $a \oplus (b \oplus c) = a \oplus (b+c+7) = a+(b+c+7)+7 = a+b+c+14$, while $(a \oplus b) \oplus c = (a+b+7) \oplus c = (a+b+7)+c+7 = a+b+c+14$. Consequently, the (closed) binary operation $\oplus$ is associative. Furthermore, $a \oplus b = a+b+7 = b+a+7 = b \oplus a$ for all $a, b \in \mathbf{Q}$, so this (closed) binary operation is also commutative.

Turning now to multiplication we find that for $a, b, c \in \mathbf{Q}$, $a \odot (b \odot c) = a \odot (b+c+(\frac{bc}{7})) = a+(b+c+(\frac{bc}{7}))+a(b+c+(\frac{bc}{7}))/7 = a+b+c+\frac{ab}{7}+\frac{ac}{7}+\frac{bc}{7}+\frac{abc}{49}$ and $(a \odot b) \odot c = (a+b+(\frac{ab}{7})) \odot c = a+b+(\frac{ab}{7})+c+(a+b+(\frac{ab}{7}))c/7 = a+b+c+(\frac{ab}{7})+(\frac{ac}{7})+(\frac{bc}{7})+(\frac{abc}{49})$, so the (closed) binary operation $\odot$ is associative.

And now we shall verify one of the distributive laws. If $a, b, c \in \mathbf{Q}$, then

$$\begin{aligned} a \odot (b \oplus c) &= a \odot (b+c+7) \\ &= a+(b+c+7)+[a(b+c+7)]/7 \\ &= a+b+c+7+(ab/7)+(ac/7)+a, \end{aligned}$$

while

$$\begin{aligned}(a \odot b) \oplus (a \odot c) &= (a \odot b) + (a \odot c) + 7 \\ &= a + b + (ab/7) + a + c + (ac/7) + 7 \\ &= a + b + c + 7 + (ab/7) + (ac/7) + a.\end{aligned}$$

Also, the rational number -7 is the zero element, and the additive inverse of each rational number a is $-14 - a$.

(b) Since $a \odot b = a + b + (ab/7) = b + a + (ba/7) = b \odot a$ for all $a, b \in \mathbf{Q}$, the ring $(\mathbf{Q}, \oplus, \odot)$ is commutative.

(c) For each $a \in \mathbf{Q}$, $a = a \odot u = a + u + (au/7) \Rightarrow u[1 + (a/7)] = 0 \Rightarrow u = 0$, because a is arbitrary. Hence the rational number 0 is the unity for this ring.
Now let $a \in \mathbf{Q}$, where $a \neq -7$, the zero element of the ring. Can we find $b \in \mathbf{Q}$ so that $a \odot b = 0$ – that is, so that $a + b + (ab/7) = 0$? It follows that $a + b + (ab/7) = 0 \Rightarrow b(1 + (a/7)) = -a \Rightarrow b = (-a)/[1 + (a/7)]$. Hence every rational number, other than -7, is a unit.

(d) From part (c) we know that $(\mathbf{Q}, \oplus, \odot)$ is a field. In order to verify that it is also an integral domain, let $a, b \in \mathbf{Q}$ with $a \odot b = -7$. Here we have $a \odot b = -7 \Rightarrow a + b + (ab/7) = -7$
$\Rightarrow a[1 + (b/7)] = -b - 7 \Rightarrow a[7 + b] = (-1)[7 + b](7)$
$\Rightarrow (a + 7)(b + 7) = 0 \Rightarrow a + 7 = 0$ or $b + 7 = 0 \Rightarrow a = -7$ or $b = -7$.
Consequently, there are no proper divisors of zero (the rational number -7) and $(\mathbf{Q}, \oplus, \odot)$ is an integral domain.

11. (a) For example, $(a+bi)+(c+di) = (a+c)+(b+d)i = (c+a)+(d+b)i = (c+di)+(a+bi)$, because addition in $\mathbf{Z}$ is commutative. In like manner, each of the other properties for R to be a commutative ring with unity follow from the corresponding property of $(\mathbf{Z}, +, \cdot)$. Finally, with respect to divisors of zero, if $(a + bi)(c + di) = (ac - bd) + (bc + ad)i = 0$ and $a + bi \neq 0$, then at least one of a, b is nonzero. Assume, without loss of generality, that $a \neq 0$. $ac - bd = 0 \Longrightarrow c = bd/a$; $bc + ad = 0 \Longrightarrow d = -bc/a$. $cd = (bd/a)(-bc/a) = (-b^2/a^2)(cd) \Longrightarrow cd(1 + (b^2/a^2)) = 0 \Longrightarrow cd(a^2 + b^2) = 0 \Longrightarrow c = 0$ or $d = 0$, since $a, b, c, d \in \mathbf{Z}$ and $a \neq 0$. $c = 0$, $d = -bc/a \Longrightarrow d = 0$. Also $d = 0$, $c = bd/a \Longrightarrow c = 0$. Hence $c + di = 0$ and R is an integral domain.

(b) $a + bi$ is a unit in R if there is an element $c + di \in R$ with $(a + bi)(c + di) = 1$. $1 = (a + bi)(c + di) = (ac - bd) + (bc + ad)i \Longrightarrow ac - bd = 1$, $bc + ad = 0 \Longrightarrow c = a/(a^2 + b^2)$, $d = -b/(a^2 + b^2)$. $c, d \in \mathbf{Z} \Longrightarrow a^2 + b^2 = 1 \Longrightarrow a = \pm 1$, $b = 0$; $a = 0$, $b = \pm 1$. Hence, the units of R are $1, -1, i$, and $-i$.

13. Given $a, b, c, d \in \mathbf{R}$, we want to find $w, x, y, z \in \mathbf{R}$ so that

$$\begin{bmatrix} a & b \\ c & d \end{bmatrix}\begin{bmatrix} w & x \\ y & z \end{bmatrix} = \begin{bmatrix} w & x \\ y & z \end{bmatrix}\begin{bmatrix} a & b \\ c & d \end{bmatrix} = \begin{bmatrix} 1 & 0 \\ 0 & 1 \end{bmatrix}.$$

From $\begin{bmatrix} a & b \\ c & d \end{bmatrix}\begin{bmatrix} w & x \\ y & z \end{bmatrix} = \begin{bmatrix} aw + by & ax + bz \\ cw + dy & cx + dz \end{bmatrix} = \begin{bmatrix} 1 & 0 \\ 0 & 1 \end{bmatrix}$

we see that

$$aw + by = 1 \qquad ax + bz = 0$$
$$cw + dy = 0 \qquad cx + dz = 1.$$

The equations involving w and y can be rewritten as

(1) $acw + bcy = c$

(2) $acw + ady = 0$.

Subtracting the second equation from the first we have

$$(bc - ad)y = c, \text{ so } y = c/(bc - ad) = -c/(ad - bc), \text{ for } ad - bc \neq 0.$$

Likewise, we find that

$$w = d/(ad - bc), \; x = -b/(ad - bc), \text{ and } z = a/(ad - bc).$$

So $\begin{bmatrix} a & b \\ c & d \end{bmatrix}^{-1} = (1/(ad - bc)) \begin{bmatrix} d & -b \\ -c & a \end{bmatrix}$, when $ad - bc \neq 0$.

[Note: The same results follow if we start with

$\begin{bmatrix} w & x \\ y & z \end{bmatrix} \begin{bmatrix} a & b \\ c & d \end{bmatrix} = \begin{bmatrix} 1 & 0 \\ 0 & 1 \end{bmatrix}$.]

15. (a) $xx = x(t + y) = xt + xy = t + y = x$
$yt = (x + t)t = xt + tt = t + t = s$
$yy = y(t + x) = yt + yx = s + s = s$
$tx = (y + x)x = yx + xx = s + x = x$
$ty = (y + x)y = yy + xy = s + y = y$

(b) Since $tx = x \neq t = xt$, this ring is not commutative.

(c) There is no unity, and, consequently, no units.

(d) The ring is neither an integral domain nor a field.

Section 14.2

1. (Theorem 14.5 (a)) Suppose that $u_1, u_2 \in R$ and that u_1, u_2 are both unity elements. Then $u_1 = u_1u_2 = u_2$. The first equality holds because u_2 is a unity element; the second equality follows since u_1 is a unity element.

(Theorem 14.5 (b)) Let $y_1, y_2 \in R$ with $xy_1 = y_1x = u = xy_2 = y_2x$, where u is the unity of R. Then $y_1 = uy_1 = (y_2x)y_1 = y_2(xy_1) = y_2u = y_2$.

(Theorem 14.10 (b)) If S is a subring of R, then $a, b \in S \Longrightarrow a+b,\ ab \in S$. Conversely, let $S = \{x_1, x_2, \ldots, x_n\}$. $T = \{x_i + x_1 | 1 \le i \le n\} \subseteq S$. $x_i + x_1 = x_j + x_1 \Longrightarrow x_i = x_j$, so $|T| = n$ and $T = S$. Hence $x_i + x_1 = x_1$ for some $1 \le i \le n$, and $x_i = z$, the zero element of R. For each $x \in S$, $x + S = \{x + x_i | 1 \le i \le n\} = S$. With $z \in S$, $x + x_j = z$ for some $x_j \in S$, so $x_j = -x \in S$. Consequently, by Theorem 14.9 S is a subring of R.

3. (a) $(ab)(b^{-1}a^{-1}) = aua^{-1} = aa^{-1} = u$ and $(b^{-1}a^{-1})(ab) = b^{-1}ub = b^{-1}b = u$, so ab is a unit. Since the multiplicative inverse of a unit is unique, it follows that $(ab)^{-1} = b^{-1}a^{-1}$.

(b) $A^{-1} = \begin{bmatrix} 2 & -7 \\ -1 & 4 \end{bmatrix}$, $B^{-1} = \begin{bmatrix} 1 & -2 \\ -2 & 5 \end{bmatrix}$, $(AB)^{-1} = \begin{bmatrix} 4 & -15 \\ -9 & 34 \end{bmatrix}$,

$(BA)^{-1} = \begin{bmatrix} 16 & -39 \\ -9 & 22 \end{bmatrix}$, $B^{-1}A^{-1} = \begin{bmatrix} 4 & -15 \\ -9 & 34 \end{bmatrix}$.

5. Since $(-a)[-(a^{-1})] = aa^{-1}$ (by Theorem 14.4 (c)) $= u$, it follows that $-a$ is a unit and, from Theorem 14.5 (b), that $(-a)^{-1} = -(a^{-1})$.

7. $z \in S, T \Longrightarrow z \in S \cap T \Longrightarrow S \cap T \ne \emptyset$. $a, b \in S \cap T \Longrightarrow a, b \in S$ and $a, b \in T \Longrightarrow$ $a+b, ab \in S$ and $a+b, ab \in T \Longrightarrow a+b, ab \in S \cap T$.
$a \in S \cap T \Longrightarrow a \in S$ and $a \in T \Longrightarrow -a \in S$ and $-a \in T \Longrightarrow -a \in S \cap T$.
So $S \cap T$ is a subring of R.

9. If not, there exist $a, b \in S$ with $a \in T_1, a \notin T_2$, and $b \in T_2, b \notin T_1$. Since S is a subring of R, it follows that $a + b \in S$. Hence $a + b \in T_1$ or $a + b \in T_2$.

Assume without loss of generality that $a + b \in T_1$. Since $a \in T_1$ we have $-a \in T_1$, so by the closure under addition in T_1 we now find that $(-a) + (a + b) = (-a + a) + b = b \in T_1$, a contradiction.

Therefore, $S \subseteq T_1 \cup T_2 \Longrightarrow S \subseteq T_1$ or $S \subseteq T_2$.

11. (a) Let $a, b \in \mathbf{Z}$ and let $\begin{bmatrix} a & 0 \\ 0 & a \end{bmatrix}, \begin{bmatrix} b & 0 \\ 0 & b \end{bmatrix} \in S$. Then $\begin{bmatrix} a & 0 \\ 0 & a \end{bmatrix} + \begin{bmatrix} b & 0 \\ 0 & b \end{bmatrix} = \begin{bmatrix} a+b & 0 \\ 0 & a+b \end{bmatrix} \in S$ and $\begin{bmatrix} a & 0 \\ 0 & a \end{bmatrix}\begin{bmatrix} b & 0 \\ 0 & b \end{bmatrix} = \begin{bmatrix} ab & 0 \\ 0 & ab \end{bmatrix} \in S$. Further, since $a \in \mathbf{Z}$ we have $-a \in \mathbf{Z}$ and $\begin{bmatrix} -a & 0 \\ 0 & -a \end{bmatrix} = -\begin{bmatrix} a & 0 \\ 0 & a \end{bmatrix} \in S$. Consequently, it follows from Theorem 14.9 that S is a subring of R.

(b) $\begin{bmatrix} 1 & 0 \\ 0 & 1 \end{bmatrix}$ (c) $\begin{bmatrix} 1 & 0 \\ 0 & 0 \end{bmatrix}$

(d) S is an integral domain while R is a noncommutative ring with unity.

(e) S is not an ideal of R – for example, $\begin{bmatrix} 1 & 1 \\ 1 & 1 \end{bmatrix}\begin{bmatrix} 1 & 0 \\ 0 & 0 \end{bmatrix} = \begin{bmatrix} 1 & 0 \\ 1 & 0 \end{bmatrix}$, and this result is not in S.

13. Since $za = z$, it follows that $z \in N(a)$ and $N(a) \neq \emptyset$. If $r_1, r_2 \in N(a)$, then $(r_1 - r_2)a = r_1a - r_2a = z - z = z$, so $r_1 - r_2 \in N(a)$. Finally, if $r \in N(a)$ and $s \in R$, then $(rs)a = (sr)a = s(ra) = sz = z$, so $rs,\ sr \in N(a)$. Hence $N(a)$ is an ideal – by Definition 14.6.

15. Two ideals: R and $\{z\}$, where z is the zero of R.

17. (a) $a = au \in aR$, so $aR \neq \emptyset$. If $ar_1, ar_2 \in aR$, then $ar_1 - ar_2 = a(r_1 - r_2) \in aR$. Also, for $ar_1 \in aR$, $r \in R$, $r(ar_1) = (ar_1)r = a(r_1r) \in aR$. Hence aR is an ideal of R.

(b) Let $a \in R$, $a \neq z$. Then $a = au \in aR$ so $aR = R$. Since $u \in R = aR$, $u = ar$ for some $r \in R$, and $r = a^{-1}$. Hence R is a field.

19. (a) $\binom{4}{2}(49)$ (b) 7^4 (d) Yes, the element (u, u, u, u). (d) 4^4

21. (a) For each $m \in \mathbf{Z}^+, (a^m)(a^1) = a^m a = a^{m+1}$ so the result is true for $n = 1$. Assume the result for $m \in \mathbf{Z}^+$ and $n = k\ (\geq 1)$. For $m \in \mathbf{Z}^+, n = k+1, (a^m)(a^n) = (a^m)(a^{k+1}) = (a^m)(a^ka) = (a^ma^k)(a) = (a^{m+k})(a) = a^{(m+k)+1} = a^{m+(k+1)} = a^{m+n}$. Consequently, by the Principle of Mathematical Induction the result is true for all $m, n \in \mathbf{Z}^+$.
In like manner, $(a^m)^n = a^{mn}$ for all $m \in \mathbf{Z}^+$ and $n = 1$. Assuming the result for $m \in \mathbf{Z}^+$ and $n = k\ (\geq 1)$, we consider the case for $m \in \mathbf{Z}^+$ and $n = k + 1$. Then $(a^m)^{(k+1)} = (a^m)^k(a^m) = (a^{mk})(a^m) = a^{mk+m}$ (from the first result) $= a^{m(k+1)} = a^{mn}$ and the result is true for all $m, n \in \mathbf{Z}^+$ by the Principle of Mathematical Induction.

(b) If R has a unity u, define $a^0 = u$, for $a \in R$, $a \neq z$. If a is a unit of R, define a^{-n} as $(a^{-1})^n$, for $n \in \mathbf{Z}^+$.

Section 14.3

1. (a) (i) $118 - 62 = 56 = 7(8)$, so $118 \equiv 62 \pmod 8$
(ii) $-237 - (-43) = -194$, but 8 does not divide -194, so -237 and -43 are *not* congruent modulo 8.

Also, $-43 \equiv 5 \pmod 8$ while $-237 \equiv 3 \pmod 8$, so -237 and -43 are *not* congruent modulo 8 .

(iii) $230 - (-90) = 320 = 40(8)$, so $230 \equiv -90 \pmod 8$.

Also, $230 = 28(8) + 6$ and $-90 = -12(8) + 6$ so $230 \equiv 6 \equiv -90 \pmod 8$.

b) (i) $243 - 76 = 167 = 18(9) + 5$ so 243 and 77 are *not* congruent modulo 9.

Also, $243 = 27(9) + 0$ while $76 = 8(9) + 4$. Since the remainders for 243 and 76 are different for division by 9, it follows that 243 and 76 are *not* congruent modulo 9.

(ii) $700 - (-137) = 837 = 93(9)$, so 700 and -137 are congruent modulo 9.

(iii) $056 - (-1199) = 1143 = 127(9)$, so -56 and -1199 are congruent modulo 9.

3. (a) $-6, 1, 8, 14$ (b) $-9, 2, 13, 24$ (c) $-7, 10, 27, 44$

5. Proof: Since $a \equiv b \pmod n$ we may write $a = b + kn$ for some $k \in \mathbf{Z}$. And $m|n \Rightarrow n = \ell m$ for some $\ell \in \mathbf{Z}$. Consequently, $a = b + kn = b + (k\ell)m$ and $a \equiv b \pmod m$.

7. Let $a = 8$, $b = 2$, $m = 6$, and $n = 2$. Then $\gcd(m, n) = \gcd(6, 2) = 2 > 1$, $a \equiv b \pmod m$ and $a \equiv b \pmod n$. But $a - b = 8 - 2 = 6 \neq k(12) = k(mn)$, for some $k \in \mathbf{Z}$. Hence $a \not\equiv b \pmod{mn}$.

9. Proof: For n odd consider the $n-1$ numbers $1, 2, 3, \ldots, n-3, n-2, n-1$ as $(n-1)/2$ pairs: 1 and $(n-1)$, 2 and $(n-2)$, 3 and $(n-3), \ldots, n - (\frac{n-1}{2}) - 1$ and $n - (\frac{n-1}{2})$. The sum of each pair is n which is congruent to 0 modulo n. Hence $\sum_{i=1}^{n-1} i \equiv 0 \pmod n$.

When n is even we consider the $n-1$ numbers $1, 2, 3, \ldots, (n/2) - 1, (n/2), (n/2) + 1, \ldots, n-3, n-2, n-1$ as $(n/2) - 1$ pairs — namely, 1 and $n-1$, 2 and $n-2$, 3 and $n-3, \ldots, (n/2) - 1$ and $(n/2) + 1$ — and the single number $(n/2)$. For each pair the sum is n, or 0 modulo n, so $\sum_{i=1}^{n-1} i \equiv (n/2) \pmod n$.

11. (a) For each $a \in \mathbf{Z}^+$ $\tau(a) = \tau(a)$, so the relation is reflexive. If $a, b \in \mathbf{Z}^+$, $\tau(a) = \tau(b) \Longrightarrow \tau(b) = \tau(a)$ so the relation is symmetric. Finally, for $a, b, c \in \mathbf{Z}^+$, $\tau(a) = \tau(b)$ and $\tau(b) = \tau(c) \Longrightarrow \tau(a) = \tau(c)$ so the relation is transitive.

(b) No, $2\mathcal{R}3$, $3\mathcal{R}5$ but $5 \not\mathcal{R} 8$. Also, $2\mathcal{R}3$, $2\mathcal{R}5$ but $4 \not\mathcal{R} 15$.

13. (a)

$$\begin{aligned} 1009 &= 59(17) + 6 & 0 < 6 < 17 \\ 17 &= 2(6) + 5 & 0 < 5 < 6 \\ 6 &= 1(5) + 1 & 0 < 1 < 5, \end{aligned}$$

so $1 = 6 - 5 = 6 - [17 - 2(6)] = 3(6) - 17 = 3[1009 - 59(17)] - 17 = 3(1009) - 178(17)$.

Hence $1 \equiv (-178)(17) \pmod{1009}$, so $[17]^{-1} = [-178] = [-178 + 1009] = [831]$.

(b)

$$\begin{aligned} 1009 &= 10(100) + 9 & 0 < 9 < 100 \\ 100 &= 11(9) + 1 & 0 < 1 < 9 \end{aligned}$$

$1 = 100 - 11(9) = 100 - 11[1009 - 10(100)] = 1009(-11) + 100(111)$.

So $1 \equiv (100)(111) \pmod{1009}$ and $[100]^{-1} = [111]$.

(c)

$$\begin{array}{rcll} 1009 &=& 1(777)+232 & 0<232<777\\ 777 &=& 3(232)+81 & 0<81<232\\ 232 &=& 2(81)+70 & 0<70<81\\ 81 &=& 1(70)+11 & 0<11<70\\ 70 &=& 6(11)+4 & 0<4<11\\ 11 &=& 2(4)+3 & 0<3<4\\ 4 &=& 1(3)+1 & 0<1<3 \end{array}$$

$$\begin{aligned} 1 &= 4-3=4-[11-2(4)]=(-1)(11)+3(4)\\ &= (-1)(11)+3[70-6(11)]=3(70)+(-19)(11)\\ &= 3(70)+(-19)[81-70]=22(70)+(-19)(81)\\ &= (-19)(81)+22[232-2(81)]=22(232)+(-63)(81)\\ &= (22)(232)+(-63)[777-3(232)]=777(-63)+211(232)\\ &= 777(-63)+211[1009-777]=1009(211)+777(-274) \end{aligned}$$

So $1 \equiv (777)(-274) \equiv (777)(1009-274) \equiv (777)(735) \pmod{1009}$ and $[777]^{-1}=[735]$.

15. (a) 16 units; 0 proper zero divisors (b) 72 units; 44 proper zero divisors
(c) 1116 units; 0 proper zero divisors.

17. $\{1,2,3,\ldots,1000\} = \{1,4,7,10,\ldots,997,1000\} \cup \{2,5,8,\ldots,995,998\} \cup \{3,6,9,\ldots,999\}$. In this partition the first cell has 334 elements while the other two cells contain 333 elements each. If three elements are selected from the same cell then their sum will be divisible by three. If one number is selected from each of the three cells then their sum is divisible by three. Consequently the probability that the sum of three elements selected from $\{1,2,3,\ldots,999\}$ is divisible by three is $[\binom{334}{3}+2\binom{333}{3}+\binom{334}{1}\binom{333}{1}^2]/\binom{1000}{3}$.

19. (a) For $n=0$ we have $10^0=1=1(-1)^0$ so $10^0 \equiv (-1)^0 \pmod{11}$. [Since $10-(-1)=11$, $10 \equiv (-1) \pmod{11}$, or $10^1 \equiv (-1)^1 \pmod{11}$. Hence the result is true for $n=0,1$.] Assume the result true for $n=k \geq 1$ and consider the case for $k+1$. Then since $10^k \equiv (-1)^k \pmod{11}$ and $10 \equiv (-1) \pmod{11}$, we have $10^{k+1}=10^k \cdot 10 \equiv (-1)^k(-1)=(-1)^{k+1} \pmod{11}$. The result now follows for all $n \in \mathbf{N}$ by the Principle of Mathematical Induction.

(b) If $x_nx_{n-1}\ldots x_2x_1x_0 = x_n \cdot 10^n + x_{n-1}\cdot 10^{n-1}+\cdots+x_2\cdot 10^2+x_1\cdot 10+x_0$ denotes an $(n+1)$-st digit integer, then
$x_nx_{n-1}\ldots x_2x_1x_0 \equiv (-1)^n x_n + (-1)^{n-1}x_{n-1}+\cdots+x_2-x_1+x_0 \pmod{11}$.

Proof: $x_nx_{n-1}\ldots x_2x_1x_0 = x_n\cdot 10^n + x_{n-1}\cdot 10^{n-1}+\cdots+x_2\cdot 10^2+x_1\cdot 10+x_0 \equiv x_n(-1)^n + x_{n-1}(-1)^{n-1}+\cdots+x_2(-1)^2+x_1(-1)+x_0 = (-1)^nx_n+(-1)^{n-1}x_{n-1}+\cdots+x_2-x_1+x_0 \pmod{11}$.

21. Let $g=\gcd(a,n)$, $h=\gcd(b,n)$. $a \equiv b \pmod{n} \Longrightarrow a=b+kn$, for some $k \in \mathbf{Z} \Longrightarrow g|b, h|a$. $g|b, g|n \Longrightarrow g|h$; $h|a, h|n \Longrightarrow h|g$. Since $g,h>0$, $g=h$.

23.

(1) Plaintext	a	ℓ	ℓ	g	a	u	ℓ	i	s	d	i	v	i	d	e	d
(2)	0	11	11	6	0	20	11	8	18	3	8	21	8	3	4	3
(3)	3	14	14	9	3	23	14	11	21	6	11	24	11	6	7	6
(4) Ciphertext	D	O	O	J	D	X	O	L	V	G	L	Y	L	G	H	G

i	n	t	o	t	h	r	e	e	p	a	r	t	s
8	13	19	14	19	7	17	4	4	15	0	17	19	18
11	16	22	17	22	10	20	7	7	18	3	20	22	21
L	Q	W	R	W	K	U	H	H	S	D	U	W	V

For each θ in row (2), the corresponding result below it in row (3) is $\theta + 3 \pmod{26}$.

25. From part (c) of Example 14.15 we know that for an alphabet of n letters there are $n \cdot \phi(n)$ affine ciphers. Here we have:
(a) $24\phi(24) = (24)[24(1-\frac{1}{2})(1-\frac{1}{3})] = (24)(8) = 192$
(b) $25\phi(25) = (25)[25(1-\frac{1}{5})] = (25)(20) = 500$
(c) $27\phi(27) = (27)[27(1-\frac{1}{3})] = (27)(18) = 486$
(d) $30\phi(30) = (30)[30(1-\frac{1}{2})(1-\frac{1}{3})(1-\frac{1}{5})] = (30)(8) = 240.$

27. (a) $x_0 = 10$

$x_1 \equiv 5(x_0) + 3 \pmod{19} \equiv \quad \pmod{19} \equiv 15 \pmod{19}$, so $x_1 = 15$.
$x_2 \equiv 5(x_1) + 3 \pmod{19} \equiv 78 \pmod{19} \equiv 2 \pmod{19}$, so $x_2 = 2$.
$x_3 \equiv 5(x_2) + 3 \pmod{19} \equiv 13 \pmod{19}$, so $x_3 = 13$.
$x_4 \equiv 5(x_3) + 3 \pmod{19} \equiv 68 \pmod{19} \equiv 11 \pmod{19}$, so $x_4 = 11$.

Further computation tells us that $x_5 = 1$, $x_6 = 8$, $x_7 = 5$, $x_8 = 9$, and $x_9 = 10$, the seed.

So this linear congruential generator produces nine distinct terms.
(b) $10, 15, 2, 13, 11, 1, 9, 5, 9, 10, 15, 2, \ldots$.

29. Proof: (By Mathematical Induction)
[Note that for $n \geq 1$, $(a^n - 1)/(a-1) = a^{n-1} + a^{n-2} + \cdots + 1$, which can be computed in the ring $(\mathbf{Z}, +, \cdot)$.]

When $n - 0$, $a^0 x_0 + c[(a^0 - 1)/(a-1)] \equiv x_0 + c[0/(a-1)] \equiv x_0 \pmod{m}$, so the formula is true in thisfirst basis ($n = 0$) case. Assuming the result for n we have

$x_n \equiv a^n x_0 + c[(a^n - 1)/(a-1)] \pmod{m}$, $0 \leq x_n < m$. Continuing to the next case we learn than

$$\begin{aligned}
x_{n+1} &\equiv ax_n + c \pmod{m} \\
&\equiv a[a^n x_0 + c[(a^n - 1)/(a-1)]] + c \pmod{m} \\
&\equiv a^{n+1} x_0 + ac[(a^n - 1)/(a-1)] + c(a-1)/(a-1) \pmod{m} \\
&\equiv a^{n+1} x_0 + c[(a6n + 1 - a + a - 1)/(a-1)] \pmod{m} \\
&\equiv a^{n+1} x_0 + c[(a^{n+1} - 1)/(a-1)] \pmod{m}
\end{aligned}$$

and we select x_{n+1} so that $0 \leq x_{n+1} < m$. It now follows by the Principle of Mathematical Induction that

$$x_n \equiv a^n x_0 + c[(a^n - 1)/(a-1)] \pmod{m},\ 0 \leq x_n < m.$$

31. Proof: Let $n, n+1$, and $n+2$ be three consecutive integers. Then $n^3+(n+1)^3+(n+2)^3 = n^3 + (n^3 + 3n^2 + 3n + 1) + (n^3 + 6n^2 + 12n + 8) = (3n^3 + 15n) + 9(n^2 + 1)$. So we consider $3n^3 + 15n = 3n(n^2 + 5)$. If $3|n$, then we are finished. If not, then $n \equiv 1 \pmod 3$ or $n \equiv 2 \pmod 3$. If $n \equiv 1 \pmod 3$, then $n^2 + 5 \equiv 1 + 5 \equiv 0 \pmod 3$, so $3|(n^2+5)$. If $n \equiv 2 \pmod 3$, then $n^2+5 \equiv 9 \equiv 0 \pmod 3$, and $3|(n^2+5)$. All cases are now covered, so we have $3|[n(n^2+5)]$. Hence $9|[3n(n^2+5)]$ and, consequently, 9 divides $(3n^3 + 15n) + 9(n^2 + 1) = n^3 + (n+1)^3 + (n+2)^3$.

33. From the presentation given in Example 14.18 it follows that for $n \in \mathbf{Z}^+$,

$$\sum_{k=0}^{n-1} p(k(n+1), n, n) = \frac{1}{n+1}\binom{2n}{n}, \text{ the } n\text{th Catalan number.}$$

35. (a) $1+2+3 = 6 \equiv 1 \pmod 5$; $0+4 = 4 \equiv 1 \pmod 3$; $2+2+7+5 = 16 \equiv 2 \pmod 7$. $h(123-04-2275) = 112$.

(b) Let $n = 112-43-8295$. Then $h(112-43-8295) = 413$.

37. (a) $h(206) = 1$ **mod** 41, since $206 = 5(41)+1$. Likewise, $h(807) = 28$ **mod** 41, $h(137) = 14$ **mod** 41, $h(444) = 34$ **mod** 41, $h(617) = 2$ **mod** 41. Since $h(330) = 2$ **mod** 41 but that parking space has been assigned, this patron is assigned to the next available space – here, it is 3. Likewise, the last two patrons are assigned to the spaces numbered $14+1 = 15$ and $3+1 = 4$.
(b) 1, 2, 3, 4, or 5.

Section 14.4

1. $s \to 0$, $t \to 1$, $v \to 2$, $w \to 3$, $x \to 4$, $y \to 5$

3. Let $(R, +, \cdot)$, $(S, \oplus, \odot)$, $(T, +', \cdot')$ be the rings. For all $a, b \in R$, $(g \circ f)(a+b) = g(f(a+b)) = g(f(a) + f(b)) = g(f(a)) +' g(f(b)) = (g \circ f)(a) +' (g \circ f)(b)$. Also, $(g \circ f)(a \cdot b) = g(f(a \cdot b)) = g(f(a) \odot f(b)) = g(f(a)) \cdot' g(f(b)) = (g \circ f)(a) \cdot' (g \circ f)(b)$. Hence, $g \circ f$ is a ring homomorphism.

5. (a) Since $f(z_R) = z_S$, it follows that $z_R \in K$ and $K \neq \emptyset$. If $x, y \in K$, then $f(x-y) = f(x+(-y)) = f(x) \oplus f(-y) = f(x) \ominus f(y) = z_S \ominus z_S = z_S$, so $x - y \in K$. Finally, if $x \in K$ and $r \in R$, then $f(rx) = f(r) \odot f(x) = f(r) \odot z_S = z_S$, and $f(xr) = f(x) \odot f(r) = z_S \odot f(r) = z_S$, so $rx, xr \in K$. Consequently, K is an ideal of R.

(b) The kernel is $\{6n | n \in \mathbf{Z}\}$.

(c) If f is one-to-one, then for each $x \in K, [f(x) = z_S = f(z_R)] \Longrightarrow [x = z_R]$, so $K = \{z_R\}$. Conversely, if $K = \{z_R\}$, let $x, y \in R$ with $f(x) = f(y)$. Then

$z_S = f(x) \ominus f(y) = f(x-y)$, so $x - y \in K = \{z_R\}$. Consequently, $x - y = z_R \Longrightarrow x = y$, and f is one-to-one.

7. (a)

x (in $\mathbf{Z}_{20}$)	$f(x)$ (in $\mathbf{Z}_4 \times \mathbf{Z}_5$)	x (in $\mathbf{Z}_{20}$)	$f(x)$ (in $\mathbf{Z}_4 \times \mathbf{Z}_5$)
0	(0,0)	10	(2,0)
1	(1,1)	11	(3,1)
2	(2,2)	12	(0,2)
3	(3,3)	13	(1,3)
4	(0,4)	14	(2,4)
5	(1,0)	15	(3,0)
6	(2,1)	16	(0,1)
7	(3,2)	17	(1,2)
8	(0,3)	18	(2,3)
9	(1,4)	19	(3,4)

(b) (i) $f((17)(19) + (12)(14)) = (1,2)(3,4) + (0,2)(2,4) = (3,3) + (0,3) = (3,1)$ and $f^{-1}(3,1) = 11$.

(ii) $f((18)(11) - (9)(15)) = (2,3)(3,1) - (1,4)(3,0) = (2,3) - (3,0) = (3,3)$ and $f^{-1}(3,3) = 3$.

9. (a) For the ring $\mathbf{Z}_8$ there are four units – namely, $1, 3, 5, 7$.
(b) The ring $\mathbf{Z}_2 \times \mathbf{Z}_2 \times \mathbf{Z}_2$ has only one unit – the element $(1,1,1)$.
(c) No

11. No, $\mathbf{Z}_4$ has two units, while the ring in Example 14.4 has only one unit.

13. Here $a_1 = 5$; $a_2 = 73$; $m_1 = 8$; $m_2 = 81$; $m = m_1 m_2 = 8 \cdot 81 = 648$; $M_1 = m/m_1 = 81$; and $M_2 = m/m_2 = 8$.
$[x_1] = [M_1]^{-1} = [10(8) + 1]^{-1} = [1]^{-1} = [1]$ in $\mathbf{Z}_8$
$[x_2] = [M_2]^{-1} = [8]^{-1} = [-10] = [71]$ in $\mathbf{Z}_{81}$
$x - a_1 M_1 x_1 + a_2 M_2 x_2 = 5 \cdot 81 \cdot 1 + 73 \cdot 8 \cdot 71 = 41869 = 64(648) + 397$.
So the smallest positive solution is 397 and all other solutions are congruent to 397 modulo 648.
Check: $397 = 48(8) + 3 = 4(81) + 73$.

15. Here $a_1 = 1$; $a_2 = 2$; $a_3 = 3$; $a_4 = 5$; $m_1 = 2$; $m_2 = 3$; $m_3 = 5$; $m_4 = 7$; $m = m_1 m_2 m_3 m_4 = 2 \cdot 3 \cdot 5 \cdot 7 = 210$; $M_1 = m/m_1 = 105$; $M_2 = m/m_2 = 70$; $M_3 = m/m_3 = 42$ and $M_4 = m/m_4 = 30$.
$[x_1] = [M_1]^{-1} = [105]^{-1} = [52(2) + 1]^{-1} = [1]^{-1} = [1]$ in $\mathbf{Z}_2$
$[x_2] = [M_2]^{-1} = [70]^{-1} = [23(3) + 1]^{-1} = [1]^{-1} = [1]$ in $\mathbf{Z}_3$
$[x_3] = [M_3]^{-1} = [42]^{-1} = [8(5) + 2]^{-1} = [2]^{-1} = [3]$ in $\mathbf{Z}_5$
$[x_4] = [M_4]^{-1} = [30]^{-1} = [4(7) + 2]^{-1} = [2]^{-1} = [4]$ in $\mathbf{Z}_7$
$x = 1 \cdot 105 \cdot 1 + 2 \cdot 70 \cdot 1 + 3 \cdot 42 \cdot 3 + 5 \cdot 30 \cdot 4 = 1223 \equiv 173 \pmod{210}$.

So $x = 173$ is the smallest positive simultaneous solution for the four congruences. Any other solution would be congruent to 173 modulo 210.
Check: $173 = 86(2) + 1 = 57(3) + 2 = 34(5) + 3 = 24(7) + 5$.

Supplementary Exercises

1. (a) False. Let $R = \mathbf{Z}$ and $S = \mathbf{Z}^+$.
(b) False. Let $R = \mathbf{Z}$ and $S = \{2x | x \in \mathbf{Z}\}$.
(c) False. Let $R = M_2(\mathbf{Z})$ and $S = \left\{ \begin{bmatrix} a & 0 \\ 0 & 0 \end{bmatrix} \mid a \in \mathbf{Z} \right\}$.
(d) True.
(e) False. $(\mathbf{Z}, +, \cdot)$ is a subring (but not a field) in $(\mathbf{Q}, +, \cdot)$.
(f) False. For each prime p, $\{a/(p^n) | a, n \in \mathbf{Z}, n \geq 0\}$ is a subring of $(\mathbf{Q}, +, \cdot)$.
(g) False. Consider the field in Table 14.6.
(h) True

3. (a) $a + a = (a + a)^2 = a^2 + a^2 + a^2 + a^2 = (a + a) + (a + a) \implies a + a = 2a = z$.

(b) For each $a \in R$, $a + a = z \implies a = -a$. For $a, b \in R$, $(a + b) = (a + b)^2 = a^2 + ab + ba + b^2 = a + ab + ba + b \implies ab + ba = z \implies ab = -ba = ba$, so R is commutative.

5. Since $az = z = za$ for all $a \in R$, we have $z \in C$ and $C \neq \emptyset$. If $x, y \in C$, then $(x + y)a = xa + ya = ax + ay = a(x + y)$, $(xy)a = x(ya) = x(ay) = (xa)y = (ax)y$, and $(-x)a = -(xa) = -(ax) = a(-x)$, for all $a \in R$, so $x + y, xy$, and $-x \in C$. Consequently, C is a subring of R.

7. (a) Since $a^3 = b^3$ and $a^5 = b^5$, it follows that $a^5 = (b^3)(b^2) = (a^3)(b^2)$. Consequently, $(a^3)(a^2) = (a^3)(b^2)$ with $a^3 \neq z$, so $a^2 = b^2$.

Now with $a^3 = b^3$ and $a^2 = b^2$ we have $(a^2)(a) = a^3 = b^3 = (b^2)(b) = (a^2)(b)$, and since $a^2 \neq z$ it follows that $a = b$.

(b) Since m, n are relatively prime we can write $1 = ms + nt$ where $s, t \in \mathbf{Z}$. With $m, n > 0$ it follows that one of s, t must be positive, and the other negative. Assume (without any loss of generality) that s is negative so that $1 - ms = nt > 0$.

Then $a^n = b^n \implies (a^n)^t = (b^n)^t \implies a^{nt} = b^{nt} \implies a^{1-ms} = b^{1-ms} \implies a(a^m)^{(-s)} = b(b^m)^{(-s)}$. But with $-s > 0$ and $a^m = b^m$, we have $(a^m)^{(-s)} = (b^m)^{(-s)}$. Consequently,

$$([(a^m)^{(-s)} = (b^m)^{(-s)} \neq z] \wedge [a(a^m)^{(-s)} = b(b^m)^{(-s)}]) \implies a = b,$$

since we may use the Cancellation Law of Multiplication in an integral domain.

9. Let $x = a_1 + b_1$, $y = a_2 + b_2$, for $a_1, a_2 \in A$, $b_1, b_2 \in B$. Then $x - y = (a_1 - a_2) + (b_1 - b_2) \in A + B$. If $r \in R$, and $a + b \in A + B$, with $a \in A$, $b \in B$, then $ra \in A$, $rb \in B$ and

$r(a+b) \in A+B$. Similarly, $(a+b)r \in A+B$, and $A+B$ is an ideal of R.

11. Consider the numbers $x_1,\ x_1+x_2,\ x_1+x_2+x_3,\ \ldots, x_1+x_2+x_3+\ldots+x_n$. If one of these numbers is congruent to 0 modulo n, the result follows. If not, there exist $1 \le i < j \le n$ with $(x_1+x_2+\ldots+x_i) \equiv (x_1+\ldots+x_i+x_{i+1}+\ldots+x_j) \pmod{n}$. Hence n divides $(x_{i+1}+\ldots+x_j)$.

13. (a) For each $t \in \mathbf{N}$,

$$\begin{array}{ll} 7^{4t+1} \equiv 7 \pmod{10} & 3^{4t+1} \equiv 3 \pmod{10} \\ 7^{4t+2} \equiv 9 \pmod{10} & 3^{4t+2} \equiv 9 \pmod{10} \\ 7^{4t+3} \equiv 3 \pmod{10} & 3^{4t+3} \equiv 7 \pmod{10} \\ 7^{4t+4} \equiv 1 \pmod{10} & 3^{4t+4} \equiv 1 \pmod{10} \end{array}$$

So in order to get the units digit of $7^m + 3^n$ as 8 we must have (i) $m \equiv 1 \pmod 4$ and $n \equiv 0 \pmod 4$, or (ii) $m \equiv 0 \pmod 4$ and $n \equiv 3 \pmod 4$, or (iii) $m \equiv 2 \pmod 4$ and $n \equiv 2 \pmod 4$.

For case (i) there are 25 choices for m (namely, $1, 5, 9, \ldots, 93, 97$) and 25 choices for n (namely, $4, 8, 12, \ldots, 96, 100$) — a total of $25^2 = 625$ choices for the pair. There are also 625 choices for the pair in each of cases (ii) and (iii). Consequently, in total, there are 625 + 625 + 625 = 1875 ways to make the selection for m, n.
(b) For case (i) there are 32 choices for m and 31 choices for n, and $32 \times 31 = 992$ choices for the pair. There are 31 choices for each of m, n, resulting in $31^2 = 961$ possible pairs, for case (ii) and case (iii). Therefore we can select m, n in this situation in 992 + 961 + 961 = 2914 ways.
(c) There are $(100)^2 = 10,000$ ways in which one can select the pair m, n.
Here we consider three cases:
(i) $m \equiv 2 \pmod 4$ and $n \equiv 1 \pmod 4$;
(ii) $m \equiv 3 \pmod 4$ and $n \equiv 2 \pmod 4$; and
(iii) $m \equiv 0 \pmod 4$ and $n \equiv 0 \pmod 4$.
For each case there are $(25)^2 = 625$ ways to select the pair m, n. Therefore, we have 1875 ways in total.
Consequently, the probability for the problem posed is $\frac{1875}{10{,}000} = 0.1875 = 3[(\frac{25}{100})(\frac{25}{100})] = 3/16$.

15. Proof: For all $n \in \mathbf{Z}$ we find that $n^2 \equiv 0 \pmod 5$ – when $5|n$ – or $n^2 \equiv 1 \pmod 5$ or $n^2 \equiv 4 \pmod 5$. Suppose that 5 does not divide any of a, b, or c. Then
(i) $a^2+b^2+c^2 \equiv 3 \pmod 5$ – when $a^2 \equiv b^2 \equiv c^2 \equiv 1 \pmod 5$;
(ii) $a^2+b^2+c^2 \equiv 1 \pmod 5$ – when each of two of a^2, b^2, c^2 is congruent to 1 modulo 5 and the other square is congruent to 4 modulo 5;
(iii) $a^2+b^2+c^2 \equiv 4 \pmod 5$ – when one of a^2, b^2, c^2 is congruent to 1 modulo 5 and each of the other two squares is congruent to 4 modulo 5; or,
(iv) $a^2+b^2+c^2 \equiv 2 \pmod 5$ – when $a^2 \equiv b^2 \equiv c^2 \equiv 4 \pmod 5$.

17. From Section 4.5 we know that $a-b$ has $(e_1+1)(e_2+1)\cdots(e_k+1)$ positive integer divisors. Consequently, there are $(e_1+1)(e_2+1)\cdots(e_k+1)-1$ possible values for n which will make $a \equiv b \pmod{n}$ true.

CHAPTER 15
BOOLEAN ALGEBRA AND SWITCHING FUNCTIONS

Section 15.1

1. (a) 1 (b) 1 (c) 1 (d) 1

3. (a) 2^n (b) $2^{(2^n)}$

5.

x	y	z	$\overline{x+y}$	$\overline{x}z$	$\overline{\overline{(x+y)}+(\overline{x}z)}$
0	0	0	1	0	0
0	0	1	1	1	0
0	1	0	0	0	1
0	1	1	0	1	0
1	0	0	0	0	1
1	0	1	0	0	1
1	1	0	0	0	1
1	1	1	0	0	1

(a) d.n.f. $xyz + x\overline{y}z + x\overline{y}\,\overline{z} + xy\overline{z} + \overline{x}y\overline{z}$

c.n.f. $(x+y+z)(x+y+\overline{z})(x+\overline{y}+\overline{z})$

(b) $f = \sum m(2,4,5,6,7) = \prod M(0,1,3)$

7. (a) 2^{64} (b) 2^6 (c) 2^6

9. $m + k = 2^n$

11. (a) $xy + (x+y)\overline{z} + y = y(x+1) + (x+y)\overline{z} = y + x\overline{z} + y\overline{z} = y(1+\overline{z}) + x\overline{z} = y + x\overline{z}$.

(b) $x + y + \overline{(\overline{x} + y + z)} = x + y + (x\overline{y}\,\overline{z}) = x(1 + \overline{y}\,\overline{z}) + y = x + y$.

(c) $yz + wz + z + [wz(xy + wz)] = z(y+1) + wx + wxyz + wz = z + wx(1+yz) + wz = z + wx + wz = z(1+w) + wx = z + wx$.

13. (a)

(i)

f	g	h	fg	$\overline{f}h$	gh	$fg+\overline{f}h+gh$	$fg+\overline{f}h$
0	0	0	0	0	0	0	0
0	0	1	0	1	0	1	1
0	1	0	0	0	0	0	0
0	1	1	0	1	1	1	1
1	0	0	0	0	0	0	0
1	0	1	0	0	0	0	0
1	1	0	1	0	0	1	1
1	1	1	1	0	1	1	1

Alternately, $fg+\overline{f}h = (fg+\overline{f})(fg+h) = (f+\overline{f})(g+\overline{f})(fg+h) = \mathbf{1}(g+\overline{f})(fg+h) = fgg+gh+\overline{f}fg+\overline{f}h = fg+gh+\mathbf{0}g+\overline{f}h = fg+gh+\overline{f}h.$

(ii) $fg+f\overline{g}+\overline{f}g+\overline{f}\overline{g} = f(g+\overline{g})+\overline{f}(g+\overline{g}) = f\cdot\mathbf{1}+\overline{f}\cdot\mathbf{1} = f+\overline{f} = \mathbf{1}$

(b) (i) $(f+g)(f+h)(g+h) = (f+g)(\overline{f}+h)$

(ii) $(f+g)(f+\overline{g})(\overline{f}+g)(\overline{f}+\overline{g}) = \mathbf{0}$

15. (a) $f\oplus f=\mathbf{0}$; $f\oplus\overline{f}=\mathbf{1}$; $f\oplus\mathbf{1}=\overline{f}$; $f\oplus\mathbf{0}=f$

(b) (i) $f\oplus g=\mathbf{0} \Leftrightarrow f\overline{g}+\overline{f}g=\mathbf{0} \Rightarrow f\overline{g}+\overline{f}g=\mathbf{0}$. $[f=1$, and $f\overline{g}=\mathbf{0}] \Rightarrow g=1$. $[f=0$ and $\overline{f}g=\mathbf{0}] \Rightarrow g=0$. Hence $f=g$.

(ii)

$$\begin{aligned}
f\oplus(g\oplus h) &= f\oplus(g\overline{h}+\overline{g}h)\\
&= f\overline{(g\overline{h}+\overline{g}h)}+\overline{f}(g\overline{h}+\overline{g}h)\\
&= f\overline{(g\overline{h})}\,\overline{(\overline{g}h)}+\overline{f}g\overline{h}+\overline{f}\overline{g}h\\
&= f(\overline{g}+h)(g+\overline{h})+\overline{f}g\overline{h}+\overline{f}\overline{g}h\\
&= f(\overline{g}g+\overline{g}\overline{h}+hg+h\overline{h})+\overline{f}g\overline{h}+\overline{f}\overline{g}h\\
&= f(\mathbf{0}+\overline{g}\overline{h}+hg+\mathbf{0})+\overline{f}g\overline{h}+\overline{f}\overline{g}h\\
&= f\overline{g}\overline{h}+fgh+\overline{f}g\overline{h}+\overline{f}\overline{g}h
\end{aligned}$$

$$\begin{aligned}
(f\oplus g)\oplus h &= (f\overline{g}+\overline{f}g)\oplus h\\
&= (f\overline{g}+\overline{f}g)\overline{h}+\overline{(f\overline{g}+\overline{f}g)}h\\
&= (f\overline{g}\overline{h}+\overline{f}g\overline{h})+\overline{(f\overline{g})}\,\overline{(\overline{f}g)}h\\
&= (f\overline{g}\overline{h}+\overline{f}g\overline{h}+(\overline{f}+g)(f+\overline{g})h\\
&= f\overline{g}\overline{h}+\overline{f}g\overline{h}+(\overline{f}f+\overline{f}\overline{g}+gf+g\overline{g})h\\
&= f\overline{g}\overline{h}+\overline{f}g\overline{h}+(\mathbf{0}+\overline{f}\overline{g}+gf+\mathbf{0})h\\
&= f\overline{g}\overline{h}+\overline{f}g\overline{h}+\overline{f}\overline{g}h+fgh
\end{aligned}$$

(iii) $\overline{f}\oplus\overline{g} = \overline{f}\overline{\overline{g}}+\overline{\overline{f}}\overline{g} = \overline{f}g+f\overline{g} = f\overline{g}+\overline{f}g = f\oplus g$

(iv) This is the only result that is not true. When f has value 1, g has value 0 and h value 1 (or g has value 1 and h value 0), then $f \oplus gh$ has value 1 but $(f \oplus g)(f \oplus h)$ has value 0.

(v) $fg \oplus fh = \overline{fg}fh + fg\overline{fh} = (\overline{f} + \overline{g})fh + fg(\overline{f} + \overline{h}) = \overline{f}fh + f\overline{g}h + f\overline{f}g + fg\overline{h} = f\overline{g}h + fg\overline{h} = f(\overline{g}h + g\overline{h}) = f(g \oplus h)$.

(vi) $\overline{f} \oplus g = \overline{f}\,\overline{g} + fg = fg + \overline{f}\,\overline{g} = f \oplus \overline{g}$.
$\overline{f \oplus g} = \overline{f\overline{g} + \overline{f}g} = (\overline{f} + g)(f + \overline{g}) = \overline{f}\,\overline{g} + fg = \overline{f} \oplus g$.

(vii) $[f \oplus g = f \oplus h] \Rightarrow [f \oplus (f \oplus g) = f \oplus (f \oplus h)] \Rightarrow [(f \oplus f) \oplus g = (f \oplus f) \oplus h] \Rightarrow [\mathbf{0} \oplus g = \mathbf{0} \oplus h] \Rightarrow [g = h]$.

Section 15.2

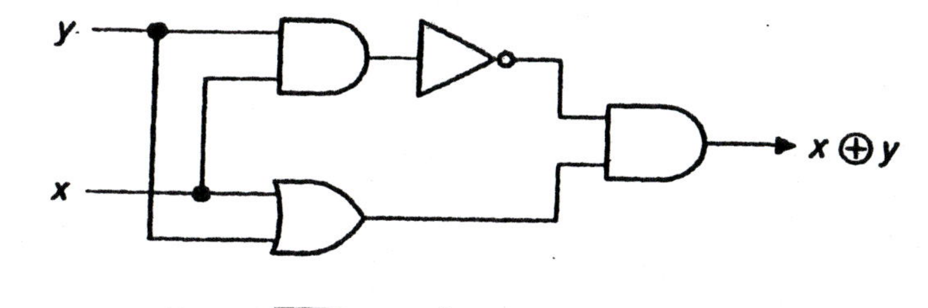

1. (a) $x \oplus y = (x + y)(\overline{xy})$

   (b) $\overline{xy}$

   (c) $\overline{x + y}$

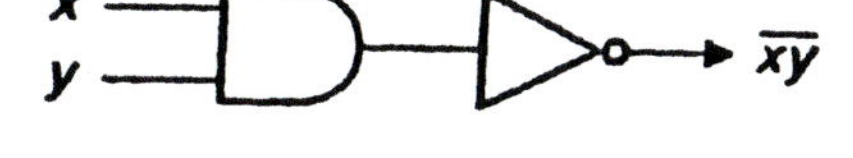

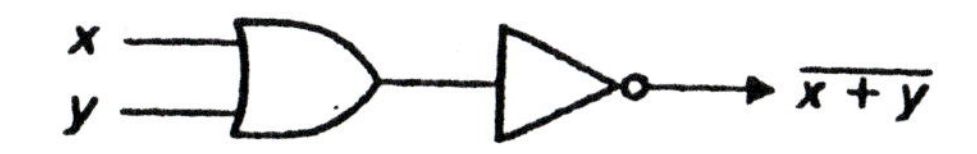

3. (a)

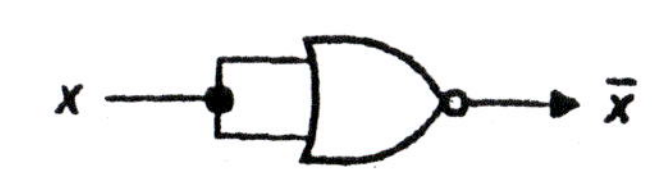

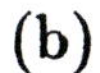

(b)

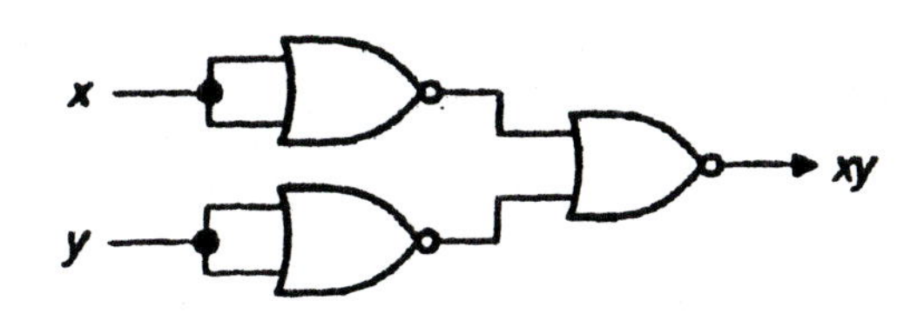

(c)

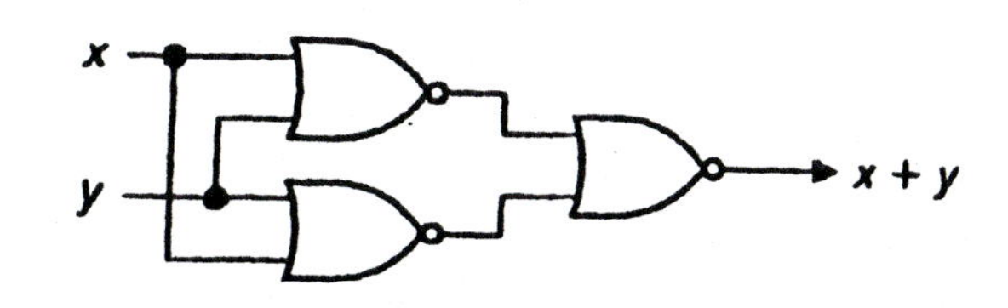

5. $f(w, x, y, z) = \overline{w}\,\overline{x}y\overline{z} + (w + x + \overline{y})z$

7. a) The output is $(x+\overline{y})(x+y)+y$. This simplifies to $x+(\overline{y}y)+y = x+0+y = x+y$ and provides us with the simpler equivalent network in part (a) of the figure.

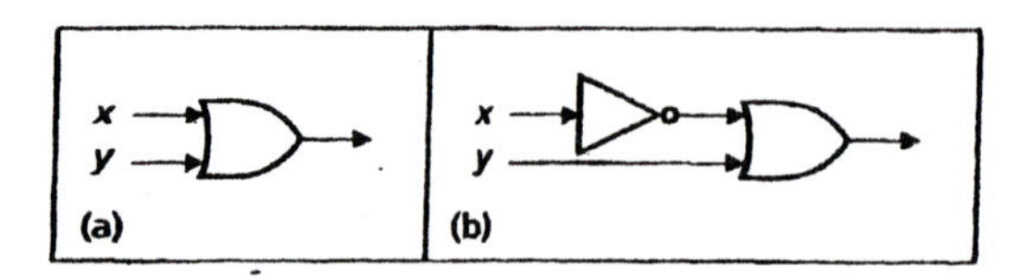

b) Here the output is $\overline{(x+\overline{y})}+(\overline{x}\,\overline{y}+y)$ which simplifies to $\overline{x}\,\overline{\overline{y}}+\overline{x}\,\overline{y}+y = \overline{x}y+\overline{x}\,\overline{y}+y = \overline{x}(y+\overline{y})+y = \overline{x}(1)+y = \overline{x}+y$. This accounts for the simpler equivalent network in part (b) of the figure.

9. (a)

$w \backslash xy$	00	01	11	10
0		1		1
1		1		1

$f(w,x,y) = \overline{x}y + x\overline{y}$

(b)

$w \backslash xy$	00	01	11	10
1	0	0		
1	0	0		

$f(w,x,y) = x$

(c)

$wx \backslash yz$	00	01	11	10
00	1			1
01		1	1	
11		1	1	
10	1			1

$f(w,x,y,z) = xz + \overline{x}\,\overline{z}$

(d)

$wx \backslash yz$	00	01	11	10
00				
01		1		1
11	1	1	1	1
10	1		1	

$f(w,x,y,z) = w\overline{y}\,\overline{z} + x\overline{y}z + wyz + xy\overline{z}$

(e)

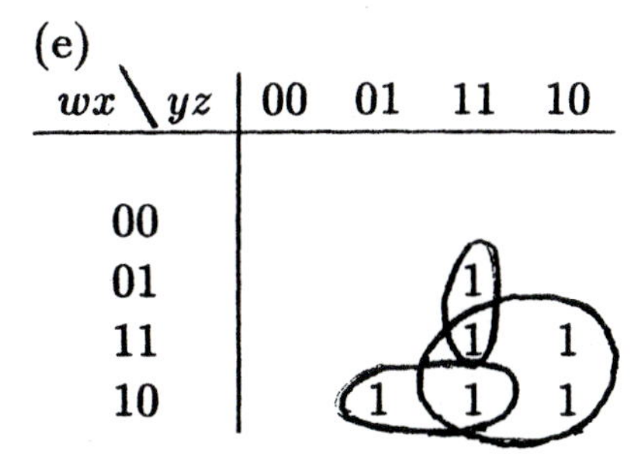

$f(w,x,y,z) = wy + xyz + w\bar{x}z$

(f)

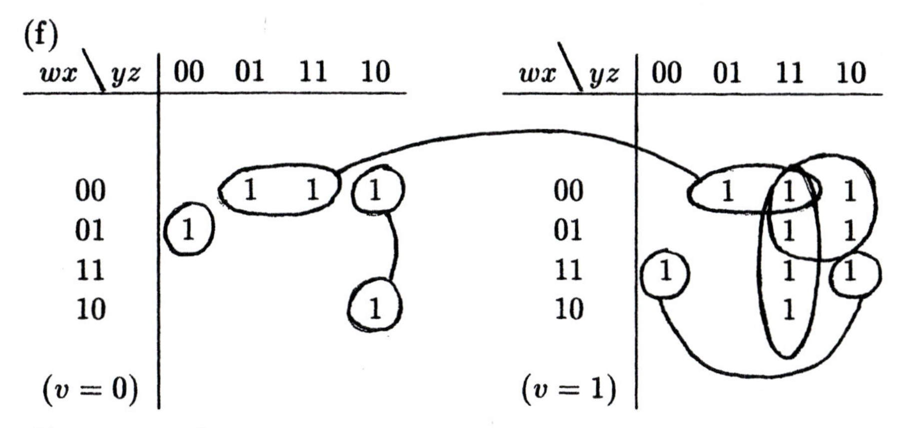

$f(v,w,x,y,z) = \bar{v}\,\bar{w}x\bar{y}\,\bar{z} + vwx\bar{z} + \bar{v}\,\bar{x}y\bar{z} + \bar{w}\,\bar{x}z + v\bar{w}y + vyz$

11. (a) 2 (b) 3 (c) 4 (d) $k+1$

13.

(a) $|f^{-1}(0)| = |f^{-1}(1)| = 8$
(b) $|f^{-1}(0)| = 12, \quad |f^{-1}(1)| = 4$
(c) $|f^{-1}(0)| = 14, \quad |f^{-1}(1)| = 2$
(d) $|f^{-1}(0)| = 4, \quad |f^{-1}(1)| = 12$
(e) $|f^{-1}(0)| = 6, \quad |f^{-1}(1)| = 10$
(f) $|f^{-1}(0)| = 7, \quad |f^{-1}(1)| = 9$

Section 15.3

1. $f(u,v,w,x,y,z) = (v+w+x+y)(u+w)(v+z)(u+y+z) =$
$(uv+uw+ux+uy+vw+w+wx+wy)(v+z)(u+y+z) =$
$(uv+ux+uy+(u+v+1+x+y)w)(v+z)(u+y+z) =$
$(uv+ux+uy+w)(uv+vy+vz+uz+yz+z) =$
$(uv+ux+uy+w)(uv+vy+z) =$
$(uv+uvx+uvy+uvw+uvy+uvxy+uvy+wvy+uvz+uxz+uyz+wz) =$
$uv+wvy+uxz+uyz+wz$

3. (a)

$wx \backslash yz$	00	01	11	10
00		1	1	
01		1	1	
11	√	√	√	√
10		1	√	√

$f(w,x,y,z) = z$

(b)

$wx \backslash yz$	00	01	11	10
00	1			
01	√	1		1
11		1		1
10	1	√	√	

$f(w,x,y,z) = \overline{x}\,\overline{y}\,\overline{z} + x\overline{y}z + xy\overline{z}$

(c)

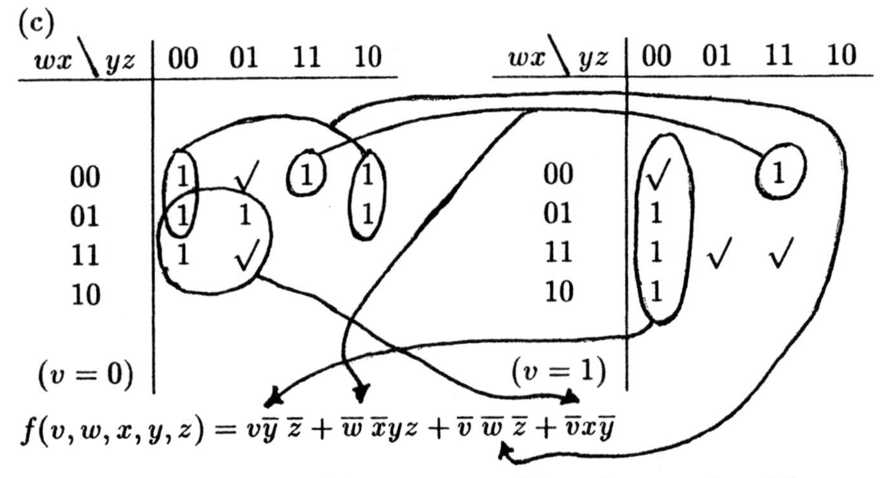

$f(v,w,x,y,z) = v\overline{y}\,\overline{z} + \overline{w}\,\overline{x}yz + \overline{v}\,\overline{w}\,\overline{z} + \overline{v}x\overline{y}$

5. (a) $(a+b+c+d+e)(a+b+c+f)(a+b+c+d+f)(a+c+d+e+g)\cdot$
$(a+d+e+g)(b+c+f+g)(d+e+f+g) = (a+b+c+d+e)\cdot$
$(a+b+c+f)(a+d+e+g)(b+c+f+g)(d+e+f+g) =$
$(a+b+c+df+ef)(a+d+e+g)(b+c+f+g)(d+e+f+g) =$
$(a+b+c+df+ef)(d+e+g+af)(b+c+f+g) = [(b+c)+(a+df+ef)(f+g)](d+e+g+af) =$
$[b+c+af+df+ag+dfg+efg](d+e+g+af) = [b+c+af+df+ef+ag](d+e+g+af) =$
$bd+cd+adf+df+def+adg+be+ce+aef+def+ef+aeg+bg+cg+afg+dfg+efg+ag+abf$
$+ acf + af + adf + aef + afg = bd + cd + df + ag + ef + be + ce + af + bg + cg.$

Section 15.4

1. (The second Distributive Law). Let $x = 2^{k_1}3^{k_2}5^{k_3}$, $y = 2^{m_1}3^{m_2}5^{m_3}$, $z = 2^{n_1}3^{n_2}5^{n_3}$ where

for $1 \le i \le 3$, $0 \le k_i, m_i, n_i \le 1$.
$\gcd(y,z) = 2^{s_1}3^{s_2}5^{s_3}$ where $s_i = \min\{m_i, n_i\}, 1 \le i \le 3$. $\text{lcm}(x, \gcd(y,z)) = 2^{t_1}3^{t_2}5^{t_3}$ where $t_i = \max\{k_i, s_i\}$, $1 \le i \le 3$. Also, $\text{lcm}(x,y) = 2^{u_1}3^{u_2}5^{u_3}$, $\text{lcm}(x,z) = 2^{v_1}3^{v_2}5^{v_3}$ where $u_i = \max\{k_i, m_i\}$, $v_i = \max\{k_i, n_i\}$, $1 \le i \le 3$, and $\gcd(\text{lcm}(x,y), \text{lcm}(x,z)) = 2^{w_1}3^{w_2}5^{w_3}$ where $w_i = \min\{u_i, v_i\}$, $1 \le i \le 3$. To prove that $\text{lcm}(x, \gcd(y,z)) = \gcd(\text{lcm}(x,y), \text{lcm}(x,z))$ we need to show that $t_i = w_i$, $1 \le i \le 3$. If $k_i = 0$, then $t_i = s_i$, $u_i = m_i$, $v_i = n_i$ and $w_i = \min\{u_i, v_i\} = \min\{m_i, n_i\} = s_i = t_i$. If $k_i = 1$, then $t_i = 1 = u_i = v_i = w_i$.

(The Identity Laws) $x + 0 =$ the lcm of x and 1 (the zero element) $= x$; $x \cdot 1 =$ the gcd of x and 30 (the one element) $= x$, since x is a divisor of 30.

(The Inverse Laws) $x + \overline{x} =$ the lcm of x and $30/x = 30$ (the one element of this Boolean algebra); $x\overline{x} =$ the gcd of x and $30/x = 1$ (the zero element of the Boolean algebra).

3. (a) $30 + 5 \cdot 7 = \text{lcm}(30, \gcd(5,7)) = \text{lcm}(30,1) = 30$
(b) $(30+5) \cdot (30+7) = \gcd(\text{lcm}(30,5), \text{lcm}(30,7)) = \gcd(30, 210) = 30$ [Note: $30 + 5 \cdot 7 = (30+5) \cdot (30+7)$ – by the Distributive Law of + over ·.]
(c) $\overline{(14+15)} = \overline{\text{lcm}(14,15)} = \overline{210} = 210/210 = 1$
(d) $21(2 + \overline{10}) = 21(2 + (\frac{210}{10})) = 21(2+21) = \gcd(21, \text{lcm}(2,21)) = \gcd(21,42) = 21$
(e) $(2+3)+5 = \text{lcm}(\text{lcm}(2,3),5) = \text{lcm}(6,5) = 30$
(f) $(6+35)(7+10) = \gcd(\text{lcm}(6,35), \text{lcm}(7,10)) = \gcd(210,70) = 70$

5. (a) $w \le 0 \Rightarrow w \cdot 0 = w$. But $w \cdot 0 = 0$, by part (a) of Theorem 15.3.
(b) $1 \le x \Rightarrow 1 \cdot x = 1$, and $1 \cdot x = x$ from our defintion of a Boolean algebra.
(c) $y \le z \Rightarrow yz = y$, and $y \le \overline{z} \Rightarrow y\overline{z} = y$. Therefore $y = yz = (y\overline{z})z = y(\overline{z}z) = y \cdot 0 = 0$.

7. $x \le y \Longleftrightarrow xy = x$. The dual of $xy = x$ is $x + y = x$.

$x + y = x \Longrightarrow xy = (x+y)y = xy + y = y(x+1) = y \cdot 1 = y$, and $xy = y \Longleftrightarrow y \le x$. Consequently, the dual of $x \le y$ is $y \le x$.

9. From Theorem 15.5(a), with x_1, x_2 distinct atoms, if $x_1, x_2 \ne 0$, then $x_1 = x_1x_2 = x_2x_1 = x_2$, a contradiction.

11. (a) Let $w \in B_2$. Then there exists a unique $x \in B_1$ such that $f(x) = w$. Consequently, $f(x) = f(x+0) = f(x) + f(0)$, or $w = w + f(0)$. Since the zero element of a Boolean algebra is unique (from Exercise 10 of this section), it follows that $f(0) = 0$.
(b) $f(0) = 0 \Rightarrow \overline{f(0)} = \overline{0} \Rightarrow f(\overline{0}) = \overline{0} \Rightarrow f(1) = 1$.
(c) $x \le y \Rightarrow xy = x \Rightarrow f(xy) = f(x) \Rightarrow f(x)f(y) = f(x) \Rightarrow f(x) \le f(y)$.
(d) Since x is an atom of $B_1, x \ne 0$ so $f(x) \ne 0$. Let $y \in B_2$ with $0 \ne y$ and $y \le f(x)$. With f an isomorphism there exists $z \in B_1$ with $f(z) = y$. Also, $f^{-1}: B_2 \longrightarrow B_1$ is an isomorphism so $f(z) \le f(x) \Longrightarrow z \le x$. With x an atom and $0 < z \le x$ we have $z = x$ so $f(z) = y = f(x)$, and $f(x)$ is an atom.

13. (a) $f(xy) = f(\overline{\overline{x}+\overline{y}}) = \overline{f(\overline{x}+\overline{y})} = \overline{f(\overline{x})+f(\overline{y})} = \overline{f(\overline{x})}\cdot\overline{f(\overline{y})} = f(\overline{\overline{x}})\cdot f(\overline{\overline{y}}) = f(x)\cdot f(y)$.
(b) Let B_1, B_2 be Boolean algebras with $f: B_1 \longrightarrow B_2$ one-to-one and onto. Then f is an isomorphism if $f(\overline{x}) = \overline{f(x)}$ and $f(xy) = f(x)f(y)$ for all $x, y \in B_1$. [Follows from part (a) by duality.]

15. For each $1 \leq i \leq n$, $(x_1+x_2+\ldots+x_n)x_i = x_1x_i + x_2x_i + \ldots + x_{i-1}x_i + x_ix_i + x_{i+1}x_i + \ldots + x_nx_i = 0+0+\ldots+0+x_i+0+\ldots+0 = x_i$, by part (b) of Theorem 15.5. Consequently, it follows from Theorem 15.7 that $(x_1+x_2+\ldots+x_n)x = x$ for all $x \in B$. Since the one element is unique (from Exercise 10) we conclude that $1 = x_1+x_2+\ldots+x_n$.

Supplementary Exercises

1. (a) When $n=2$, x_1+x_2 denotes the Boolean sum of x_1 and x_2. For $n \geq 2$, we define $x_1+x_2+\ldots+x_n+x_{n+1}$ recursively by $(x_1+x_2+\ldots+x_n)+x_{n+1}$. [A similar definition can be given for the Boolean product.]
For $n=2$, $\overline{x_1+x_2} = \overline{x}_1\overline{x}_2$ is true, for this is one of the DeMorgan Laws. Assume the result for $n=k\ (\geq 2)$ and consider the case of $n=k+1$. $\overline{(x_1+x_2+\ldots+x_k+x_{k+1})} = \overline{(x_1+x_2+\ldots+x_k)+x_{k+1}} = \overline{(x_1+x_2+\ldots+x_k)}\cdot\overline{x_{k+1}} = \overline{x}_1\overline{x}_2\cdots\overline{x}_k\overline{x}_{k+1}$. Consequently, the result follows for all $n \geq 2$ by the Principle of Mathematical Induction.
(b) Follows from part (a) by duality.

3. Let v, w, x, y, z indicate that Eileeen invites Margaret, Joan, Kathleen, Nettie, and Cathy, respectively. The conditons in (a) – (e) can then be expressed as

(a) $(v \rightarrow w) \Longleftrightarrow (\overline{v}+w)$ (b) $(x \rightarrow vy) \Longleftrightarrow (\overline{x}+vy)$

(c) $\overline{w}z + w\overline{z}$ (d) $yz + \overline{y}\,\overline{z}$ (e) $x+y+xy \Longleftrightarrow x+y$

$(\overline{v}+w)(\overline{x}+vy)(\overline{w}z+w\overline{z})(yz+\overline{y}\,\overline{z})(x+y) \Longleftrightarrow (\overline{v}+w)(\overline{x}y+vy)(\overline{w}z+w\overline{z})(yz+\overline{y}\,\overline{z})$
$\Longleftrightarrow (\overline{v}+w)(\overline{x}y+vy)(\overline{w}yz+w\overline{y}\,\overline{z}) \Longleftrightarrow (\overline{v}+w)(\overline{w}\,\overline{x}yz+\overline{w}vyz) \Longleftrightarrow \overline{v}\,\overline{w}\,\overline{x}yz$

Consequently, the only way Eileen can have her party and satisfy conditions (a) – (e) is to invite only Nettie and Cathy out of this group of five of her friends.

5. Proof: If $x \leq z$ and $y \leq z$ then from Exercise 6(b) of Section 15.4 we have $x+y \leq z+z$. And by the idempotent law we have $z+z=z$.
Conversely, suppose that $x+y \leq z$. We find that $x \leq x+y$, because $x(x+y) = x+xy$ (by the idempotent law) $= x$ (by the absorption law). Since $x \leq x+y$ and $x+y \leq z$ we have $x \leq z$, because a partial order is transitive. [The proof that $y \leq z$ follows in a similar way.]

7. Proof:
(a) $x \leq y \Rightarrow x+\overline{x} \leq y+\overline{x} \Rightarrow 1 \leq y+\overline{x} \Rightarrow y+\overline{x} = \overline{x}+y = 1$. Conversely, $\overline{x}+y = 1 \Rightarrow x(\overline{x}+y) = x\cdot 1 \Rightarrow x\overline{x}(=0)+xy = x \Rightarrow xy = x \Rightarrow x \leq y$.
(b) $x \leq \overline{y} \Rightarrow x\overline{y} = x \Rightarrow xy = (x\overline{y})y = x(\overline{y}y) = x\cdot 0 = 0$. Conversely, $xy = 0 \Rightarrow x = x\cdot 1 =$

$x(y + \overline{y}) = xy + x\overline{y} = x\overline{y}$, and $x = x\overline{y} \Rightarrow x \leq \overline{y}$.

9. (a)

$wx \backslash yz$	00	01	11	10
00	1	1	1	1
01			1	1
11			1	1
10				

$f(w, x, y, z) = \overline{w}\,\overline{x} + xy$

(b)

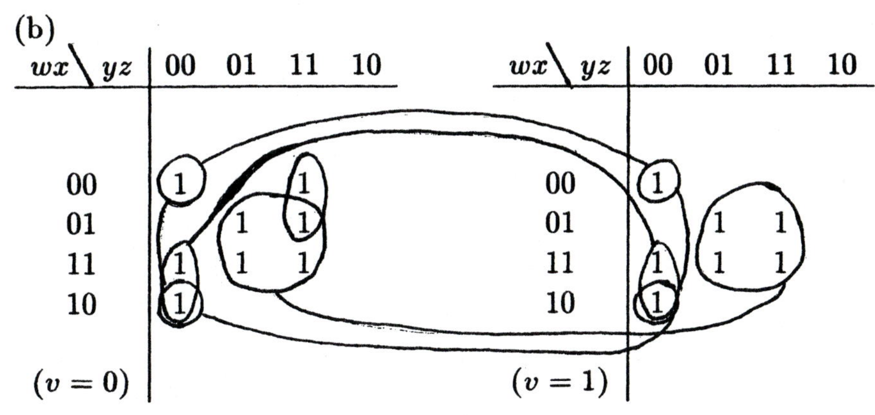

$g(v, w, x, y, z) = \overline{v}\,\overline{w}yz + xz + w\overline{y}\,\overline{z} + \overline{x}\,\overline{y}\,\overline{z}$

11. (a) $2^{(2^{n-1})}$ (b) 2^4; 2^{n+1}

13. (a) $60 = 2^2 \cdot 3 \cdot 5$ so there are 12 divisors of 60. Since 12 is not a power of 2 these divisors cannot yield a Boolean algebra.
(b) $120 = 2^3 \cdot 3 \cdot 5$ and there are 16 divisors of 60. Let $x = 4$. Then $\overline{x} = 120/4 = 30$ and $x \cdot \overline{x} =$ gcd of x and $\overline{x} = \gcd(4, 30) = 2$, not 1. Hence although $16 = 2^4$ the divisors of 120 do not yield a Boolean algebra.

CHAPTER 16
GROUPS, CODING THEORY, AND POLYA'S METHOD OF ENUMERATION

Section 16.1

1. (a) Yes. The identity is 1 and each element is its own inverse.
(b) No. The set is not closed under addition and there is no identity.
(c) No. The set is not closed under addition.
(d) Yes. The identity is 0; the inverse of $10n$ is $10(-n)$ or $-10n$.
(e) Yes. The identity is 1_A and the inverse of $g: A \rightarrow A$ is $g^{-1}: A \rightarrow A$.
(f) Yes. The identity is 0; the inverse of $a/(2^n)$ is $(-a)/(2^n)$.

3. Subtraction is not an associative (closed) binary operation – e.g., $(3-2)-4 = -3 \neq 5 = 3-(2-4)$.

5. Since $x, y \in \mathbf{Z} \Longrightarrow x+y+1 \in \mathbf{Z}$, the operation is a (closed) binary operation (or $\mathbf{Z}$ is closed under $\circ$). For all $w, x, y \in \mathbf{Z}$, $w \circ (x \circ y) = w \circ (x+y+1) = w+(x+y+1)+1 = (w+x+1)+y+1 = (w \circ x) \circ y$, so the (closed) binary operation is associative. Furthermore, $x \circ y = x+y+1 = y+x+1 = y \circ x$, for all $x, y \in \mathbf{Z}$, so $\circ$ is also commutative. If $x \in \mathbf{Z}$ then $x \circ (-1) = x+(-1)+1 = x[= (-1) \circ x]$, so -1 is the identity element for $\circ$. And finally, for each $x \in \mathbf{Z}$, we have $-x-2 \in \mathbf{Z}$ and $x \circ (-x-2) = x+(-x-2)+1 = -1[= (-x-2)+x]$, so $-x-2$ is the inverse for x under $\circ$. Consequently, $(\mathbf{Z}, \circ)$ is an abelian group.

7. $U_{20} = \{x | x \in \mathbf{Z}_{20}, \gcd(x, 20) = 1\} = \{1, 3, 7, 9, 11, 13, 17, 19\}$

$U_{24} = \{x | x \in \mathbf{Z}_{24}, \gcd(x, 24) = 1\} = \{1, 5, 7, 11, 13, 17, 19, 23\}$

9. (a) The result follows from Theorem 16.1(b) since both $(a^{-1})^{-1}$ and a are inverses of a^{-1}.
(b) $(b^{-1}a^{-1})(ab) = b^{-1}(a^{-1}a)b = b^{-1}(e)b = b^{-1}b = e$ and $(ab)(b^{-1}a^{-1}) = a(bb^{-1})a^{-1} = a(e)a^{-1} = aa^{-1} = e$. So $b^{-1}a^{-1}$ is an inverse of ab, and by Theorem 16.1(b), $(ab)^{-1} = b^{-1}a^{-1}$.

11. (a) $\{0\}$; $\{0,6\}$; $\{0,4,8\}$; $\{0,3,6,9\}$; $\{0,2,4,6,8,10\}$; $\mathbf{Z}_{12}$.
(b) $\{1\}$; $\{1,10\}$; $\{1,3,4,5,9\}$; $\mathbf{Z}_{11}^*$.
(c) $\{\pi_0\}$; $\{\pi_0, \pi_1, \pi_2\}$; $\{\pi_0, r_1\}$; $\{\pi_0, r_2\}$; $\{\pi_0, r_3\}$; S_3

13. (a) There are 10: five rotations through $i(72^\circ), 0 \leq i \leq 4$, and five reflections about lines containing a vertex and the midpoint of the opposite side.

(b) For a regular n-gon ($n \geq 3$) there are $2n$ rigid motions. There are the n rotations through $i(360°/n)$, $0 \leq i \leq n-1$. There are n reflections. For n odd each reflection is about a line through a vertex and the midpoint of the opposite side. For n even, there are $n/2$ reflections about lines through opposite vertices and $n/2$ reflections about lines through the midpoints of opposite sides.

15. Since $eg = ge$ for all $g \in G$, it follows that $e \in H$ and $H \neq \emptyset$. If $x, y \in H$, then $xg = gx$ and $yg = gy$ for all $g \in G$. Consequently, $(xy)g = x(yg) = x(gy) = (xg)y = (gx)y = g(xy)$ for all $g \in G$, and we have $xy \in H$. Finally, for all $x \in H$ and $g \in G$, $xg^{-1} = g^{-1}x$. So $(xg^{-1})^{-1} = (g^{-1}x)^{-1}$, or $gx^{-1} = x^{-1}g$, and $x^{-1} \in H$. Therefore H is a subgroup of G.

17. (a) Let $(g_1, h_1), (g_2, h_2) \in G \times H$. Then $(g_1, h_1) \cdot (g_2, h_2) = (g_1 \circ g_2, h_1 * h_2)$, where $g_1 \circ g_2 \in G$, $h_1 * h_2 \in H$, since $(G, \circ)$ and $(H, *)$ are closed. Hence $G \times H$ is closed. For $(g_1, h_1), (g_2, h_2), (g_3, h_3) \in G \times H$, $[(g_1, h_1) \cdot (g_2, h_2)] \cdot (g_3, h_3) = (g_1 \circ g_2, h_1 * h_2) \cdot (g_3, h_3) = ((g_1 \circ g_2) \circ g_3, (h_1 * h_2) * h_3) = (g_1 \circ (g_2 \circ g_3), h_1 * (h_2 * h_3)) = (g_1, h_1) \cdot (g_2 \circ g_3, h_2 * h_3) = (g_1, h_1) \cdot [(g_2, h_2) \cdot (g_3, h_3)]$, since the operations in G and H are associative. Hence, $G \times H$ is associative under $\cdot$.
Let e_G, e_H denote the identities for G, H, respectively. Then (e_G, e_H) is the identity in $G \times H$.
Finally, let $(g, h) \in G \times H$. If g^{-1} is the inverse of g in G and h^{-1} is the inverse of h in H, then (g^{-1}, h^{-1}) is the inverse of (g, h) in $G \times H$.

(b) (i) $|\mathbf{Z}_6| \times |\mathbf{Z}_6| \times |\mathbf{Z}_6| = 6 \times 6 \times 6 = 216$
(ii) $H_1 = \{(x, 0, 0) | x \in \mathbf{Z}_6\}$ is a subgroup of order 6; $H_2 = \{(x, y, 0) | x, y \in \mathbf{Z}_6,\ y = 0, 3\}$ is a subgroup of order 12; $H_3 = \{(x, y, 0) | x, y \in \mathbf{Z}_6\}$ has order 36.
(iii) $-(2,3,4) = (4,3,2)$; $-(4,0,2) = (2,0,4)$; $-(5,1,2) = (1,5,4)$.

19. (a) $x = 1$, $x = 4$ (b) $x = 1$, $x = 10$

(c) $x = x^{-1} \Rightarrow x^2 \equiv 1 \pmod p \Rightarrow x^2 - 1 \equiv 0 \pmod p \Rightarrow (x-1)(x+1) \equiv 0 \pmod p \Rightarrow x - 1 \equiv 0 \pmod p$ or $x + 1 \equiv 0 \pmod p \Rightarrow x \equiv 1 \pmod p$ or $x \equiv -1 \equiv p - 1 \pmod p$.

(d) The result is true for $p = 2$, since $(2-1)! = 1! \equiv -1 \pmod 2$. For $p \geq 3$, consider the elements $1, 2, \ldots, p-1$ in $(\mathbf{Z}_p^*, \cdot)$, The elements $2, 3, \ldots, p-2$ yield $(p-3)/2$ pairs of the form x, x^{-1}. [For example, when $p = 11$ we find that $2, 3, 4, \ldots, 9$ yield the four pairs 2,6; 3,4; 5,9; 7,8.] Consequently, $(p-1)! \equiv (1)(1)^{(p-3)/2}(p-1) \equiv p - 1 \equiv -1 \pmod p$.

Section 16.2

1. (b) $f(a) * f(a)^{-1} = e_H = f(e_G) = f(a \circ a^{-1}) = f(a) * f(a^{-1})$. By left cancellation in the group $(H, *)$, it follows that $f(a)^{-1} = f(a^{-1})$.
(c) If $n = 0$, the result follows from part (a) of Theorem 16.5. So consider $n \in \mathbf{Z}^+$.

For $n = 1$, $f(a^n) = f(a^1) = f(a) = [f(a)]^1 = [f(a)]^n$, so the result follows for $n = 1$. Now assume the result true for $n = k$ (≥ 1) and consider $n = k + 1$. Then $f(a^n) = f(a^{k+1}) = f(a^k \cdot a) = f(a^k) \cdot f(a) = [f(a)]^k \cdot f(a) = [f(a)]^{k+1} = [f(a)]^n$. So by the Principle of Mathematical Induction, the result is true for all $n \geq 1$.

For $n \geq 1$, we have $a^{-n} = (a^{-1})^n$ – as defined in the material following Theorem 16.1. So $f(a^{-n}) = f[(a^{-1})^n] = [f(a^{-1})]^n$ by our previous work. Then $[f(a^{-1})]^n = [(f(a))^{-1}]^n = [f(a)]^{-n}$ – by part (b) of Theorem 16.1. Hence $f(a^{-n}) = [f(a)]^{-n}$.

Consequently, $f(a^n) = [f(a)]^n$, for all $a \in G$ and all $n \in \mathbf{Z}$.

3. $f(0) = (0,0)$ $\quad f(1) = (1,1)$ $\quad f(2) = (2,0)$
$f(3) = (0,1)$ $\quad f(4) = (1,0)$ $\quad f(5) = (2,1)$

5. We need to express the element $(4,6)$ of $\mathbf{Z} \times \mathbf{Z}$ in terms of the elements $(1,3)$ and $(3,7)$, so let us write

$$(4,6) = a(1,3) \oplus b(3,7), \quad \text{where } a, b \in \mathbf{Z}.$$

Then $f(4,6) = f(a(1,3) \oplus b(3,7)) = f(a(1,3)) + f(b(3,7)) = af(1,3) + bf(3,7)$.
With $(4,6) = a(1,3) \oplus b(3,7)$ we have $4 = a + 3b$ and $6 = 3a + 7b$, from which it follows that $a = -5$ and $b = 3$.
Consequently, $f(4,6) = -5g_1 + 3g_2$.

7. (a) $o(\pi_0) = 1$, $o(\pi_1) = o(\pi_2) = 3$, $o(r_1) = o(r_2) = o(r_3) = 2$.

(b) (See Fig. 16.6) $o(\pi_0) = 1$, $o(\pi_1) = o(\pi_3) = 4$, $o(\pi_2) = o(r_1) = o(r_2) = o(r_3) = o(r_4) = 2$.

9. (a) The elements of order 10 are 4, 12, 28, and 36.

(b) The elements of order 10 are a^4, a^{12}, a^{28}, and a^{36}.

11. $\mathbf{Z}_5^* = \langle 2 \rangle = \langle 3 \rangle$; $\quad \mathbf{Z}_7^* = \langle 3 \rangle = \langle 5 \rangle$; $\quad \mathbf{Z}_{11}^* = \langle 2 \rangle = \langle 6 \rangle = \langle 7 \rangle = \langle 8 \rangle$.

13. Let $(G,+)$, $(H,*)$, $(K,\cdot)$ be the given groups. For any $x, y \in G$, $(g \circ f)(x + y) = g(f(x + y)) = g(f(x) * f(y)) = (g(f(x)) \cdot (g(f(y)) = ((g \circ f)(x)) \cdot ((g \circ f)(y))$, since f, g are homomorphisms. Hence, $g \circ f : G \to K$ is a group homomorphism.

15. (a) $(\mathbf{Z}_{12}, +) = \langle 1 \rangle = \langle 7 \rangle = \langle 11 \rangle$
$(\mathbf{Z}_{16}, +) = \langle 1 \rangle = \langle 3 \rangle = \langle 5 \rangle = \langle 7 \rangle = \langle 9 \rangle = \langle 11 \rangle = \langle 13 \rangle = \langle 15 \rangle$
$(\mathbf{Z}_{24}, +) = \langle 1 \rangle = \langle 5 \rangle = \langle 7 \rangle = \langle 11 \rangle = \langle 13 \rangle = \langle 17 \rangle = \langle 19 \rangle = \langle 23 \rangle$

(b) Let $G = \langle a^k \rangle$. Since $G = \langle a \rangle$, $a = (a^k)^s$ for some $s \in \mathbf{Z}$. Then $a^{1-ks} = e$, so $1 - ks = tn$ since $o(a) = n$. $1 - ks = tn \Longrightarrow 1 = ks + tn \Longrightarrow \gcd(k,n) = 1$. Conversely, let $G = \langle a \rangle$ where $a^k \in G$ and $\gcd(k,n) = 1$. Then $\langle a^k \rangle \subseteq G$. $\gcd(k,n) = 1 \Longrightarrow 1 = ks + tn$, for some $s, t \in \mathbf{Z} \Longrightarrow a = a^1 = a^{ks+nt} = (a^k)^s (a^n)^t = (a^k)^s (e)^t = (a^k)^s \in \langle a^k \rangle$. Hence $G \subseteq \langle a^k \rangle$. So $G = \langle a^k \rangle$, or a^k generates G.

(c) $\phi(n)$.

Section 16.3

1. (a) $\left\{\begin{pmatrix}1&2&3&4\\2&3&4&1\end{pmatrix},\begin{pmatrix}1&2&3&4\\3&4&1&2\end{pmatrix},\begin{pmatrix}1&2&3&4\\4&1&2&3\end{pmatrix},\begin{pmatrix}1&2&3&4\\1&2&3&4\end{pmatrix}\right\}$

(b)

$$\begin{pmatrix}1&2&3&4\\2&1&4&3\end{pmatrix}H=\left\{\begin{pmatrix}1&2&3&4\\3&2&1&4\end{pmatrix},\begin{pmatrix}1&2&3&4\\4&3&2&1\end{pmatrix},\begin{pmatrix}1&2&3&4\\1&4&3&2\end{pmatrix},\begin{pmatrix}1&2&3&4\\2&1&4&3\end{pmatrix}\right\}$$

$$\begin{pmatrix}1&2&3&4\\4&2&3&1\end{pmatrix}H=\left\{\begin{pmatrix}1&2&3&4\\1&3&4&2\end{pmatrix},\begin{pmatrix}1&2&3&4\\2&4&1&3\end{pmatrix},\begin{pmatrix}1&2&3&4\\3&1&2&4\end{pmatrix},\begin{pmatrix}1&2&3&4\\4&2&3&1\end{pmatrix}\right\}$$

$$\begin{pmatrix}1&2&3&4\\1&2&4&3\end{pmatrix}H=\left\{\begin{pmatrix}1&2&3&4\\2&3&1&4\end{pmatrix},\begin{pmatrix}1&2&3&4\\3&4&2&1\end{pmatrix},\begin{pmatrix}1&2&3&4\\4&1&3&2\end{pmatrix},\begin{pmatrix}1&2&3&4\\1&2&4&3\end{pmatrix}\right\}$$

$$\begin{pmatrix}1&2&3&4\\1&3&2&4\end{pmatrix}H=\left\{\begin{pmatrix}1&2&3&4\\2&4&3&1\end{pmatrix},\begin{pmatrix}1&2&3&4\\3&1&4&2\end{pmatrix},\begin{pmatrix}1&2&3&4\\4&2&1&3\end{pmatrix},\begin{pmatrix}1&2&3&4\\1&3&2&4\end{pmatrix}\right\}$$

$$\begin{pmatrix}1&2&3&4\\1&4&2&3\end{pmatrix}H=\left\{\begin{pmatrix}1&2&3&4\\2&1&3&4\end{pmatrix},\begin{pmatrix}1&2&3&4\\3&2&4&1\end{pmatrix},\begin{pmatrix}1&2&3&4\\4&3&1&2\end{pmatrix},\begin{pmatrix}1&2&3&4\\1&4&2&3\end{pmatrix}\right\}$$

$$\begin{pmatrix}1&2&3&4\\1&2&3&4\end{pmatrix}H=H$$

3. Since the order of γ is 2, $|<\gamma>|=2$ and there are $|S_4|/|<\gamma>|=24/2=12$ cosets.

5. From Lagrange's Theorem we know that $|K|=66(=2\cdot3\cdot11)$ divides $|H|$ and that $|H|$ divides $|G|=660(=2^2\cdot3\cdot5\cdot11)$. Consequently, since $K\neq H$ and $H\neq G$, it follows that $|H|$ is $2(2\cdot3\cdot11)=132$ or $5(2\cdot3\cdot11)=330$.

7. (a)

$\cdot$	(1)(2)(3)(4)	(12)(34)	(13)(24)	(14)(23)
(1)(2)(3)(4)	(1)(2)(3)(4)	(12)(34)	(13)(24)	(14)(23)
(12)(34)	(12)(34)	(1)(2)(3)(4)	(14)(23)	(13)(24)
(13)(24)	(13)(24)	(14)(23)	(1)(2)(3)(4)	(12)(34)
(14)(23)	(14)(23)	(13)(24)	(12)(34)	(1)(2)(3)(4)

It follows from Theorem 16.3 that H is a subgroup of G. And since the entries in the above table are symmetric about the diagonal from the upper left to the lower right, we have H an abelian subgroup of G.

(b) Since $|G|=4!=24$ and $|H|=4$, there are $24/4=6$ left cosets of H in G.

(c) Consider the function $f:\ H\to\mathbf{Z}_2\times\mathbf{Z}_2$ defined by

$f:\ (1)(2)(3)(4)\to(0,0)$, $f:\ (12)(34)\to(1,0)$,
$f:\ (13)(24)\to(0,1)$, $f:\ (14)(23)\to(1,1)$.

This function f is one-to-one and onto, and for all $x, y \in H$ we find that

$$f(x \cdot y) = f(x) \oplus f(y).$$

Consequently, f is an isomorphism.
[Note: There are other possible answers that can be given here. In fact, there are six possible isomorphisms that one can define here.]

9. (a) If H is a proper subgroup of G, then by Lagrange's Theorem $|H|$ is 2 or p. If $|H| = 2$, then $H = \{e, x\}$ where $x^2 = e$, so $H = \langle x \rangle$. If $|H| = p$, let $y \in H$, where $y \neq e$. Then $o(y) = p$, so $H = \langle y \rangle$.

(b) Let $x \in G$, $x \neq e$. Then $o(x) = p$ or $o(x) = p^2$. If $o(x) = p$, then $|\langle x \rangle| = p$. If $o(x) = p^2$, then $G = \langle x \rangle$ and $\langle x^p \rangle$ is a subgroup of G of order p.

11. (a) Let $x \in H \cap K$. $x \in H \Longrightarrow o(x)|10 \Longrightarrow o(x) = 1, 2, 5$, or 10. $x \in K \Longrightarrow o(x)|21 \Longrightarrow o(x) = 1, 3, 7$, or 21. Hence $o(x) = 1$ and $x = e$.

(b) Since $\gcd(m, n) = 1$, there exists $a, b \in \mathbf{Z}$ with $ma + nb = 1$. Let $x \in H \cap K$. Then $x \in H \Rightarrow o(x)|m$. Likewise, $x \in K \Rightarrow o(x)|n$. With $o(x)|m$, $o(x)|n$, and $ma + nb = 1$, it follows that $o(x)|1$. So $x = e$.

13. (a) In $(\mathbf{Z}_p^*, \cdot)$ there are $p - 1$ elements, so by Exercise 8, for each $[x] \in (\mathbf{Z}_p^*, \cdot)$, $[x]^{p-1} = [1]$, or $x^{p-1} \equiv 1 \pmod p$, or $x^p \equiv x \pmod p$. For all $a \in \mathbf{Z}$, if $p \mid a$ then $a \equiv 0 \pmod p$ and $a^p \equiv 0 \equiv a \pmod p$. If $p \nmid a$, then $a \equiv b \pmod p$, $1 \leq b \leq p - 1$ and $a^p \equiv b^p \equiv b \equiv a \pmod p$.

(b) In the group G of units of $\mathbf{Z}_n$ there are $\phi(n)$ units. If $a \in \mathbf{Z}$ and $\gcd(a, n) = 1$ then $[a] \in G$ and $[a]^{\phi(n)} = [1]$ or $a^{\phi(n)} \equiv 1 \pmod n$
(c) and (d) These results follow from Exercises 6 and 8. They are special cases of Exercise 8.

Section 16.4

1. Here $n = 2573$ and $e = 7$.
The assignment for the given plaintext is:

IN	VE	ST	IN	ST	OC	KS
0813	2104	1819	0813	1819	1402	1018

Since

$(0813)^7$ **mod** $2573 = 0462$
$(2104)^7$ **mod** $2573 = 0170$
$(1819)^7$ **mod** $2573 = 1809$
$(0813)^7$ **mod** $2573 = 0462$
$(1819)^7$ **mod** $2573 = 1809$
$(1402)^7$ **mod** $2573 = 1981$
$(1018)^7$ **mod** $2573 = 0305$,

the ciphertext is
0462 0170 1809 0462 1809 1981 0305

3. Here $n = 2501 = (41)(61)$, so $r = \phi(n) = (40)(60) = 2400$. Further, $e = 11$ is a unit in $\mathbf{Z}_{2400}$ and $d = e^{-1} = 1091$.

Since the encrypted ciphertext is
1418 1436 2370 1102 1805 0250,
we calculate the following:

$(1418)^{1091} \bmod 2501 = 0317$ $\quad$ $(1102)^{1091} \bmod 2501 = 0005$
$(1436)^{1091} \bmod 2501 = 0821$ $\quad$ $(1805)^{1091} \bmod 2501 = 0411$
$(2370)^{1091} \bmod 2501 = 0418$ $\quad$ $(0250)^{1091} \bmod 2501 = 2423$

Consequently, the assignment for the original message is
0317 0821 0418 0005 0411 2423
and this reveals the message as

DRIVE SAFELYX.

5. Here $n = pq = 121,361$ and $r = \phi(n) = 120,432$.

Since $p + q = n - r + 1 = 930$ and $p - q = \sqrt{(n - r + 1)^2 - 4n} = \sqrt{864,900 - 485,444} = \sqrt{379,456} = 616$, it follows that

$$p = 157 \text{ and } q = 773.$$

Section 16.5

1. (a) $e = 0001001$ $\quad$ (b) $r = 1111011$ $\quad$ (c) $c = 0101000$

3. (a) (i) $D(111101100) = 101$ $\quad$ (ii) $D(000100011) = 000$
(iii) $D(010011111) = 011$

(b) 000000000, 000000001, 100000000

(c) 64

Sections 16.6 and 16.7

1. $S(101010, 1) = \{101010, 001010, 111010, 100010, 101110, 101000, 101011\}$

$S(111111, 1) = \{111111, 011111, 101111, 110111, 111011, 111101, 111110\}$

3. (a) $|S(x,1)| = 11$; $\quad |S(x,2)| = 56$; $\quad |S(x,3)| = 176$

(b) $|S(x,k)| = 1 + \binom{n}{1} + \binom{n}{2} + \ldots + \binom{n}{k} = \sum_{i=0}^{k} \binom{n}{i}$

5. (a) The minimum distance between code words is 3. The code can detect all errors of weight ≤ 2 or correct all single errors.

(b) The minimum distance between code words is 5. The code can detect all errors of weight ≤ 4 or correct all errors of weight ≤ 2.
(c) The minimum distance between code words is 2. The code detects all single errors but has no correction capability.
(d) The minimum distance between code words is 3. The code can detect all errors of weight ≤ 2 or correct all single errors.

7. (a) $C = \{00000, 10110, 01011, 11101\}$. The minimum distance between code words is 3, so the code can detect all errors of weight ≤ 2 or correct all single errors.

(b) $H = \begin{bmatrix} 1 & 0 & 1 & 0 & 0 \\ 1 & 1 & 0 & 1 & 0 \\ 0 & 1 & 0 & 0 & 1 \end{bmatrix}$

(c) (i) 01 (ii) 11 (v) 11 (vi) 10
For (iii) and (iv) the syndrome is $(111)^{tr}$ which is not a column of H. Assuming a double error, if $(111)^{tr} = (110)^{tr} + (001)^{tr}$, then the decoded received word is 01 (for (iii)) and 10 (for (iv)). If $(111)^{tr} = (011)^{tr} + (100)^{tr}$, we get 10 (for (iii)) and 01 (for (iv)).

9. $G = [I_8|A]$ where I_8 is the 8×8 multiplicative identity matrix and A is a column of eight 1's. $H = [A^{tr}|1] = [11111111|1]$.

11. Compare the generator (parity-check) matrix in Exercise 9 with the parity-check (generator) matrix in Exercise 10.

Sections 16.8 and 16.9

1. $\binom{256}{2}$ calculations are needed to find the minimum distance between code words. (A calculation here determines the distance between a pair of code words.) If E is a group homomorphism we need to calculate the weights of the 255 nonzero code words.

3. (a)

Syndrome	Coset Leader			
000	00000	10110	01011	11101
110	10000	00110	11011	01101
011	01000	11110	00011	10101
100	00100	10010	01111	11001
010	00010	10100	01001	11111
001	00001	10111	01010	11100
101	11000	01110	10011	00101
111	01100	11010	00111	10001

[The last two rows are not unique.]

(b)

Received Word	Code Word	Decoded Message
11110	10110	10
11101	11101	11
11011	01011	01
10100	10110	10
10011	01011	01
10101	11101	11
11111	11101	11
01100	00000	00

5. (a) G is 57×63; H is 6×63
(b) The rate is 57/63.

7. (a) The Hamming (7,4) code corrects all single errors in transmission, so the probability of the correct decoding of 1011 is $(0.99)^7 + \binom{7}{1}(0.99)^6(0.01)$

(b) $[(0.99)^7 + \binom{7}{1}(0.99)^6(0.01)]^5$

Section 16.10

1. (a) $\pi_2^* = \begin{pmatrix} C_1C_2C_3C_4C_5C_6C_7C_8C_9C_{10}C_{11}C_{12}C_{13}C_{14}C_{15}C_{16} \\ C_1C_4C_5C_2C_3C_8C_9C_6C_7C_{10}C_{11}C_{14}C_{15}C_{12}C_{13}C_{16} \end{pmatrix}$

$$\pi_3^* = \begin{pmatrix} C_1C_2C_3C_4C_5C_6C_7C_8C_9C_{10}C_{11}C_{12}C_{13}C_{14}C_{15}C_{16} \\ C_1C_5C_2C_3C_4C_9C_6C_7C_8C_{11}C_{10}C_{15}C_{12}C_{13}C_{14}C_{16} \end{pmatrix}$$

$$r_2^* = \begin{pmatrix} C_1C_2C_3C_4C_5C_6C_7C_8C_9C_{10}C_{11}C_{12}C_{13}C_{14}C_{15}C_{16} \\ C_1C_5C_4C_3C_2C_6C_9C_8C_7C_{11}C_{10}C_{13}C_{12}C_{15}C_{14}C_{16} \end{pmatrix}$$

$$r_4^* = \begin{pmatrix} C_1C_2C_3C_4C_5C_6C_7C_8C_9C_{10}C_{11}C_{12}C_{13}C_{14}C_{15}C_{16} \\ C_1C_4C_3C_2C_5C_9C_8C_7C_6C_{10}C_{11}C_{12}C_{15}C_{14}C_{13}C_{16} \end{pmatrix}$$

(b) $r_3^{-1} = r_3$

$$r_3^* = (r_3^{-1})^* = \begin{pmatrix} C_1C_2C_3C_4C_5C_6C_7C_8C_9C_{10}C_{11}C_{12}C_{13}C_{14}C_{15}C_{16} \\ C_1C_2C_5C_4C_3C_7C_6C_9C_8C_{10}C_{11}C_{14}C_{13}C_{12}C_{15}C_{16} \end{pmatrix}$$

$$= (r_3^*)^{-1}$$

$$(\pi_1^{-1})^* = \pi_3^* = \begin{pmatrix} C_1C_2C_3C_4C_5C_6C_7C_8C_9C_{10}C_{11}C_{12}C_{13}C_{14}C_{15}C_{16} \\ C_1C_5C_2C_3C_4C_9C_6C_7C_8C_{11}C_{10}C_{15}C_{12}C_{13}C_{14}C_{16} \end{pmatrix}$$

$$(\pi_1^*)^{-1} = \begin{pmatrix} C_1C_2C_3C_4C_5C_6C_7C_8C_9C_{10}C_{11}C_{12}C_{13}C_{14}C_{15}C_{16} \\ C_1C_3C_4C_5C_2C_7C_8C_9C_6C_{11}C_{10}C_{13}C_{14}C_{15}C_{12}C_{16} \end{pmatrix}^{-1}$$

$$= \begin{pmatrix} C_1C_2C_3C_4C_5C_6C_7C_8C_9C_{10}C_{11}C_{12}C_{13}C_{14}C_{15}C_{16} \\ C_1C_5C_2C_3C_4C_9C_6C_7C_8C_{11}C_{10}C_{15}C_{12}C_{13}C_{14}C_{16} \end{pmatrix}$$

(c) $\pi_3^* r_4^* = \begin{pmatrix} C_1 C_2 C_3 C_4 C_5 C_6 C_7 C_8 C_9 C_{10} C_{11} C_{12} C_{13} C_{14} C_{15} C_{16} \\ C_1 C_5 C_4 C_3 C_2 C_6 C_9 C_8 C_7 C_{11} C_{10} C_{13} C_{12} C_{15} C_{14} C_{16} \end{pmatrix}$

$= r_2^* = (\pi_3 r_4)^*.$

$\pi_1^* r_1^* = \begin{pmatrix} C_1 C_2 C_3 C_4 C_5 C_6 C_7 C_8 C_9 C_{10} C_{11} C_{12} C_{13} C_{14} C_{15} C_{16} \\ C_1 C_2 C_5 C_4 C_3 C_7 C_6 C_9 C_8 C_{10} C_{11} C_{14} C_{13} C_{12} C_{15} C_{16} \end{pmatrix}$

$= r_3^* = (\pi_1 r_1)^*.$

3. (a) $o(\alpha) = 7;$ $o(\beta) = 12;$ $o(\gamma) = 3;$ $o(\delta) = 6.$

(b) Let $\alpha \in S_n$, with $\alpha = c_1 c_2 \ldots c_k$, a product of disjoint cycles. Then $o(\alpha)$ is the lcm of $\ell(c_1), \ell(c_2), \ldots, \ell(c_k)$, where $\ell(c_i) =$ length of c_i, $1 \leq i \leq k$.

5. For $0 \leq i \leq 4$, let π_i denote a clockwise rotation through $i(72°)$. Also, there are five reflections r_i, $1 \leq i \leq 5$, each about a line through a vertex and the midpoint of the opposite side. Here $|G| = 10$.
(a) $\Psi(\pi_0^*) = 2^5$ $\quad \Psi(\pi_i^*) = 2,\ \ 2 \leq i \leq 4$
$\Psi(r_i^*) = 2^3,\ \ 1 \leq i \leq 5.$

The number of distinct configurations is $(1/10)[2^5 + 4(2) + 5(2^3)] = 8.$

(b) $(1/10)[3^5 + 4(3) + 5(3^3)] = 39$

7. (a) $G = \{\pi_i | 0 \leq i \leq 3\}$, where π_i is a clockwise rotation through $i \cdot 90°$. The number of distinct bracelets is $(1/4)[4^4 + 4 + 4^2 + 4] = 70.$

(b) $G = \{\pi_i | 0 \leq i \leq 3\} \cup \{r_i | 1 \leq i \leq 4\}$, where each r_i, $1 \leq i \leq 4$, is one of the two reflections about a line through two opposite beads of the midpoints of two opposite lengths of wire. Then the number of distinct bracelets is $(1/8)[4^4+4+4^2+4+4^3+4^3+4^2+4^2] = 55.$

9. Triangular Figure:
(a) $G = \{\pi_0, \pi_1, \pi_2\}$ $\quad (1/3)[2^4 + 2^2 + 2^2] = 8$
(b) $G = \{\pi_0, \pi_1, \pi_2, r_1, r_2, r_3\}$ $\quad (1/6)[2^4 + 2^2 + 2^2 + 3(2^3)] = 8$

Square Figure:
(a) $G = \{\pi_0, \pi_1, \pi_2 \pi_3\}$ $\quad (1/4)[2^5 + 2(2^2) + 2^3] = 12$
(b) $G = \{\pi_0, \pi_1, \pi_2, \pi_3, r_1, r_2, r_3, r_4\}$ $\quad (1/8)[2^5 + 2(2^2) + 2^3 + 2(2^3) + 2(2^4)] = 12$

11. (a) $(1/4)[2^9 + 2^3 + 2^5 + 2^3] = 140$
(b) $(1/8)[2^9 + 2^3 + 2^5 + 2^3 + 2^6 + 2^6 + 2^6 + 2^6] = 102$

13. $G = \{\pi_i | 0 \leq i \leq 6\}$, where π_i is the (clockwise) rotation through $i \cdot (360°/7)$.
$(1/7)[3^7 + 6(3)] = 315$

Section 16.11

1. (a) $(1/4)[5^4 + 5^2 + 2(5)] = 165$
(b) $(1/8)[5^4 + 5^2 + 2(5) + 2(5^2) + 2(5^3)] = 120$

3. (Triangular Figure):
(a) $G = \{\pi_0, \pi_1, \pi_2\}$ $\quad (1/3)[4^4 + 2(4^2)] = 96$
(b) $G = \{\pi_0, \pi_1, \pi_2, r_1, r_2, r_3\}$ $\quad (1/6)[4^4 + 2(4^2) + 3(4^3)] = 80$

(Square Figure):
(a) $G = \{\pi_0, \pi_1, \pi_2, \pi_3\}$, $\quad (1/4)[4^5 + 2(4^2) + 4^3] = 280$
(b) $G = \{\pi_0, \pi_1, \pi_2, \pi_3, r_1, r_2, r_3, r_4\}$ $\quad (1/8)[4^5 + 2(4^2) + 4^3 + 2(4^3) + 2(4^4)] = 220$

(Hexagonal Figure):
(a) $G = \{\pi_0, \pi_1\}$ where π_i is the rotation through $i \cdot 180°$, $i = 0, 1$.
$(1/2)[4^9 + 4^5] = 131,584$
(b) $G = \{\pi_0, \pi_1, r_1, r_2\}$ where $r_1(r_2)$ is the vertical (horizontal) reflection.
$(1/4)[4^9 + 4^5 + 4^5 + 4^7] = 70,144$

5. (a) $(1/6)[5^6 + 2(5) + 2(5^2) + 5^3] = 2635$
(b) $(1/12)[5^6 + 2(5) + 2(5^2) + 4(5^3) + 3(5^4)] = 1505$
(c)

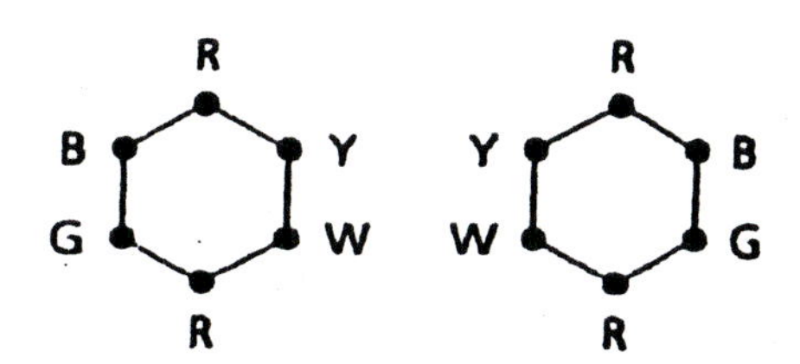

7. (a) $(1/8)[3^4 + 2(3) + 3^2 + 2(3^3) + 2(3^2)] = 21$
(b) $(1/8)[3^8 + 2(3^2) + 3^4 + 2(3^5) + 2(3^5)] = 954$
(c) No, $k = 21$, $m = 21$, so $km = 441 \neq 954 = n$. Here the location of a certain edge must be considered relative to the location of the vertices.

For example, [first square] is not equivalent to [second square] even though [third square] is equivalent to [fourth square] and [fifth square] is equivalent to [sixth square].

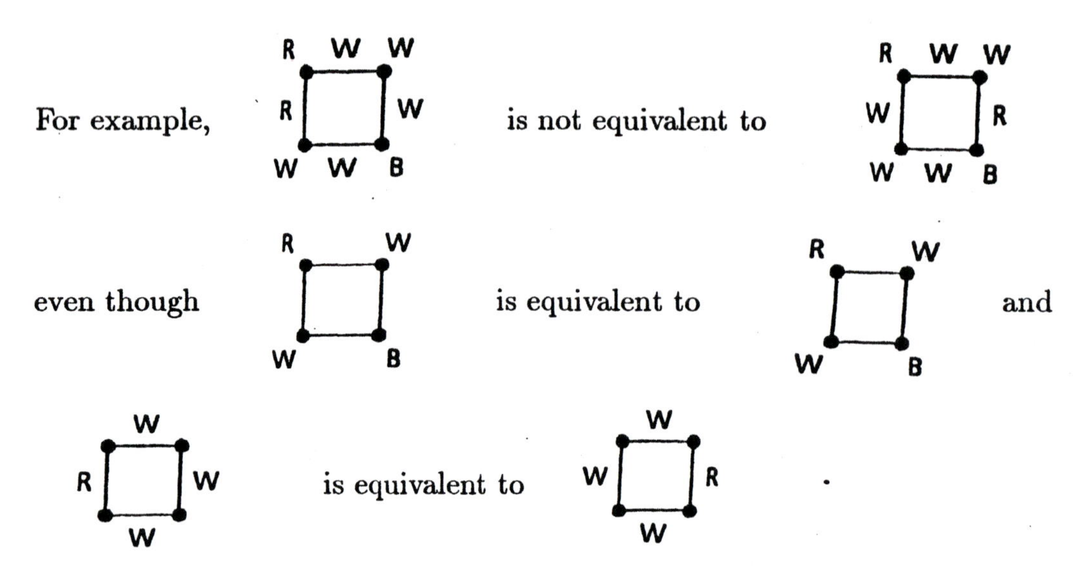

Section 16.12

1. (a) (i) $(1/4)[(r + w)^4 + 2(r^4 + w^4) + (r^2 + w^2)^2] = r^4 + w^4 + r^3w + 2r^2w^2 + rw^3$
(ii) $(1/8)[(r + w)^4 + 2(r^4 + w^4) + 3(r^2 + w^2)^2 + 2(r + w)^2(r^2 + w^2)] = r^4 + w^4 + r^3w + 2r^2w^2 + rw^3$

(b) (i) $(1/4)[(r + b + w)^4 + 2(r^4 + b^4 + w^4) + (r^2 + b^2 + w^2)^2]$
(ii) $(1/8)[(r + b + w)^4 + 2(r^4 + b^4 + w^4) + 3(r^2 + b^2 + w^2)^2 + 2(r + b + w)^2(r^2 + b^2 + w^2)]$

3. (a) (See Example 16.35)

	Rigid Motion	Cycle Structure Representation
(1)	Identity	x_1^6
(2)	Rotation through 90°	$x_1^2x_4$
	Rotation through 180°	$x_1^2x_2^2$
	Rotation through 270°	$x_1^2x_4$
(3)	Rotations of 180°	x_2^3
(4)	Rotations of 120°	x_3^2

There are then $(1/24)[2^6 + 6(2^3) + 3(2^4) + 6(2^3) + 8(2^2)] = 10$ distinct 2-colorings of the faces of the cube.

(b) $(1/24)[(r+w)^6+6(r+w)^2(r^4+w^4)+3(r+w)^2(r^2+w^2)^2+6(r^2+w^2)^3+8(r^3+w^3)^2]$

(c) For three red and three white faces we consider the coefficients of the summands that involve r^3w^3 :

$(r+w)^6$:	$\binom{6}{3}=20$
$3(r+w)^2(r^2+w^2)^2$:	12
$8(r^3+w^3)^2$:	16

The answer is $(1/24)[20+12+16]=2$

5. Let g denote green and y gold.

(Triangular Figure): $(1/6)[(g+y)^4+2(g+y)(g^3+y^3)+3(g+y)^2(g^2+y^2)]$

(Square Figure): $(1/8)[(g+y)^5+2(g+y)(g^4+y^4)+(g+y)(g^2+y^2)^2+2(g+y)(g^2+y^2)^2+2(g+y)^3(g^2+y^2)]$

(Hexagonal Figure): $(1/4)[(g+y)^9+(g+y)(g^2+y^2)^4+(g+y)(g^2+y^2)^4+(g+y)^5(g^2+y^2)^2]$.

7. (a)

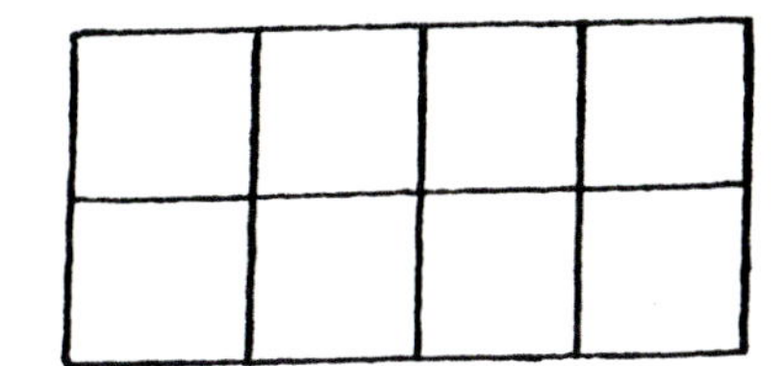

Here $G=\{\pi_0,\pi_1\}$, where π_1 denotes the $180°$ rotation.

$(1/2)[2^8+2^4]=136$ distinct ways to 2-color the squares of the chessboard.

(b) $(1/2)[(r+w)^8+(r^2+w^2)^4]$

(c) Four red and four white faces: $(1/2)[\binom{8}{4}+\binom{4}{2}]=38$
Six red and two white faces: $(1/2)[\binom{8}{6}+\binom{4}{1}]=16$

9. Let $c_1,c_2,\dots,c_m$ denote the m colors. Since the term $(c_1+c_2+\dots+c_m)^n$ is involved in the pattern inventory, there are $\binom{m+n-1}{n}$ distinct summands.

Supplementary Exercises

1. (a) Since $f(e_G)=e_H$, it follows that $e_G\in K$ and $K\neq\emptyset$. If x, $y\in K$, then $f(x)=f(y)=e_H$ and $f(xy)=f(x)f(y)=e_He_H=e_H$, so $xy\in K$. Also, for $x\in K$, $f(x^{-1})=[f(x)]^{-1}=e_H^{-1}=e_H$, so $x^{-1}\in K$. Hence K is a subgroup of G.

(b) If $x\in K$, then $f(x)=e_H$. For all $g\in G$, $f(gxg^{-1})=f(g)f(x)f(g^{-1})=f(g)e_Hf(g^{-1})=f(g)f(g^{-1})=f(gg^{-1})=f(e_G)=e_H$.
Hence, for all $x\in K, g\in G$, we find that $gxg^{-1}\in K$.

3. Let $a,\ b \in G$. Then $a^2b^2 = ee = e = (ab)^2 = abab$. But $a^2b^2 = abab \Longrightarrow aabb = abab \Longrightarrow ab = ba$, so G is abelian.

5. Let $G = < g >$ and let $h = f(g)$. If $h_1 \in H$, then $h_1 = f(g^n)$ for some $n \in \mathbf{Z}$, since f is onto. Therefore, $h_1 = f(g^n) = [f(g)]^n = h^n$, and $H = \langle h \rangle$.

7. Proof: For all $a, b \in G$,

$$(a \circ a^{-1}) \circ b^{-1} \circ b = b \circ b^{-1} \circ (a^{-1} \circ a) \Longrightarrow$$

$$a \circ a^{-1} \circ b = b \circ a^{-1} \circ a \Longrightarrow a \circ b = b \circ a,$$

and so it follows that $(G, \circ)$ is an abelian group.

9. (a) Consider a permutation σ that is counted in $P(n+1, k)$. If $(n+1)$ is a cycle (of length 1) in σ, then σ (restricted to $\{1, 2, \ldots, n\}$) is counted in $P(n, k-1)$. Otherwise, consider any permutation τ that is counted in $P(n, k)$. For each cycle of τ, say $(a_1 a_2 \ldots a_r)$, there are r locations in which to place $n+1$ - (1) Between a_1 and a_2; (2) Between a_2 and $a_3; \ldots; (r-1)$ Between a_{r-1} and a_r; and (r) Between a_r and a_1. Hence there are n locations, in total, to locate $n+1$ in τ. Consequently, $P(n+1, k) = P(n, k-1) + nP(n, k)$.

(b) $\sum_{k=1}^{n} P(n, k)$ counts all of the permutations in S_n, which has $n!$ elements.

11. (a) Suppose that n is composite. We consider two cases.
(1) $n = m \cdot r$, where $1 < m < r < n$: Here $(n-1)! = 1 \cdot 2 \cdots (m-1) \cdot m \cdot (m+1) \cdots (r-1) \cdot r \cdot (r+1) \cdots (n-1) \equiv 0 \pmod{n}$. Hence $(n-1)! \not\equiv -1 \pmod{n}$.
(2) $n = q^2$, where q is a prime: If $(n-1)! \equiv -1 \pmod{n}$ then $0 \equiv q(n-1)! \equiv q(-1) \equiv n - q \not\equiv 0 \pmod{n}$. So in this case we also have $(n-1)! \not\equiv -1 \pmod{n}$.

(b) From Wilson's Theorem, when p is an odd prime, we find that

$$-1 \equiv (p-1)! \equiv (p-3)!(p-2)(p-1) \equiv (p-3)!(p^2 - 3p + 2) \equiv 2(p-3)! \pmod{p}.$$

CHAPTER 17
FINITE FIELDS AND COMBINATORIAL DESIGNS

Section 17.1

1. $f(x)+g(x) = 2x^4+(2+3)x^3+(3+5)x^2+(1+6)x+(4+1) = 2x^4+5x^3+8x^2+7x+5 = 2x^4+5x^3+x^2+5$

$f(x)-g(x) = 2x^4+(2-3)x^3+(3-5)x^2+(1-6)x+(4-1) = 2x^4+(-1)x^3+(-2)x^2+(-5)x+3 = 2x^4+6x^3+5x^2+2x+3$

$f(x)g(x) = (2)(3)x^7+[(2)(5)+(2)(3)]x^6+[(2)(6)+(2)(5)+(3)(3)]x^5+[(2)(1)+(2)(6)+(3)(5)+(1)(3)]x^4+[(2)(1)+(3)(6)+(1)(5)+(4)(3)]x^3+[(3)(1)+(1)(6)+(4)(5)]x^2+[(1)(1)+(4)(6)]x+4 = 6x^7+16x^6+31x^5+32x^4+37x^3+29x^2+25x+4 = 6x^7+2x^6+3x^5+4x^4+2x^3+x^2+4x+4.$

3. $(10)(11)^2$; $(10)(11)^3$; $(10)(11)^4$; $(10)(11)^n$

5. (Theorem 17.1) We shall prove one of the distributive laws. Let $f(x)=\sum_{i=0}^{n} a_i x^i$, $g(x)=\sum_{j=0}^{m} b_j x^j$, $h(x)=\sum_{k=0}^{p} c_k x^k$, where $m \geq p$. For $0 \leq t \leq m+n$, the coefficient of x^t in $f(x)[g(x)+h(x)]$ is $\sum a_i(b_j+c_j)$ where the sum is taken over all $0 \leq i \leq n, 0 \leq j \leq m$ with $i+j=t$. But this is the same as $(\sum a_i b_j)+(\sum a_i c_j)$, for $0 \leq i \leq n$, $0 \leq j \leq m$, $i+j=t$, because $a_i(b_j+c_j)=a_i b_j + a_i c_j$ in ring R, and this is the coefficient of x^t in $f(x)g(x)+f(x)h(x)$.

(Corollary 17.1)

(a) Let $f(x)=\sum_{i=0}^{n} a_i x^i$, $g(x)=\sum_{j=0}^{m} b_j x^j$. For all $0 \leq t \leq m+n$ the coefficient of x^t in $f(x)g(x)$ is $\sum_{i+j=t} a_i b_j = \sum_{i+j=t} b_j a_i$ (since R is commutative), and this last summation is the coefficient of x^t in $g(x)f(x)$. Hence, $f(x)g(x)=g(x)f(x)$ and $R[x]$ is commutative.

(b) Let 1 denote the unity of R. Then 1 or $1x^0$ is the unity in $R[x]$.

(c) Let R be an integral domain and let $f(x)=\sum_{i=0}^{n} a_i x^i$, $g(x)=\sum_{j=0}^{m} b_j x^j$ with $a_n \neq 0$, $b_m \neq 0$. If $f(x)g(x)=0$, then $a_n b_m = 0$ contradicting R as an integral domain. Conversely, if $R[x]$ is an integral domain and $a, b \in R$ with $a \neq 0$ and $b \neq 0$, then $ab=(ax^0)(bx^0) \neq 0$ and R is an integral domain.

7. (a) and (b) $f(x)=(x^2+4)(x-2)(x+2)$; the roots are ± 2.

(c) $f(x) = (x+2i)(x-2i)(x-2)(x+2)$; the roots are $\pm 2, \pm 2i$

(d) (a) $f(x) = (x^2-5)(x^2+5)$; no rational roots
(b) $f(x) = (x-\sqrt{5})(x+\sqrt{5})(x^2+5)$; the roots are $\pm\sqrt{5}$
(c) $f(x) = (x-\sqrt{5})(x+\sqrt{5})(x-\sqrt{5}i)(x+\sqrt{5}i)$; the roots are $\pm\sqrt{5}, \pm i\sqrt{5}$

9. Using the Remainder Theorem (Theorem 17.4) we find
(a) $f(3) = 8060$ (b) $f(1) = 1$ (c) $f(-9) = f(2) = 6$

11. 4; 6; $p-1$

13. Let $f(x) = \sum_{i=0}^{m} a_i x^i$ and $h(x) = \sum_{i=0}^{k} b_i x^i$, where $a_i \in R$ for $0 \le i \le m$, and $b_i \in R$ for $0 \le i \le k$, and $m \le k$. Then $f(x)+h(x) = \sum_{i=0}^{k}(a_i+b_i)x^i$, where $a_{m+1} = a_{m+2} = \ldots = a_k = z$, the zero of R, so $G(f(x)+h(x)) = G(\sum_{i=0}^{k}(a_i+b_i)x^i) = \sum_{i=0}^{k} g(a_i+b_i)x^i = \sum_{i=0}^{k}[g(a_i)+g(b_i)]x^i = \sum_{i=0}^{k} g(a_i)x^i + \sum_{i=0}^{k} g(b_i)x^i = G(f(x))+G(h(x))$.

Also, $f(x)h(x) = \sum_{i=0}^{m+k} c_i x^i$, where $c_i = a_i b_0 + a_{i-1}b_1 + \ldots + a_1 b_{i-1} + a_0 b_i$, and

$$G(f(x)h(x)) = G(\sum_{i=0}^{m+k} c_i x^i) = \sum_{i=0}^{m+k} g(c_i)x^i.$$

Since $g(c_i) = g(a_i)g(b_0) + g(a_{i-1})g(b_1) + \ldots + g(a_1)g(b_{i-1}) + g(a_0)g(b_i)$,

$$\sum_{i=0}^{m+k} g(c_i)x^i = (\sum_{i=0}^{m} g(a_i)x^i)(\sum_{i=0}^{k} g(b_i)x^i) = G(f(x)) \cdot G(h(x)).$$

Consequently, $G: R[x] \longrightarrow S[x]$ is a ring homomorphism.

15. In $\mathbf{Z}_4[x]$, $(2x+1)(2x+1) = 1$, so $(2x+1)$ is a unit. This does not contradict Exercise 14 because $(\mathbf{Z}_4, +, \cdot)$ is not an integral domain.

17. First note that for $f(x) = a_n x^n + a_{n-1}x^{n-1} + \cdots + a_2x^2 + a_1x + a_0$, we have $a_n + a_{n-1} + \cdots + a_2 + a_1 + a_0 = 0$ if and only if $f(1) = 0$. Since the zero polynomial is in S, the set S is not empty. With $f(x)$ as given here, let $g(x) = b_m x^m + b_{m-1}x^{m-1} + \cdots + b_2x^2 + b_1x + b_0 \in S$. (Here $m \le n$, and for $m < n$ we have $b_{m+1} = b_{m+2} = \cdots = b_n = 0$.) Then $f(1) - g(1) = 0 - 0 = 0$ so $f(x) - g(x) \in S$.

Now consider $h(x) = \sum_{i=0}^{k} r_i x^i \in F[x]$. Here $h(x)f(x) \in F[x]$ and $h(1)f(1) = h(1) \cdot 0 = 0$, so $h(x)f(x) \in S$.

Consequently, S is an ideal in $F[x]$.

Section 17.2

1. (a) x^2+3x-1 is irreducible over $\mathbf{Q}$. Over $\mathbf{R}$, $\mathbf{C}$,

$$x^2+3x-1 = [x - ((-3+\sqrt{13})/2)][x - ((-3-\sqrt{13})/2)].$$

(b) $x^4 - 2$ is irreducible over $\mathbf{Q}$.
Over $\mathbf{R}$, $x^4 - 2 = (x - \sqrt[4]{2})(x + \sqrt[4]{2})(x^2 + \sqrt{2})$;
$x^4 - 2 = (x - \sqrt[4]{2})(x + \sqrt[4]{2})(x - \sqrt[4]{2}i)(x + \sqrt[4]{2}i)$ over $\mathbf{C}$

(c) $x^2 + x + 1 = (x + 2)(x + 2)$ over $\mathbf{Z}_3$. Over $\mathbf{Z}_5, x^2 + x + 1$ is irreducible;
$x^2 + x + 1 = (x + 5)(x + 3)$ over $\mathbf{Z}_7$.

(d) $x^4 + x^3 + 1$ is irreducible over $\mathbf{Z}_2$.

(e) $x^3 + 3x^2 - x + 1$ is irreducible over $\mathbf{Z}_5$.

3. Degree 1: x; $x + 1$
Degree 2: $x^2 + x + 1$
Degree 3: $x^3 + x^2 + 1$; $x^3 + x + 1$

5. 7^5

7. (a) Yes, since the coefficients of the polynomials are from a field.

(b) $h(x)|f(x), g(x) \Longrightarrow f(x) = h(x)u(x)$, $g(x) = h(x)v(x)$, for some $u(x), v(x) \in F[x]$.
$m(x) = s(x)f(x) + t(x)g(x)$ for some $s(x), t(x) \in F[x]$, so
$m(x) = h(x)[s(x)u(x) + t(x)v(x)]$ and $h(x)|m(x)$.

(c) If $m(x) \not| \; f(x)$, then $f(x) = q(x)m(x) + r(x)$ where $r(x) \neq 0$ and $0 \leq \deg r(x) < \deg m(x)$. $m(x) = s(x)f(x) + t(x)g(x)$ so $r(x) = f(x) - q(x)[s(x)f(x) + t(x)g(x)] = (1 - q(x))s(x)f(x) - q(x)t(x)g(x)$, so $r(x) \in S$. With $\deg r(x) < \deg m(x)$ we contradict the choice of $m(x)$. Hence $r(x) = 0$ and $m(x)|f(x)$.

9. (a) By the long division of polynomials we have
$x^5 - x^4 + x^3 + x^2 - x - 1 = (x^3 - 2x^2 + 5x - 8)(x^2 + x - 2) + (17x - 17)$
$x^2 + x - 2 = (17x - 17)[(1/17)x + (2/17)x]$,
so the gcd of $f(x), g(x)$ is
$(x - 1) = (1/17)(x^5 - x^4 + x^3 + x^2 - x - 1) - (1/17)(x^2 + x - 2)(x^3 - 2x^2 + 5x - 8)$.

(b) The gcd is $1 = (x + 1)(x^4 + x^3 + 1) + (x^3 + x^2 + x)(x^2 + x + 1)$

(c) The gcd is $x^2 + 2x + 1 = (x^4 + 2x^2 + 2x + 2) + (x + 2)(2x^3 + 2x^2 + x + 1)$

11. $f(x) = x^3 + 2x^2 + ax - b$
$g(x) = x^3 + x^2 - bx + a$
From the Division Algorithm for polynomials we find that $f(x) = g(x) + r(x)$, where $r(x) = x^2 + (a + b)x - (a + b)$, a polynomial of degree 2.
Since we want $r(x)$ to be the gcd, we must have $r(x)$ dividing $f(x)$.
Since $f(x) = r(x)[x + (2 - a - b)] + [(2a + b) - (2 - a - b)(a + b)]x + [-b + (a + b)(2 - a - b)]$, we must have $(2a + b) - (2 - a - b)(a + b) = 0$ and $-b + (a + b)(2 - a - b) = 0$. Consequently,
$0 = (2a + b) - (2 - a - b)(a + b) = a^2 + b^2 + 2ab - b$
$0 = -b + (a + b)(2 - a - b) = -a^2 - b^2 - 2ab + 2a + b$. So $2a = 0$, or $a = 0$ and $b^2 - b = 0$.
There are two solutions:

$a = 0, b = 0; \quad a = 0, b = 1.$

13. (a) $f(x) \equiv f_1(x) \pmod{s(x)} \Longrightarrow f(x) = f_1(x) + h(x)s(x)$;
$g(x) \equiv g_1(x) \pmod{s(x)} \Longrightarrow g(x) = g_1(x) + k(x)s(x)$

Hence $f(x) + g(x) = f_1(x) + g_1(x) + (h(x) + k(x))s(x)$, so $f(x) + g(x) \equiv f_1(x) + g_1(x) \pmod{s(x)}$, and $f(x)g(x) = f_1(x)g_1(x) + (f_1(x)k(x) + g_1(x)h(x) + h(x)k(x)s(x))s(x)$, so $f(x)g(x) \equiv f_1(x)g_1(x) \pmod{s(x)}$.

(b) These properties follow from the corresponding properties for $F[x]$. For example, for the distributive law,

$$\begin{aligned}[f(x)]([g(x)] + [h(x)]) &= [f(x)][g(x) + h(x)] = [f(x)(g(x) + h(x))] \\ &= [f(x)g(x) + f(x)h(x)] = [f(x)g(x)] + [f(x)h(x)] \\ &= [f(x)][g(x)] + [f(x)][h(x)]\end{aligned}$$

(c) If not, there exists $g(x) \in F[x]$ where $\deg g(x) > 0$ and $g(x) | f(x), s(x)$. But then $s(x)$ would be reducible.

(d) A nonzero element of $F[x]/(s(x))$ has the form $[f(x)]$ where $f(x) \neq 0$ and $\deg f(x) < \deg s(x)$. With $f(x)$, $s(x)$ relatively prime, there exist $r(x)$, $t(x)$ with $1 = f(x)r(x) + s(x)t(x)$, so $1 \equiv f(x)r(x) \pmod{s(x)}$ or $[1] = [f(x)][r(x)]$. Hence $[r(x)] = [f(x)]^{-1}$.

(e) q^n

15. (a) $[x+2][2x+2]+[x+1] = [2x^2+1]+[x+1] = [2x^2+x+2] = [4x+2+x+2] = [2x+1]$ (Note: With $x^2 + x + 2 \equiv 0 \pmod{s(x)}$, it follows that $x^2 \equiv -x - 2 \equiv 2x + 1 \pmod{s(x)}$.)

(b) $[2x+1]^2[x+2] = [x^2+x+1][x+2] = [x^3+2] = [x(2x+1)+2] = [2x^2+x+2] = [2x+1]$

(c) Find $a, b \in \mathbf{Z}_3$ so that $[2x + 2][ax + b] = [1]$.
$[2ax^2 + (2a + 2b)x + 2b] = [1]$
$[2a(-x - 2) + (2a + 2b)x + 2b] = [2bx + (2b + 2a)] = [1]$
$2b \equiv 0 \pmod 3$, $(2b + 2a) \equiv 1 \pmod 3 \Longrightarrow b \equiv 0 \pmod 3$,
$a \equiv 2 \pmod 3$, so $(22)^{-1} = [2x + 2]^{-1} = [2x]$.

17. (a) $\mathbf{Z}_p[x]/(s(x)) = \{a_0 + a_1x + a_2x^2 + \ldots + a_{n-1}x^{n-1} | a_0, a_1, a_2, \ldots, a_{n-1} \in \mathbf{Z}_p\}$ which has order p^n.

(b) The multiplicative group of nonzero elements of this field is a cyclic group of order $p^n - 1$, so it has $\phi(p^n - 1)$ generators.

19. (a) 6 (b) 12 (c) 12 (d) $\text{lcm}(m, n)$ (e) 0

21. Here n must be a power of a prime. So the possible values of n are 101, 103, 107, 109, 113, 121, 125, 127, 128, 131, 137, 139, 149.

23. For $s(x) = x^3 + x^2 + x + 2 \in \mathbf{Z}_3[x]$ one finds that $s(0) = 2$, $s(1) = 2$, and $s(2) = 1$. It then follows from part (b) of Theorem 17.7 and parts (b) and (c) of Theorem 17.11 that $\mathbf{Z}_3[x]/(s(x))$ is a finite field with $3^3 = 27$ elements.

25. (a) Since $0 = 0 + 0\sqrt{2} \in \mathbf{Q}[\sqrt{2}]$, the set $\mathbf{Q}[2]$ is nonempty. For $a + b\sqrt{2}$, $c + d\sqrt{2} \in \mathbf{Q}[\sqrt{2}]$, we have
$(a + b\sqrt{2}) - (c + d\sqrt{2}) = (a - c) + (b - d)\sqrt{2}$, with $(a - c), (b - d) \in \mathbf{Q}$; and
$(a + b\sqrt{2})(c + d\sqrt{2}) = (ac + 2bd) + (ad + bc)\sqrt{2}$, with $ac + 2bd, ad + bc \in \mathbf{Q}$.
Consequently, it follows from part (a) of Theorem 14.10 that $\mathbf{Q}[\sqrt{2}]$ is a subring of $\mathbf{R}$.

(b) In order to show that $\mathbf{Q}[\sqrt{2}]$ is a subfield of $\mathbf{R}$ we need to find in $\mathbf{Q}[\sqrt{2}]$ a multiplicative inverse for each nonzero element in $\mathbf{Q}[\sqrt{2}]$.
Let $a + b\sqrt{2} \in \mathbf{Q}[\sqrt{2}]$ with $a + b\sqrt{2} \neq 0$. If $b = 0$, then $a \neq 0$ and $a^{-1} \in \mathbf{Q}$ — and $a^{-1} + 0 \cdot \sqrt{2} \in \mathbf{Q}[\sqrt{2}]$. For $b \neq 0$, we need to find $c + d\sqrt{2} \in \mathbf{Q}[\sqrt{2}]$ so that

$$(a + b\sqrt{2})(c + d\sqrt{2}) = 1$$

Now $(a + b\sqrt{2})(c + d\sqrt{2}) = 1 \Rightarrow (ac + 2bd) + (ad + bc)\sqrt{2} = 1 \Rightarrow ac + 2bd = 1$ and $ad + bc = 0 \Rightarrow c = -ad/b$ and $a(-ad/b) + 2bd = 1 \Rightarrow -a^2d + 2b^2d = b \Rightarrow d = b/(2b^2 - a^2)$ and $c = -a/(2b^2 - a^2)$. [Note: $2b^2 - a^2 \neq 0$ because $\sqrt{2}$ is irrational.]
Consequently, $(a + b\sqrt{2})^{-1} = [-a/(2b^2 - a^2)] + [b/(2b^2 - a^2)]\sqrt{2}$, with $[-a/(2b^2 - a^2)]$, $[b/(2b^2 - a^2)] \in \mathbf{Q}$. So $\mathbf{Q}[\sqrt{2}]$ is a subfield of $\mathbf{R}$.

Since $s(x) = x^2 - 2$ is irreducible over $\mathbf{Q}$ we know from part (b) of Theorem 17.11 that $\mathbf{Q}[x]/(x^2 - 2)$ is a field.
Define the correspondence $f : \mathbf{Q}[x]/(x^2 - 2) \to \mathbf{Q}[2]$ by

$$f([a + bx]) = a + b\sqrt{2}.$$

By an argument similar to the one given in Example 17.10 and part (a) of Exercise 24 it follows that f is an isomorphism.

Section 17.3

1.

(a)

1	2	3	4
2	1	4	3
4	3	2	1
3	4	1	2

(b)

1	2	3	4
3	4	1	2
2	1	4	3
4	3	2	1

(c)

1	3	4	2
4	2	1	3
3	1	2	4
2	4	3	1

3. $a_{ri}^{(k)} = a_{rj}^{(k)} \Longrightarrow f_k f_r + f_i = f_k f_r + f_j \Longrightarrow f_i = f_j \Longrightarrow i = j.$

5.

$$L_3: \begin{array}{ccccc} 4 & 5 & 1 & 2 & 3 \\ 2 & 3 & 4 & 5 & 1 \\ 5 & 1 & 2 & 3 & 4 \\ 3 & 4 & 5 & 1 & 2 \\ 1 & 2 & 3 & 4 & 5 \end{array} \qquad L_4: \begin{array}{ccccc} 5 & 1 & 2 & 3 & 4 \\ 4 & 5 & 1 & 2 & 3 \\ 3 & 4 & 5 & 1 & 2 \\ 2 & 3 & 4 & 5 & 1 \\ 1 & 2 & 3 & 4 & 5 \end{array}$$

In standard form the Latin squares L_i, $1 \leq i \leq 4$, become

$$L_1': \begin{array}{ccccc} 1 & 2 & 3 & 4 & 5 \\ 2 & 3 & 4 & 5 & 1 \\ 3 & 4 & 5 & 1 & 2 \\ 4 & 5 & 1 & 2 & 3 \\ 5 & 1 & 2 & 3 & 4 \end{array} \qquad L_2': \begin{array}{ccccc} 1 & 2 & 3 & 4 & 5 \\ 3 & 4 & 5 & 1 & 5 \\ 5 & 1 & 2 & 3 & 4 \\ 2 & 3 & 4 & 5 & 1 \\ 4 & 5 & 1 & 2 & 3 \end{array}$$

$$L_3': \begin{array}{ccccc} 1 & 2 & 3 & 4 & 5 \\ 4 & 5 & 1 & 2 & 3 \\ 2 & 3 & 4 & 5 & 1 \\ 5 & 1 & 2 & 3 & 4 \\ 3 & 4 & 5 & 1 & 2 \end{array} \qquad L_4': \begin{array}{ccccc} 1 & 2 & 3 & 4 & 5 \\ 5 & 1 & 2 & 3 & 4 \\ 4 & 5 & 1 & 2 & 3 \\ 3 & 4 & 5 & 1 & 2 \\ 2 & 3 & 4 & 5 & 1 \end{array}$$

7. Introduce a third factor such as four types of transmission fluid or four types of tires.

Section 17.4

1.

Field	Number of Points	Number of Lines	Number of Points on a Line	Number of Lines on a Point
$GF(5)$	25	30	5	6
$GF(3^2)$	81	90	9	10
$GF(7)$	49	56	7	8
$GF(2^4)$	256	272	16	17
$GF(31)$	961	992	31	32

3. There are nine points and 12 lines. These lines fall into four parallel classes.

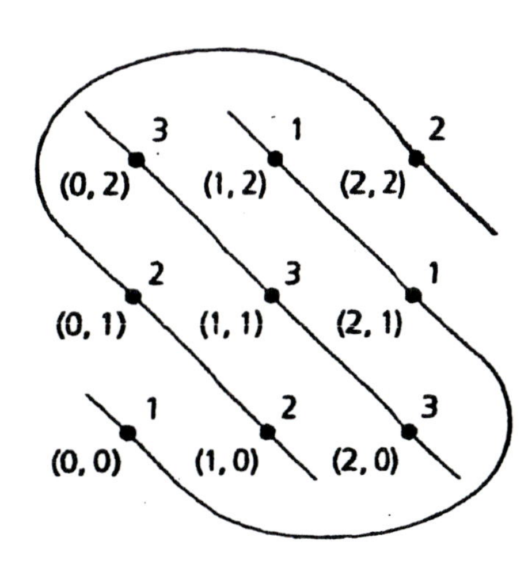

(i) Slope of 0.

$y = 0;\ y = 1;\ y = 2$

(ii) Infinite slope

$x = 0;\ x = 1;\ x = 2$

(iii) Slope 1

$y = x;\ y = x + 1;\ y = x + 2$

(iv) Slope 2 (as shown in the figure).

(1) $y = 2x$

(2) $y = 2x + 1$

(3) $y + 2x + 2$

The Latin square corresponding to the fourth parallel class is

3	1	2
2	3	1
1	2	3

5. (a) $y = 4x + 1$ (b) $y = 3x + 10$ or $2x + 3y + 3 = 0$
(c) $y = 10x$ or $10y = 11x$

7. (a) Vertical line: $x = c$. The line $y = mx + b$ intersects this vertical line at the unique point $(c, mc + b)$. As b takes on the values of F, there are no two column entries (on the line $x = c$) that are the same.

Horizontal line: $y = c$. The line $y = mx + b$ intersects this horizontal line at the unique point $(m^{-1}(c - b), c)$. As b takes on the values of F, no two row entries (on the line $y = c$) are the same.

(b) Let L_i be the Latin square for the parallel class of slope m_i, $i = 1, 2$, $m_i \neq 0$, m_i finite. If an ordered pair (j, k) appears more than once when L_1, L_2 are superimposed, then there are two pairs of lines: (1) $y = m_1x + b_1$, $y = m_2x + b_2$; and (2) $y = m_1x + b_1'$, $y = m_2x + b_2'$ which both intersect at (j, k). But then $b_1 = k - m_1j = b_1'$ and $b_2 = k - m_2j = b_2'$.

Section 17.5

1. $v = 9,\ b = 12,\ r = 4,\ k = 3,\ \lambda = 1.$

3. $\lambda = 2$

1 2 3 4	1 3 5 7	2 3 6 7	3 4 5 6
1 2 5 6	1 4 6 7	2 4 5 7	

5. (a) $vr = bk \Longrightarrow 4v = 28(3) \Longrightarrow v = 21$; $\lambda(v-1) = r(k-1) \Longrightarrow 20\lambda = 4(2) \Longrightarrow \lambda \notin \mathbf{Z}^+$, so no such design can exist.
(b) $vr = bk \Longrightarrow (17)(8) = 5b \Longrightarrow b \notin \mathbf{Z}^+$, so no such design can exist in this case either.

7. (a) $\lambda(v-1) = r(k-1) = 2r \Longrightarrow \lambda(v-1)$ is even. $\lambda v(v-1) = vr(k-1) = bk(k-1) = b(3)(2) \Longrightarrow 6|\lambda v(v-1)$.

(b) Here $\lambda = 1$. By part (a) $6|v(v-1) \Longrightarrow 3|v(v-1) \Longrightarrow 3|v$ or $3|(v-1)$, since 3 is prime. Also, by part (a) $\lambda(v-1) = (v-1)$ is even, so v is odd.
(i) $3|v \Longrightarrow v = 3t$, t odd $\Longrightarrow v = 3(2s+1) = 6s+3$ and $v \equiv 3 \pmod 6$.
(ii) $3|(v-1) \Longrightarrow v-1 = 3t$, t even $\Longrightarrow v-1 = 6x \Longrightarrow v = 6x+1$ and $v \equiv 1 \pmod 6$.

9. $k = 3$, $\lambda = 1$, $b = 12 \Longrightarrow vr = 36, v-1 = 2r \Longrightarrow (2r+1)r = 36 \Longrightarrow 2r^2 + r - 36 = 0 \Longrightarrow (2r+9)(r-4) = 0 \Longrightarrow v = 9$, $r = 4$.

11. $v = 15$, $k = 5$, $\lambda = 2 \Longrightarrow 15r = 5b$, $2(14) = 4r \Longrightarrow$ (a) $b = 21$; (b) $r = 7$.

13. There are λ blocks that contain both x and y. Since r is the replication number of the design, it follows that $r - \lambda$ blocks contain x but not y. Likewise there are $r - \lambda$ blocks containing y but not x. Consequently, the number of blocks in the design that contain x or y is $(r-\lambda) + (r-\lambda) + \lambda = 2r - \lambda$.

15. (a) $n+1 = 6 \Longrightarrow n = 5$, so there are $n^2+n+1 = 31$ points in this projective plane.
(b) $n^2+n+1 = 57 \Longrightarrow n = 7$, so there are $n+1 = 8$ points on each line of this plane.

17. (a) $v = b = 31$; $r = k = 6$; $\lambda = 1$
(b) $v = b = 57$; $r = k = 8$; $\lambda = 1$
(c) $v = b = 73$; $r = k = 9$; $\lambda = 1$

Supplementary Exercises

1. $n = 9$

3. (a) Let $a, b \in \mathbf{Z}$ with $(x-a)(x+b) = x^2 + x - n$. Then $x^2 + (b-a)x - ab = x^2 + x - n \Longrightarrow b - a = 1$ and $ab = n$. For $1 \leq a \leq 31$ and $b = a+1$, $n = ab \leq 992$. Hence there are 31 values of n, namely $a(a+1)$ for $1 \leq a \leq 31$.

(b) Here $(x-a)(x+b) = x^2 + 2x - n \Longrightarrow b - a = 2$ and $ab = n$. When $1 \leq a \leq 30$ and $b = a+2$ we find that $n = ab \leq 960$, so there are 30 such values of n in this case.

(c) In this case there are 29 values of n. Each n has the form $a(a+5)$ for $1 \leq a \leq 29$.

(d) If $(x-a)(x+b) = g(x)$, then $b-a=k$ and $ab=n$. When $k=1000$, $b=a+1000$ and $ab = a^2+1000a > 1000$. For $k=999$, with $a=1$ and $b=1000$ we have $n=ab=1000$ and $x^2+999x-1000 = (x+1000)(x-1)$. In fact, for each $1 \leq k \leq 999$, let $a=1$. Then $b=k+1$ and $n=ab=k+1$, and it follows that $x^2+kx-n = x^2+kx-(k+1) = [x+(k+1)](x-1)$. Hence the smallest positive integer k for which $g(x)$ cannot be so factored is $k=1000$.

5. For all $a \in \mathbf{Z}_p$, $a^p = a$ (See part (a) of Exercise 13 at the end of Section 16.3), so a is a root of $x^p - x$ and $x-a$ is a factor of $x^p - x$. Since $(\mathbf{Z}_p, +, \cdot)$ is a field, the polynomial $x^p - x$ can have at most p roots. Therefore $x^p - x = \prod_{a \in \mathbf{Z}_p}(x-a)$.

7. $\{1,2,4\}$, $\{2,3,5\}$, $\{4,5,7\}$

9. (a) $n^2+n+1 = 73 \Longrightarrow n = 8 \Longrightarrow n+1 = 9$, the number of points on each line.

(b) $n+1 = 10 \Longrightarrow n = 9 \Longrightarrow n^2+n+1 = 91$, the number of lines in this projective plane.

11. (a) r 1's in each row; k 1's in each column.

(b) $A \cdot J_b$ is a $v \times b$ matrix whose (i,j) entry is r, since there are r 1's in each row of A and every entry in J_b is 1. Hence $A \cdot J_b = rJ_{v \times b}$. Likewise, $J_v \cdot A$ is a $v \times b$ matrix whose (i,j) entry is k, since there are k 1's in each column of A and every entry in J_v is 1. Hence $J_v \cdot A = kJ_{v \times b}$.

(c) The (i,j) entry in $A \cdot A^{tr}$ is obtained from the componentwise multiplication of rows i and j of A. If $i = j$ this results in the number of 1's in row i, which is r. For $i \neq j$, the number of 1's is the number of times x_i and x_j appear in the same block – this is given by λ. Hence $A \cdot A^{tr} = (r-\lambda)I_v + \lambda J_v$.

(d)

$$\begin{vmatrix} r & \lambda & \lambda & \lambda & \cdots & \lambda \\ \lambda & r & \lambda & \lambda & \cdots & \lambda \\ \lambda & \lambda & r & \lambda & \cdots & \lambda \\ \lambda & \lambda & \lambda & r & \cdots & \lambda \\ \cdots & \cdots & \cdots & \cdots & \cdots & \cdots \\ \lambda & \lambda & \lambda & \lambda & \cdots & r \end{vmatrix} \overset{(1)}{=} \begin{vmatrix} r & \lambda - r & \lambda - r & \lambda - r & \cdots & \lambda - r \\ \lambda & r-\lambda & 0 & 0 & \cdots & 0 \\ \lambda & 0 & r-\lambda & 0 & \cdots & 0 \\ \lambda & 0 & 0 & r-\lambda & \cdots & 0 \\ \cdots & \cdots & \cdots & \cdots & \cdots & \cdots \\ \lambda & 0 & 0 & 0 & \cdots & r-\lambda \end{vmatrix}$$

$$\overset{(2)}{=} \begin{vmatrix} r+(v-1)\lambda & 0 & 0 & 0 & \cdots & 0 \\ \lambda & r-\lambda & 0 & 0 & \cdots & 0 \\ \lambda & 0 & r-\lambda & 0 & \cdots & 0 \\ \lambda & 0 & 0 & r-\lambda & \cdots & 0 \\ \cdots & \cdots & \cdots & \cdots & \cdots & \cdots \\ \lambda & 0 & 0 & 0 & \cdots & r-\lambda \end{vmatrix}$$

$[r+(v-1)\lambda](r-\lambda)^{v-1} = (r-\lambda)^{v-1}[r+r(k-1)] = rk(r-\lambda)^{v-1}$

(1) Multiply column 1 by -1 and add it to the other $v-1$ columns.
(2) Add rows 2 through v to row 1.

THE

APPENDICES

APPENDIX 1
EXPONENTIAL AND LOGARITHMIC FUNCTIONS

1. (a) $\sqrt{xy^3} = x^{1/2}y^{3/2}$ (b) $\sqrt[4]{81x^{-5}y^3} = 3x^{-5/4}y^{3/4} = \dfrac{3y^{3/4}}{x^{5/4}}$

(c) $5\sqrt[3]{8x^9y^{-5}} = 5(8^{1/3}x^{9/3}y^{-5/3}) = 5(2x^3y^{-5/3}) = \dfrac{10x^3}{y^{5/3}}$

3. (a) $(5^{3/4})(5^{13/4}) = 5^{[(3/4)+(13/4)]} = 5^{16/4} = 5^4 = 625$

(b) $(7^{3/5})/(7^{18/5}) = 7^{[(3/5)-(18/5)]} = 7^{(3-18)/5} = 7^{-15/5} = 7^{-3} = 1/7^3 = 1/343$

(c) $(5^{1/2})(20^{1/2}) = (5^{1/2})(4 \cdot 5)^{1/2} = (5^{1/2})(4^{1/2})(5^{1/2}) = 2(5^{1/2})^2 = 2(5) = 10$

5.

(a) $\log_2 128 = 7$ (b) $\log_{125} 5 = 1/3$
(c) $\log_{10} 1/10,000 = -4$ (d) $\log_2 b = a$

7.

(a) $x^5 = 243 \Rightarrow x = 3$ (b) $x = 3^{-3} = 1/27$
(c) $10^x = 1000 \Rightarrow x = 3$ (d) $x^{5/2} = 32 \Rightarrow x = 32^{2/5} \Rightarrow x = 4$

9. (a) Proof (by Mathematical Induction):
For $n = 1$ the statement is $\log_b r^1 = 1 \cdot \log_b r$, so the result is true for this first case. Assuming the result for $n = k\ (\geq 1)$ we have: $\log_b r^k = k \log_b r$. Now for the case where $n = k+1$ we find that $\log_b r^{k+1} = \log_b(r \cdot r^k) = \log_b r + \log_b r^k$ (by part (a) of Theorem A1.2) $= \log_b r + k \log_b r$ (by the induction hypothesis) $= (1+k)\log_b r = (k+1)\log_b r$. Therefore the result follows for all $n \in \mathbf{Z}^+$ by the Principle of Mathematical Induction.

(b) For all $n \in \mathbf{Z}^+$, $\log_b r^{-n} = \log_b(1/r^n) = \log_b 1 - \log_b r^n$ (by part (b) of Theorem A1.2) $= 0 - n \log_b r$ (by part (a) above) $= (-n)\log_b r$.

11. (a) Let $x = \log_2 3$. Then $2^x = 3$ and $x(\ln 2) = \ln 2^x = \ln 3$, so $\log_2 3 = x = \ln 3/\ln 2 = 1.0986/0.6931 \doteq 1.5851$.

(b) $\log_5 2 = \ln 2/\ln 5 = 0.6931/1.6094 \doteq 0.4307$

(c) $\log_3 5 = \ln 5/\ln 3 = 1.6094/1.0986 \doteq 1.4650$

13. (a) $1 = \log_{10} x + \log_{10} 6 = \log_{10} 6x \Longrightarrow 6x = 10^1 = 10 \Longrightarrow x = 10/6 = 5/3$.

(b) $\ln(x/(x-1)) = \ln 3 \Longrightarrow x/(x-1) = 3 \Longrightarrow x = 3(x-1) \Longrightarrow x = 3x - 3 \Longrightarrow -2x = -3 \Longrightarrow x = 3/2$.

(c) $2 = \log_3(x^2 + 4x + 4) - \log_3(2x-5) = \log_3[(x^2 + 4x + 4)/(2x-5)] \Longrightarrow 3^2 = 9 = (x^2 + 4x + 4)/(2x-5) \Longrightarrow 9(2x-5) = x^2 + 4x + 4 \Longrightarrow 18x - 45 = x^2 + 4x + 4 \Longrightarrow x^2 - 14x + 49 = 0 \Longrightarrow (x-7)^2 = 0 \Longrightarrow x = 7$.

15. Proof: Let $x = a^{\log_b c}$ and $y = c^{\log_a b}$. Then
$x = a^{\log_b c} \Longrightarrow \log_b x = \log_b[a^{\log_b c}] = (\log_b c)(\log_b a)$, and
$y = c^{\log_a b} \Longrightarrow \log_b y = \log_b[c^{\log_b a}] = (\log_b a)(\log_b c)$.
Consequently, we find that $\log_b x = \log_b y$ from which it follows that $x = y$.

APPENDIX 2
PROPERTIES OF MATRICES

1.

(a) $A + B = \begin{bmatrix} 3 & 2 & 5 \\ 0 & 2 & 7 \end{bmatrix}$

(b) $(A + B) + C = \begin{bmatrix} 3 & 3 & 7 \\ 5 & 6 & 4 \end{bmatrix}$

(c) $B + C = \begin{bmatrix} 1 & 2 & 3 \\ 6 & 6 & 1 \end{bmatrix}$

(d) $A + (B + C) = \begin{bmatrix} 3 & 3 & 7 \\ 5 & 6 & 4 \end{bmatrix}$

(e) $2A = \begin{bmatrix} 4 & 2 & 8 \\ -2 & 0 & 6 \end{bmatrix}$

(f) $2A + 3B = \begin{bmatrix} 7 & 5 & 11 \\ 1 & 6 & 18 \end{bmatrix}$

(g) $2C + 3C = \begin{bmatrix} 0 & 5 & 10 \\ 25 & 20 & -15 \end{bmatrix}$

(h) $5C = \begin{bmatrix} 0 & 5 & 10 \\ 25 & 20 & -15 \end{bmatrix}$

(i) $2B - 4C = \begin{bmatrix} 2 & -2 & -6 \\ -18 & -12 & 20 \end{bmatrix}$

(j) $A + 2B - 3C = \begin{bmatrix} 4 & 0 & 0 \\ -14 & -8 & 20 \end{bmatrix}$

(k) $2(3B) = \begin{bmatrix} 6 & 6 & 6 \\ 6 & 12 & 24 \end{bmatrix}$

(l) $(2 \cdot 3)B = \begin{bmatrix} 6 & 6 & 6 \\ 6 & 12 & 24 \end{bmatrix}$

3.

(a) [12], or 12

(b) $\begin{bmatrix} 9 & 21 \\ 12 & 27 \end{bmatrix}$

(c) $\begin{bmatrix} -10 & -10 \\ 18 & 24 \end{bmatrix}$

(d) $\begin{bmatrix} -5 & -7 & 8 \\ 29 & 21 & 2 \\ -23 & -35 & 6 \end{bmatrix}$

(e) $\begin{bmatrix} a & b & c \\ d & e & f \\ 3g & 3h & 3i \end{bmatrix}$

(f) $\begin{bmatrix} a & b & c \\ 3g & 3h & 3i \\ d & e & f \end{bmatrix}$

5. (a) $(-1/5)\begin{bmatrix} 1 & -2 \\ -3 & 1 \end{bmatrix}$

(b) $\begin{bmatrix} 0 & 1 \\ 1 & 0 \end{bmatrix}$

(c) The inverse does not exist, since $\begin{vmatrix} -3 & 1 \\ -6 & 2 \end{vmatrix} = -6-(-6)=0.$

(d) $\begin{bmatrix} 1 & 3 \\ 2 & 7 \end{bmatrix}$

7. (a) $A^{-1} = (1/2)\begin{bmatrix} 2 & -1 \\ 0 & 1 \end{bmatrix}$ (b) $B^{-1} = (1/5)\begin{bmatrix} 1 & -2 \\ 3 & -1 \end{bmatrix}$ (c) $AB = \begin{bmatrix} -4 & 3 \\ -6 & 2 \end{bmatrix}$

(d) $(AB)^{-1} = (1/10)\begin{bmatrix} 2 & -3 \\ 6 & -4 \end{bmatrix}$ (e) $B^{-1}A^{-1} = (1/10)\begin{bmatrix} 2 & -3 \\ 6 & -4 \end{bmatrix}$

9. (a) $\begin{bmatrix} 3 & -2 \\ 4 & -3 \end{bmatrix}\begin{bmatrix} x \\ y \end{bmatrix} = \begin{bmatrix} 5 \\ 6 \end{bmatrix}$

$$\begin{bmatrix} x \\ y \end{bmatrix} = \begin{bmatrix} 3 & -2 \\ 4 & -3 \end{bmatrix}^{-1}\begin{bmatrix} 5 \\ 6 \end{bmatrix} = (-1)\begin{bmatrix} -3 & 2 \\ -4 & 3 \end{bmatrix}\begin{bmatrix} 5 \\ 6 \end{bmatrix} = \begin{bmatrix} 3 \\ 2 \end{bmatrix}$$

(b) $\begin{bmatrix} 5 & 3 \\ 3 & -2 \end{bmatrix}\begin{bmatrix} x \\ y \end{bmatrix} = \begin{bmatrix} 35 \\ 2 \end{bmatrix}$

$$\begin{bmatrix} x \\ y \end{bmatrix} = \begin{bmatrix} 5 & 3 \\ 3 & -2 \end{bmatrix}^{-1}\begin{bmatrix} 35 \\ 2 \end{bmatrix} = (-1/19)\begin{bmatrix} -2 & -3 \\ -3 & 5 \end{bmatrix}\begin{bmatrix} 35 \\ 2 \end{bmatrix} = \begin{bmatrix} 4 \\ 5 \end{bmatrix}$$

11. $\det(2A) = 2^2(31) = 124, \ \det(5A) = 5^2(31) = 775$

13. (a) $\begin{vmatrix} 1 & 0 & 2 \\ 6 & -2 & 1 \\ 4 & 3 & 2 \end{vmatrix} = (1)(-1)^{1+1}\begin{vmatrix} -2 & 1 \\ 3 & 2 \end{vmatrix} + 2(-1)^{1+3}\begin{vmatrix} 6 & -2 \\ 4 & 3 \end{vmatrix} =$

$(-4-3)+2(18+8) = -7+52 = 45$

(b) $\begin{vmatrix} 4 & 7 & 0 \\ 4 & 2 & 0 \\ 3 & 6 & 2 \end{vmatrix} = (2)(-1)^{3+3}\begin{vmatrix} 4 & 7 \\ 4 & 2 \end{vmatrix} = (2)(8-28) = -40$

(c) $\begin{vmatrix} 1 & 2 & -4 \\ 0 & 1 & 0 \\ 3 & 3 & 2 \end{vmatrix} = (1)(-1)^{1+1}\begin{vmatrix} 1 & -4 \\ 3 & 2 \end{vmatrix} = 2+12 = 14$

15. (a) (i) $\begin{vmatrix} 1 & 2 & 1 \\ 0 & -1 & -1 \\ 2 & 3 & 0 \end{vmatrix} = 2(-1)^{3+1}\begin{vmatrix} 2 & 1 \\ -1 & -1 \end{vmatrix} + 3(-1)^{3+2}\begin{vmatrix} 1 & 1 \\ 0 & -1 \end{vmatrix}$

$= 2(-2-(-1)) - 3(-1) = 2(-1) + 3 = 1.$

(ii) 5 (iii) 25

(b) (i) 51 (ii) 306 (iii) 510

APPENDIX 3
COUNTABLE AND UNCOUNTABLE SETS

1. (a) True (b) False (c) True
(d) True (e) True (f) True
(g) False: Let $A = \mathbf{Z}^+ \cup (0,1]$ and $B = (0,1]$. Then A, B are both uncountable, but $A - B = \{2,3,4,\ldots\}$ is countable.

3. If B were countable, then by Theorem A3.3 it would follow that A is countable. This leads us to a contradiction since we are given that A is uncountable.

5. Since S, T are countably infinite, we know from Theorem A3.2 that we can write $S = \{s_1, s_2, s_3, \ldots\}$ and $T = \{t_1, t_2, t_3, \ldots\}$ — two (infinite) sequences of distinct terms. Define the function

$$f : S \times T \to \mathbf{Z}^+$$

by $f(s_i, t_j) = 2^i 3^j$, for all $i, j \in \mathbf{Z}^+$. If $i, j, k, \ell \in \mathbf{Z}^+$ with $f(s_i, t_j) = f(s_k, t_\ell)$, then $f(s_i, t_j) = f(s_k, t_\ell) \Rightarrow 2^i 3^j = 2^k 3^\ell \Rightarrow i = k, j = \ell$ (By the Fundamental Theorem of Arithmetic) $\Rightarrow s_i = s_k$ and $t_j = t_\ell \Rightarrow (s_i, t_j) = (s_k, t_\ell)$. Therefore f is a one-to-one function and $S \times T \sim f(S \times T) \subset \mathbf{Z}^+$. So from Theorem A3.3 we know that $S \times T$ is countable.

7. The function $f : (\mathbf{Z} - \{0\}) \times \mathbf{Z} \times \mathbf{Z} \to \mathbf{Q}$ given by $f(a, b, c) = 2^a 3^b 5^c$ is one-to-one. (Verify this!) So by Theorems A3.3 and A3.8 it follows that $(\mathbf{Z} - \{0\}) \times \mathbf{Z} \times \mathbf{Z}$ is countable. Now for all $(a, b, c) \in (\mathbf{Z} - \{0\}) \times \mathbf{Z} \times \mathbf{Z}$ there are at most two (distinct) real solutions for the quadratic equation $ax^2 + bx + c = 0$. From Theorem A3.9 it then follows that the set of all real solutions of the quadratic equations $ax^2 + bx + c = 0$, where $a, b, c \in \mathbf{Z}$ and $a \neq 0$, is countable.